Comparative Anatomy of the
VERTEBRATES

Other Related Books

A Manual of Practical Entomology, 2nd ed.	*Trigunayat, M.M.*
Agricultural Insect Pests and Their Control	*Awasthi, V.B.*
Animal Biodiversity: Patterns and Processes	*Ananthakrishnan,T.N*
Beekeeping : A Compressive Guide to Bees and Beekeeping	*Abrol, D.P.*
Dynamics of Insect Behaviour	*Ananthakrishnan, T.N*
Ecodynamics of Insect Communities	*Ananthakrishnan,T.N*
Ecological Entomology: Insect Life in Odd Environments	*Ananthakrishnan,T.N*
Freshwater Aquaculture, 3rd Ed.	*Rath, R.K.*
Freshwater Fish Pond Culture & Management	*Chakroff, M.*
Insect Biodiversity: Functional Dynamics and Ecological Perspectives	*Ananthakrishnan,T.N*
Insect Neuroendocrines	*Awasthi, V.B.*
Insect, Mite and Vertebrate Pests and their Management in Horticultural Crops	*Reddy, P.P.*
Integrated Pest Management	*Dhawan, A.K*
Introduction to General and Applied Entomology, 3rd Edition	*Awasthi, V.B.*
Medical Entomology	*Tyagi, B.K.*
Medical Physics for Human Health Care	*Bhatnagar, P.K.*
Principles of Insect Behaviour, 2nd Ed.	*Awasthi, V.B.*
Termite Life and Termite Control in Tropical South Asia	*Roonwal, M.L.*
The Chemistry of Pesticides their Metabolism, Mode of Action and Uses in Crop Protection	*Hassall, K.A.*
The Invincible Deadly Mosquitoes	*Tyagi, B.K.*
Theory and Practice of Integrated Pest Management	*Arora, R.*
Vector-Borne Diseases: Epidemiology & Control	*Tyagi, B.K.*
Vertebrate Pests in Agriculture- The Indian Scenario	*Sridhara, S.*

Comparative Anatomy *of the* VERTEBRATES

Theodore H. Eaton, Jr.

Professor of Zoology
East Carolina College
Greenville, North Carolina

United Book Prints

1st Reprint in India

Published by:

United Book Prints
4806/24, Ansari Road,
Daryaganj,
New Delhi - 110 002
Tel.: 011-41511055
E-mail: unitedbookprints@gmail.com

Print : 2023

Print: 2016

ISBN: 978-93-83692-28-6

Printed in India

CONTENTS

PREFACE

Students should come to Comparative Anatomy prepared by one or more substantial courses in zoology, and should be in their upper college years. They can then use the course effectively to equip them for medical school, for advanced study in zoology, or for teaching. Yet Comparative Anatomy is much more than a professional training course. Its values and meanings are as rich as those of all biology. It is basic to an understanding of man's origin and relationships; it is at the heart of our knowledge of evolution. By acquainting us with the changing structure of a great host of animals, the study of Comparative Anatomy gives a framework in which we can properly place our ideas of function and of embryonic development.

In one course or one book it is possible only to introduce the student to Comparative Anatomy. But that introduction ought to contain elements of all the wider meanings of the subject, and should take recent knowledge into account particularly. Therefore I have included a chapter on embryology, and presented some of the important ideas of those who work with prehistoric animals, so far as these clarify our picture of the relationships of the vertebrate classes. It is necessary always to be aware of the "time dimension," knowing that today's life is an outgrowth of different life in past ages, and that we can trace quite accurately the steps by which man and other modern forms have become what they are.

My conviction that it helps, in studying anatomy, to know the names, appearance, and ways of life of the animals concerned has led to a fairly comprehensive first chapter on the kinds of chordates, illustrated as fully as space would permit. Since structure without function has little meaning, I try to lay emphasis on the action of organs, tissues, and mechanisms of the body. A panoramic view in the last chapter may help in coordinating parts of the whole picture.

In order that bulk may not discourage full use of this text during a semester or a quarter, I have written concisely, but hope that it may be found full and clear enough to satisfy the more energetic student as well as the critical instructor.

It has been necessary to use, by permission, many illustrations from different sources, and I am grateful for them, but the majority are either new or redrawn by the author. The bibliography at the end lists books and papers which have been particularly useful to me, including those referred to in the text, but it is far from being either complete or well balanced. The sources cited there will lead the student to other work, however, in which he may be interested.

Here I should like to acknowledge my great obligation to Professor Charles L. Camp, of the University of California, under whom I received my first training in this field, and who gave me a vigorous enthusiasm for it; and to Professor Alfred S. Romer, of Harvard, with whom I have had most fruitful opportunities to work. Both men have read the manuscript and have given me the benefit of much critical comment. To Professor William K. Gregory, of the American Museum of Natural History, I owe my thanks for valuable suggestions and encouragement on a number of occasions. Finally I wish to inscribe this book to my mother and father, whose manifold faith and help have made the task easier.

THEODORE H. EATON, JR.

April, 1951

Comparative Anatomy of the VERTEBRATES

CHAPTER I

The Kinds of Chordates

To find meaning in a survey of comparative anatomy we must have at least a nodding acquaintance with the vertebrate animals and their poor relations, the protochordates. Together these comprise one great phylum of animals, including types as diverse as fishes, frogs, birds, and men. It would not be enough to study comparative anatomy by laying out sets of disembodied parts and noting merely that our own muscles, our hearts, our bones are the result of transformation of more primitive muscles, hearts, and bones. This we might do, but we should lose sight of the ***whole*** in which the parts live. So we must know the chordates first as living beings, must be conscious of the ability of organs to work together in a body and to change, sometimes, when faced with new circumstances. The background for comparative anatomy, therefore, is a clear-cut picture of the kinds of chordates and their ways of life.

This picture will fit together most easily if we start with a few traits by which we tell a chordate from anything else (Fig. 1). It is an animal with a ***notochord,*** a gristlelike rod running lengthwise in the body, during some part of its life. Usually this comes early and then disappears, to be replaced by a column of vertebrae, but in some of the less progressive kinds it remains throughout life. Of course a chordate ordinarily has a head and tail end (in anterior and posterior directions respectively), a right and left side which are mirror images of each other, and a back and belly surface (dorsal and ventral). Running along the dorsal side is a ***central nervous system*** which is a hollow tube, not far below the surface, usually well protected by hard parts. A chordate also has, in the walls of its throat, paired slits opening to the outside, or, if it is far enough up in the scale, they will show only as grooves during its early development and will then vanish. These are the ***gill clefts,*** for when they are present they allow water to stream out past the gills, providing oxygen for breathing. Finally, the blood circulates in a system of closed tubes, so arranged that on the ventral side of the body a main vessel propels it

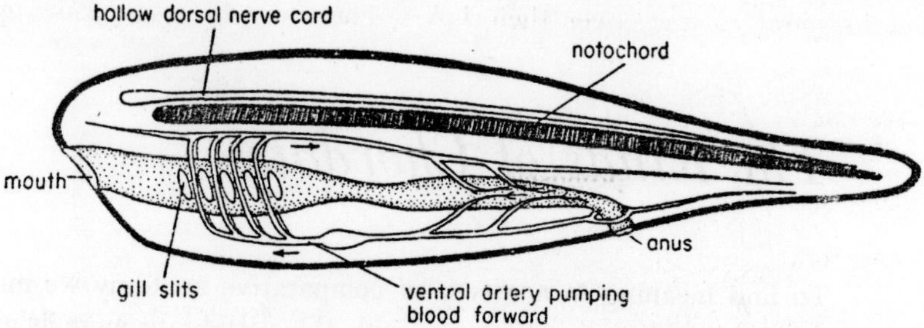

Fig. 1. Diagram of primary chordate characteristics.

forward, while along the dorsal side an artery connected with this carries it toward the posterior end.

Separation of chordates into subphyla, classes, orders, and so on is done by looking for certain traits that may be common to a particular group but are not found in other groups. The more numerous and profound, the better these "characters" serve our purpose, for we want to arrange chordates in a system that we can understand, and that will stand up under criticism. That is not quite all we want, for it has long been realized that no such scheme will endure unless it is built on actual rela-

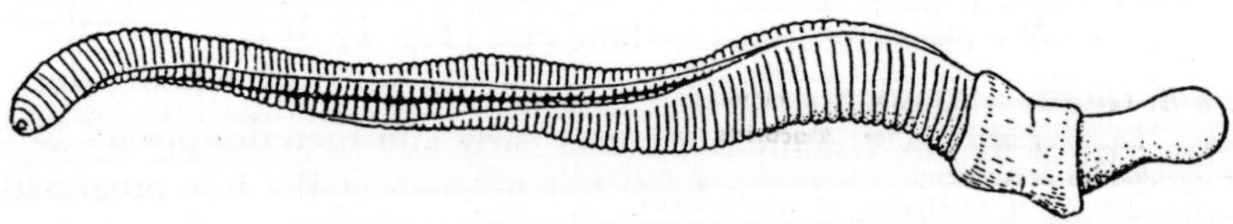

Fig. 2. A wormlike hemichordate, *Ptychodera*.

tionships, that is, the amount of divergence from a common ancestry. This is where classification (systematics) and anatomy meet. Each is indispensable to the other, and new facts may change the picture at any time. When they come, they are to be welcomed.

In this chapter we shall outline the anatomy of protochordates, and then summarize the classes of vertebrates.

THE HEMICHORDATES

Hemichordata are the first of three subphyla of small marine animals, commonly called protochordates because they seem to show beginnings of characters familiar to us in the vertebrates. Most hemichordates are worms, burrowing in mud which they swallow and then emit in spiral castings over their holes. These are the class Enterop-

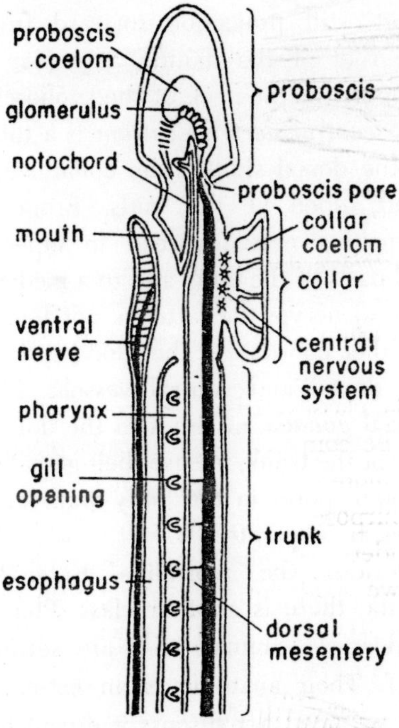

Fig. 3. Longitudinal section of a hemichordate, *Balanoglossus.* (After MacBride in *Encyclopedia Britannica.*)

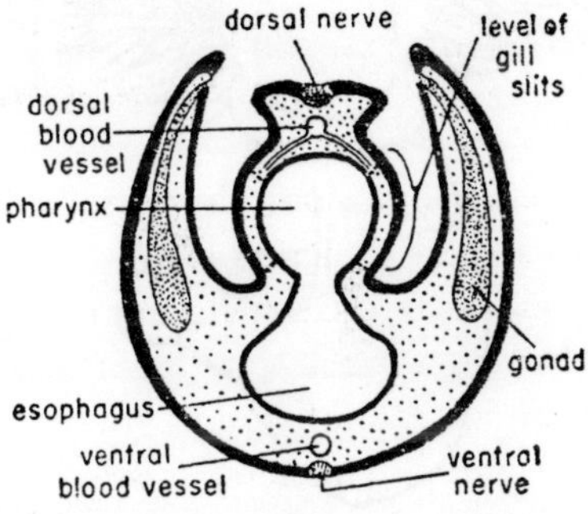

Fig. 4. Cross section of a hemichordate through pharynx, diagrammatic.

neusta; *Balanoglossus* and *Ptychodera* (Fig. 2) are familiar kinds. The body is not segmented, for they have no relationship to annelid worms, but it has three obvious divisions: a distensible ***proboscis,*** a short, thick ***collar,*** and a ***trunk.*** The ***mouth*** is under the base of the proboscis (Figs. 3, 4). It opens through a short ***oral cavity*** into the ***pharynx,*** which runs back into the first part of the trunk. Here a long series of paired U-shaped ***gill slits*** open dorsally in the sides of a groove. The oral cavity and pharynx, as well as the outer surfaces of proboscis and collar, are covered with *cilia* that sweep a current of water into the mouth and out the gill slits. This is primarily a feeding, not a breathing, mechanism, for mucus secreted by cells on the proboscis entangles microscopic food and it is thus carried to the hind end of the pharynx in little lumps that presently drop into the intestine. The animal uses this way of feeding when it protrudes from its burrow and explores with its proboscis; no doubt it digests some organic matter too from the mud swallowed during burrowing.

Fig. 5. A sessile hemichordate, *Cephalodiscus*. (After Parker and Haswell.)

The ***notochord*** is nothing but a short, stiff projection forward from the roof of the mouth, acting as a support for the neck of the proboscis. The ***central nervous system*** is a tube in the dorsal side of the collar, very short, open at the ends. From it branching nerves spread to superficial parts of the body and to a median ventral nerve; the latter is not found in other chordates. The blood system has dorsal and ventral vessels. The paired ***gonads***, in series in the dorsal side of the trunk, release their gametes through pores in the body wall. The sexes are separate.

Besides the wormlike Enteropneusta, there is another class, Pterobranchia, containing two genera of deep-sea animals that are sessile (fixed to the substratum, like plants). Their anatomy is an extreme travesty on that of *Balanoglossus,* and we need not discuss it except to say that they feed by a ciliary-mucus mechanism carried on tentacles, that the notochord is represented by a pair of stiffened patches of epithelial cells that serve to fit the shape of the pharynx to the current produced by the cilia, and that the central nervous system is represented by a neural plate on the dorsal surface. In one genus, *Cephalodiscus* (Fig. 5), there are only two gill slits, while the other, *Rhabdopleura,* has none; they are simply an outlet for water, not a respiratory device.

THE CEPHALOCHORDATES

Two genera, *Branchiostoma* (Amphioxus, Fig. 6) and *Asymmetron,* make the subphylum Cephalochordata. The first has been studied in great detail because its structure and development illuminate many points in chordate evolution. Like *Balanoglossus* it hides in the sea sand and feeds by cilia, but there the resemblance stops, for Amphioxus is not wormlike, but swims with a rapid irregular darting, and has a streamlined body pointed at both ends, with a small fin on the tail. It is not over four inches long, and is common in warm shallow seas.

The ***notochord*** runs from the projecting anterior tip (***rostrum***) to the tail end, hence the name of the subphylum "head cord." V-shaped muscle segments (***myomeres***) run in series, alternately instead of opposite, along both sides of the notochord. Their contraction on one side and then the other makes the notochord and the body bend in rhythmic S-shaped curves, and the flattened tail fin therefore thrusts the animal forward. The ***dorsal nerve cord*** runs lengthwise above the the notochord, and from it two nerves in each segment run out on each side, one dorsal, the other ventral. These correspond to the dorsal and ventral roots of the spinal nerves in vertebrates. There is no brain, but anteriorly a nerve leads to a small pit opening to the left of the middorsal line, supposedly an ***olfactory organ;*** and along the nerve cord itself are small spots of black pigment, thought to be sensitive to light.

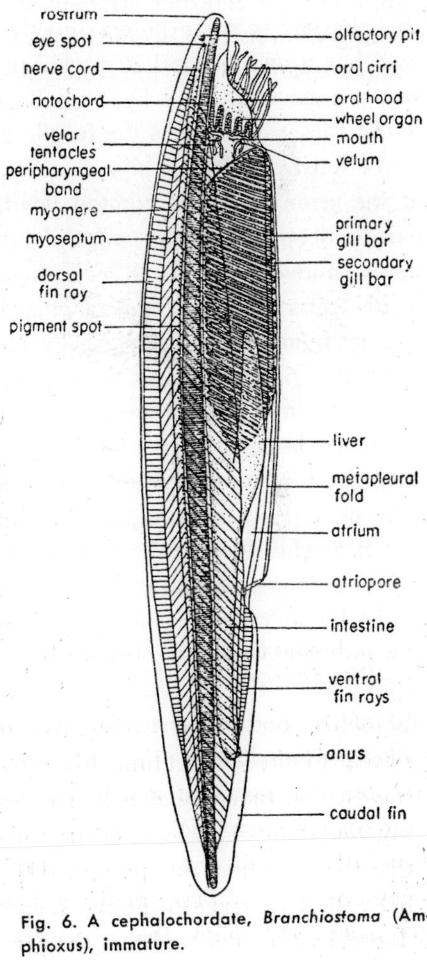

Fig. 6. A cephalochordate, *Branchiostoma* (Amphioxus), immature.

The feeding apparatus is of most interest, for none more elaborate occurs in any chordate. It is essentially a ciliary funnel and sieve. The cuplike ***oral hood,*** fringed by fingerlike ***cirri,*** admits a current of water into the ***mouth,*** which is a hole in a circular membrane, the ***velum.*** This in turn has a lobed fringe, the ***wheel organ,*** covered with beating cilia that draw the current into the *pharynx,* a long deep chamber. The sides of this are perforated by a great many (up to 200) pairs of oblique narrow ***gill slits,*** and the delicate ***gill bars*** between are braced by small bridges called ***synapticulae.*** Through the slits, of course, the current runs out, leaving food trapped inside, to be carried into the intestine. But the pharynx is not exposed directly to the outside world, for an outer wall has grown around it, enclosing it in a space, the ***atrium,*** and the water must then leave by a small posterior opening, the ***atriopore.*** No doubt this wall protects the fragile gill basket.

We must crawl back into the pharynx for a moment, however, to look at the arrangement of ***ciliated bands*** that catch and carry the food. A narrow deep groove, the ***endostyle,*** runs the whole length of the pharynx in the midventral line; it secretes mucus, which cilia then propel forward. Anteriorly this band of cilia divides, just behind the velum, into two ***peripharyngeal bands.*** They (with the help of cilia sweeping up along the gill bars) carry the mucus and food entangled in it up to the middorsal line, where they unite in a ciliated ***epipharyngeal groove,*** and this is the last lap, for it leads back to the opening of the ***esophagus.*** The latter is much smaller in diameter than the pharynx, and continues with little change to the straight ***intestine.*** From this a finger-shaped blind pouch reaches forward for a variable distance along the right side of the pharynx. It is probably equivalent to the ***liver*** of vertebrates.

The blood system has a ventral artery with a pulsating enlargement in it, a rudimentary ***heart,*** so that it pumps blood forward and then through capillaries up the gill bars to the dorsal artery, where it flows back to the rest of the body. Above the pharynx lie microscopic excretory organs called ***protonephridia*** (Fig. 264), the functioning units of which are ***solenocytes*** or flame cells, so called because of the flickering cilia that wash excretory products out of their ducts. No such organs are known in other chordates, but they do occur in several invertebrate phyla; hence *Branchiostoma* is doubtless primitive in this character at least. The ***gonads*** are paired sacs below the myomeres, bulging into the atrium, and they burst to release gametes, which then escape by the atriopore. A related genus, *Asymmetron,* has only a single row of gonads, and differs in other minor ways.

THE *TUNICATES,* OR *SEA SQUIRTS*

This subphylum, the *Urochordata* (*Tunicata*), numbers several hundred kinds, most of which have not the faintest outward resemblance to any other chordates. The simple tunicates live attached to a solid substratum, and produce buds as well as gametes. The compound tunicates may be attached or may float in chains or clusters in the open sea; sometimes they are transparent, and some are luminous at night. We may take for an example, one of the simple sessile tunicates (Fig. 7).

Each individual is a sac with two openings at its upper end, the ***mouth*** and ***atriopore.*** The body is covered by a ***tunic*** of about the same compo-

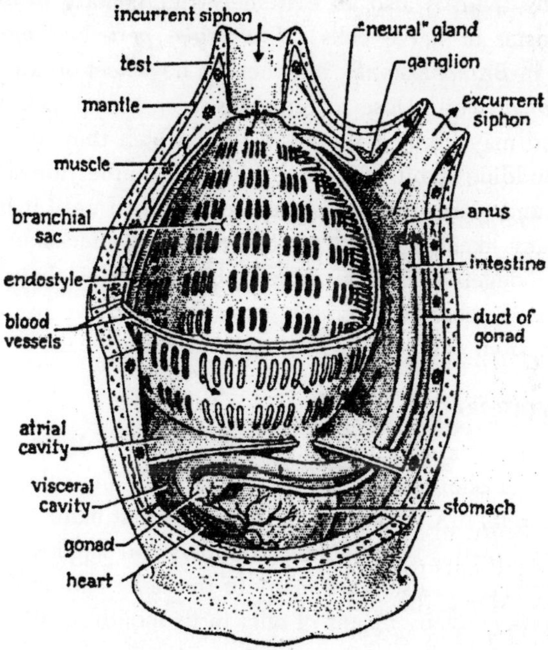

Fig. 7. Diagram of structure of a simple ascidian. (From *General Zoology* by Storer.)

sition as plant cellulose; through its oral and atrial siphons water will squirt when the animal contracts suddenly. This accounts for two of the names used for this subphylum. "Urochord" means "tail rod," for they have a swimming larval stage in which there is a ***notochord*** in the tail, as well as muscles and a tubular nerve cord; the tail disappears when the larva transforms into the stay-at-home adult.

The ***pharynx*** and feeding mechanism are much the same as in *Branchiostoma.* Imagine the latter fixed with its head up, and with the atriopore brought forward dorsally to lie not far from the mouth. Water now pours into the mouth, diffuses through the pharyngeal basket and out into the atrium. The intestine is expanded into a pouchlike ***stomach,*** and the ***anus*** enters the atrium. Here too enter ducts from the ***gonads*** (***ovary*** and ***testis***). Tunicates are ***hermaphroditic,*** having both sexes in the same individual.

There is a ventral ***heart,*** a ventral artery that parallels the endostyle and sends branches into the gill bars, and a dorsal artery that receives these. The blood flow is reversible, an exception to the rule among chordates. Dorsally there is also an extremely rudimentary ***brain,*** simply a ganglion (cluster of nerve cells). ***Endostyle, peri-*** and ***epipharyngeal bands*** are as in *Branchiostoma,* but there is no wheel organ.

Branchiostoma, while close to tunicates, seems more primitive in certain ways, and may be near the stock from which they arose. After the sessile and budding habit was established in simple tunicates they diverged more and more from the structure described until it becomes difficult to see any likeness whatever. The tunicates offer some marvellous examples of "degenerative" evolution. Undoubtedly they are a side branch from early chordates, which was able to survive by withdrawing quite adroitly from competition with most other animals.

THE *VERTEBRATES*

Presence of a skull may not be the greatest distinction between higher and lower chordates, but it is the basis for the subphylum ***Vertebrata*** (= ***Craniata***). A backbone, to which the first name refers, is by no means universal in this subphylum. Lampreys and hagfishes are without complete vertebrae, and so were many of the older extinct fishes. But no adult vertebrate feeds by means of cilia in its mouth or pharynx, a character of all protochordates.

In our discussion of each class of vertebrates we shall say something about kinds that are no longer living. This is because the characters of living kinds are often hopelessly puzzling until we find out how they came to be as they are. Not only are some extinct groups tremendously fascinating in themselves (dinosaurs, for instance), but they unfold before us the beginnings of all that live today. Progress toward mammals and toward man has been exceedingly long,—more than half a billion years of vertebrate evolution. We shall use, therefore, a timetable of this progress, just as in a study of human society we would not get far without some knowledge of what happened in the Middle Ages, the Renaissance, and the eighteenth century. An acquaintance with certain crucial prehistoric periods—the Devonian and Carboniferous, Triassic, Cretaceous, Eocene, among others—is directly and fundamentally a part of our knowledge of vertebrate anatomy. It was not always so, but it is so now, for the field is no longer limited to what happens to be preserved in our alcohol jars. See Chart of the Relationships of Vertebrate Groups, end leaves.

Branch Agnatha: Primitive Fishes Without Jaws

The subphylum separates into two fundamental and very unequal divisions, of which this, without jaws, is the more primitive. The

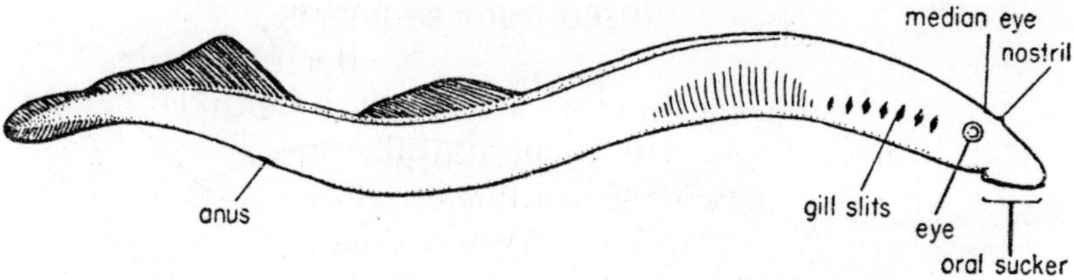

Fig. 8. Lamprey, *Entosphenus tridentatus.*

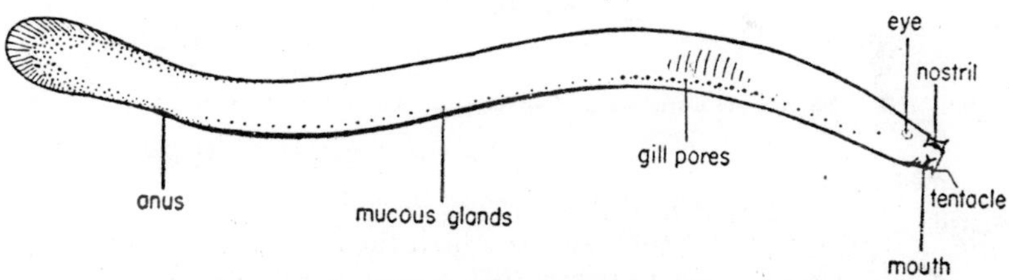

Fig. 9. A hagfish, *Bdellostoma.*

jawless fishes living today are two peculiar eel-shaped orders, the lampreys (*Petromyzontia,* Fig. 8) and hagfishes (*Myxinoidea,* Fig. 9). Both are parasites on respectable fishes, both have funnel-shaped sucking mouths (hence the name of the class, Cyclostomata, round mouthed), soft cartilaginous skeletons, and no paired fins; they are without scales and have only one nostril. The gills are in muscular pouches, opening by a row of pores along each side. Lampreys come up rivers during the spring spawning season, a few living only in fresh water, but hagfishes are wholly marine.

The larval lamprey ("Ammocoetes") is a remarkable minnowlike animal that burrows in the sand and feeds by the same mechanism of ciliated mucus bands that we found in the protochordates. It is distantly related to Amphioxus, but when it becomes an adult it puts away childish things.

Lampreys and hagfishes are descendants of the most ancient known group of bony-armored fishes, the ostracoderms, which were not parasites although they lacked jaws, and which frequently had paired fins (Figs. 10, 11). The first signs of them come in the Ordovician, and they had vanished by the close of the Devonian, some 350 million years ago.

Branch Gnathostomata: Fishes with Jaws, and Tetrapods

With the coming of jaws, by modification of the first gill arch of some very early ostracoderm, we enter upon a bewildering array of new

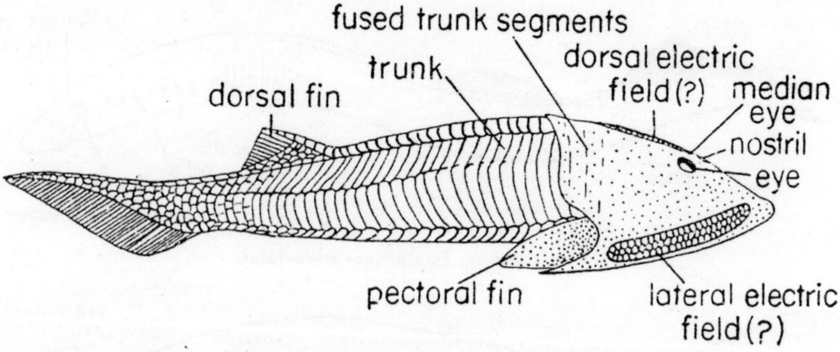

Fig. 10. An ostracoderm, *Cephalaspis*. (Modified after Abel.)

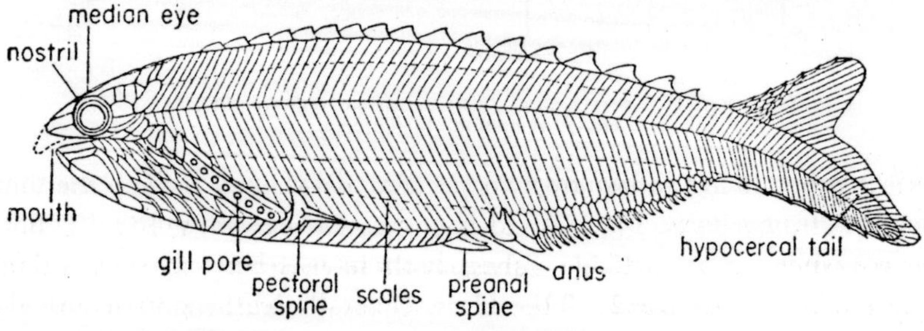

Fig. 11. An ostracoderm, *Rhyncholepis*. (After Kiaer.)

forms and habits. Even among fishes proper there are enough profound differences to warrant recognizing three or four distinct classes rather than the outmoded single group Pisces.

One class, ***Placodermi,*** consists of extinct fishes which (in acanthodians, at least) had a full-sized pair of gills on the rear margin of their jaws; the first gill arch, in other words, had not yet forsaken its early function although it had taken on the additional function of jaws, and bore teeth (Fig. 12). Placoderms ("plated skin") were armored fishes, either with small diamond-shaped scales, as in acanthodians, or with large plates of bone, as in antiarchs and arthrodires. The latter group were monsters of their day (the Devonian), ranging in length up to 30 feet. Their heads moved up and down against a pair of pivots in the body armor, hence the name jointed neck (arthrodire).

Then in the middle Devonian there arose a diverse class that has many well-known representatives today, the sharks, rays, and chimaeras (***Chon-***

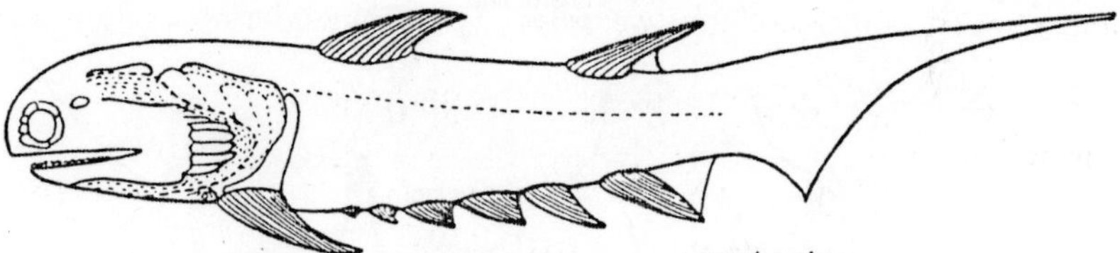

Fig. 12. An acanthodian, *Climatius.* (After Watson.)

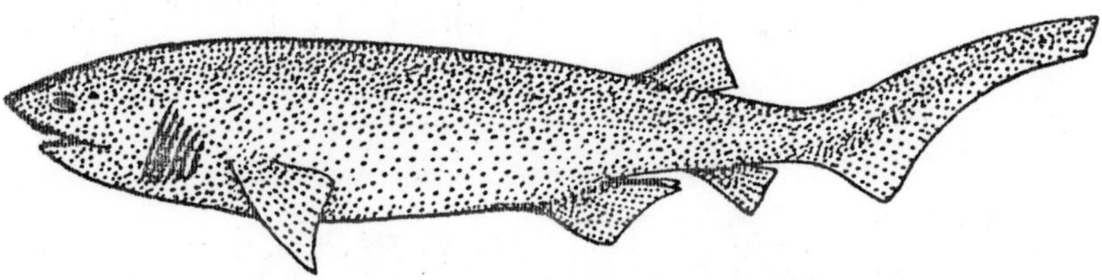

Fig. 13. *Hexanchus griseus,* a shark. (Redrawn from Jordan, *Fishes,* Appleton.)

drichthyes, or ***Elasmobranchii***). Some placoderm may have been their ancestor, but the connections are not known yet. The skeleton in modern elasmobranchs is of cartilage only, but there was bone in some of the early kinds (Romer, 1942). The skin is generally leathery with fine embedded scales whose points project slightly. Some kinds lay very large eggs, but many sharks are viviparous. Sharks (Fig. 13) have five to seven gill clefts on the sides of the throat, and also, as a rule, a vestigial pore-like cleft, the spiracle, just behind and above the jaws. A ray (Fig. 14) is like a shark but flat, much spread out, and with a pavement of flat teeth in its mouth. The chimaeras or ghost fishes (Holocephali) number only half a dozen living kinds, all but one of them (Fig. 15) in the deep sea, and with a skin flap (operculum) covering the gills on each side; their weird shape accounts for the name.

The next two classes (often lumped into one, the *Osteichthyes*) con-

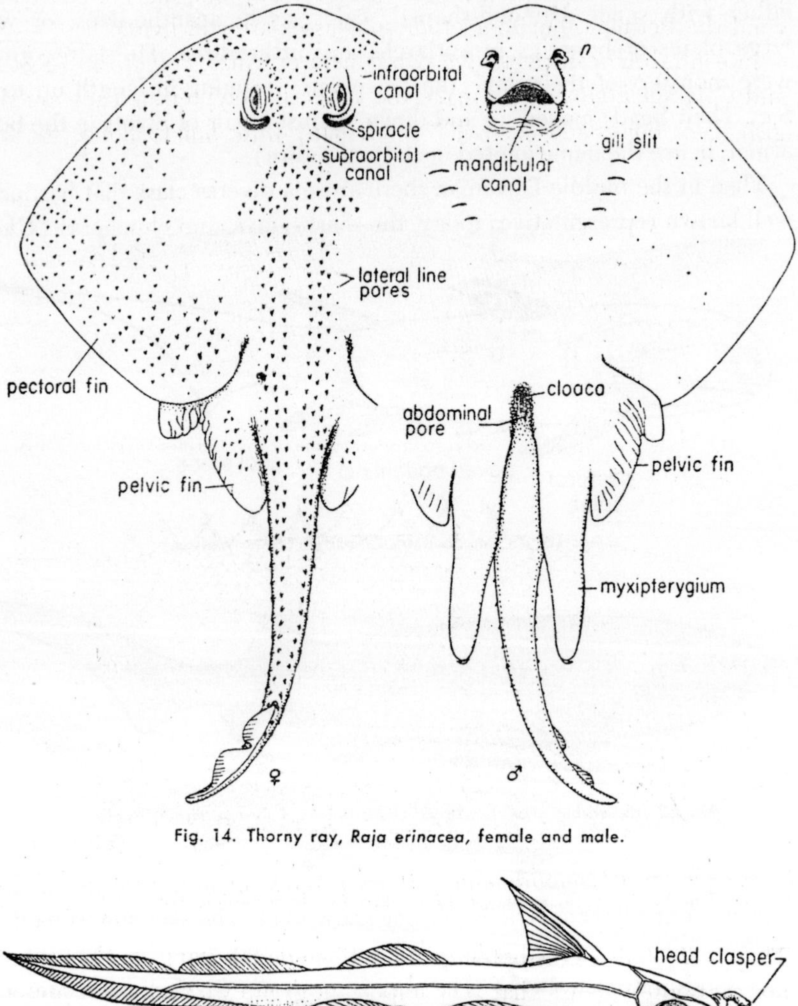

Fig. 14. Thorny ray, *Raja erinacea*, female and male.

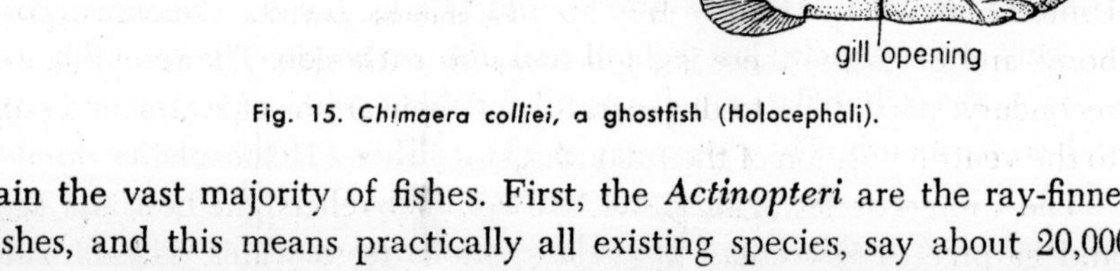

Fig. 15. *Chimaera colliei*, a ghostfish (Holocephali).

tain the vast majority of fishes. First, the *Actinopteri* are the ray-finned fishes, and this means practically all existing species, say about 20,000. For convenience they have to be split into five subclasses, a couple of dozen orders, and scores of families. The stem group, the paleoniscids,

(Fig. 16), which appeared in the Middle Devonian, did not survive beyond the Permian, but four other subclasses are still living.

In central African rivers live two genera of fishes, *Polypterus* (Fig. 17) and *Calamoichthys,* which make the subclass Polypterini. They are sluggish, very bony, and carry a row of many little finlets along the back.

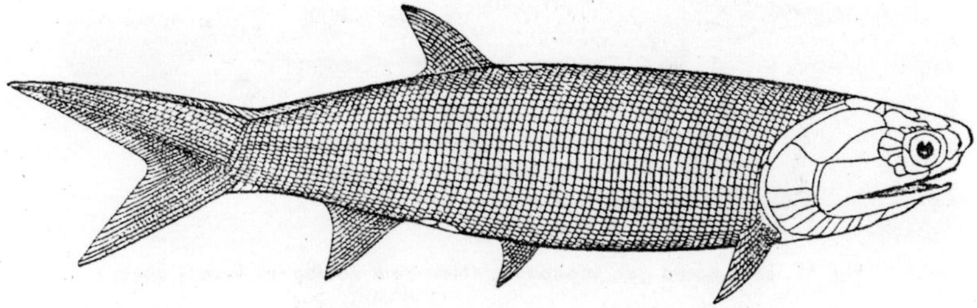

Fig. 16. A paleoniscid, *Paleoniscus macropomus,* one of the primitive ray-finned fishes.

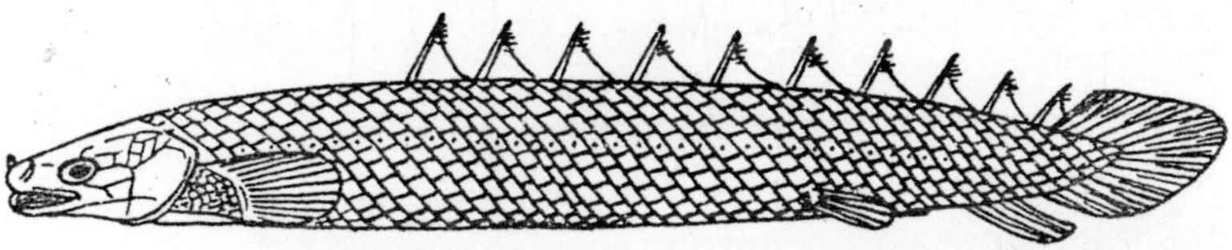

Fig. 17. *Polypterus bichir,* a primitive bony fish in the rivers of Africa. (After Parker and Haswell.)

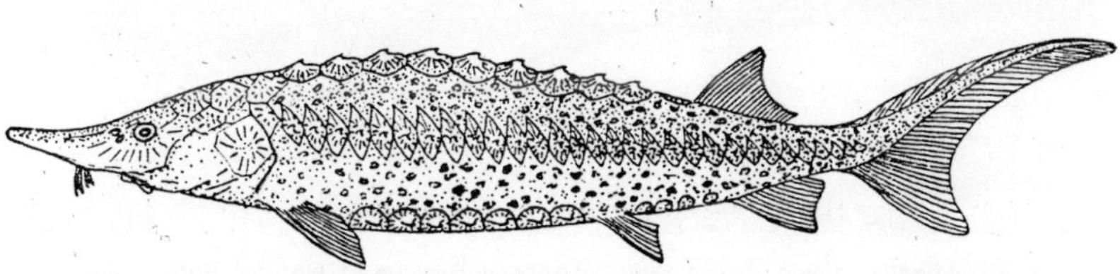

Fig. 18. Sturgeon, *Acipenser sturio.* (Redrawn from Jordan, *Fishes,* Appleton.)

They were called crossopterygians until some 20 years ago, but they are *not* crossopterygians! That very important group we shall meet presently.

Then there is a small subclass Chondrostei, the sturgeons (Fig. 18) and paddlefishes. They occur in the rivers of the middle and western United States. Superficially they are like sharks, but the skeleton is partly bony, and of course there is a gill cover on each side. The resemblance is secondary, partly due to degeneration of their armor of scales, and partly to the ventral position of their mouths.

The United States again provides two other relicts, the bowfin (*Amia*) and garpike (*Lepisosteus,* Fig. 19), both in fresh water. They are remnants of the big subclass Holostei, the bony ganoids.

Fig. 19. Long-nosed gar, *Lepisosteus*. (New York Zoological Society photo.)

Fig. 20. Atlantic salmon, *Salmo salar*. (American Museum of Natural History photo.)

Finally the great swarming proletarian masses of fish are the subclass Teleostei, which, incidentally, is almost impossible to separate from the preceding. Here are the herring, salmon, trout, catfish, minnow, carp, pickerel, cod, mullet, bass, tuna, and most of the others you can call to mind (Figs. 20, 21, 22).

Lungfishes (***Dipnoi***) and lobefins (***Crossopterygii***) together comprise a small but tremendously important class, ***Choanichthyes***, meaning fish with nostrils that open into the mouth. Their importance lies in their evolutionary position, on the line leading to land animals (tetrapods). Both subclasses appeared first in the Lower Devonian along with paleoniscids, and both have survived, although just barely so, to the present day.

Existing lungfishes are of three genera, *Epiceratodus* (Fig. 23) in Aus-

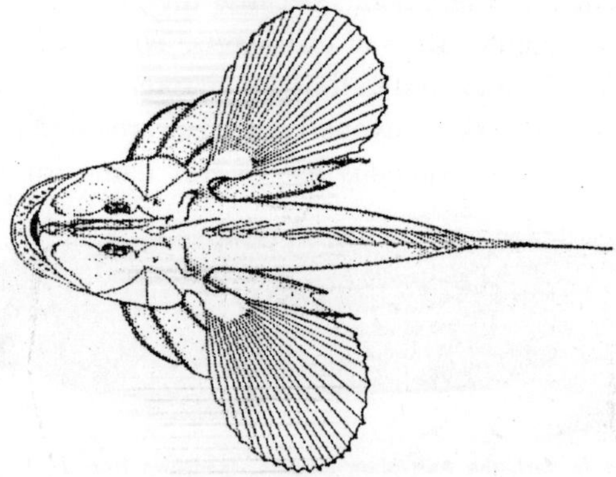

Fig. 21. Young angler fish, *Lophius piscatorius.*

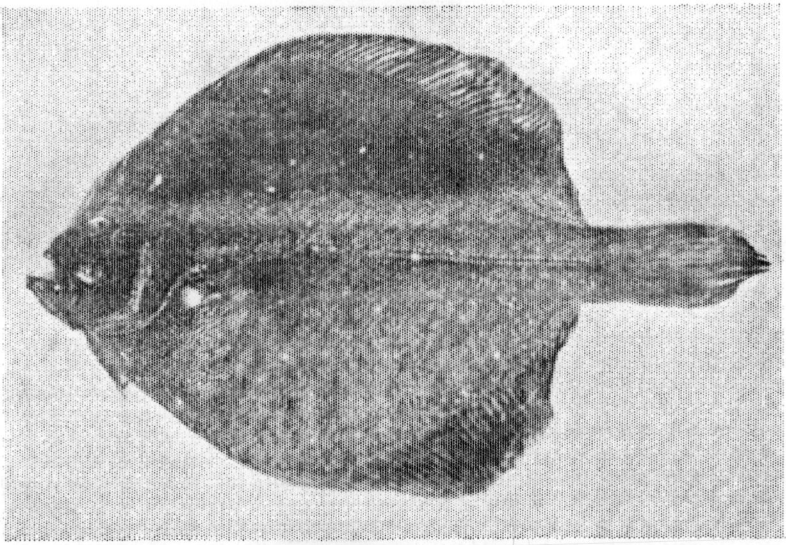

Fig. 22. A flounder, *Lophopsetta maculata.* (American Museum of Natural History photo.)

tralia, *Protopterus* in Africa, *Lepidosiren* in South America, all living in fresh water which is subject to seasonal stagnation or drying up. *Protopterus,* at least, lives then by burrowing into the bottom mud and hibernating in a cell for months, if necessary, until water comes again. Although they have gills like other fishes, they rise to the surface occasionally for a breath of air, since they are provided with lungs. Their normal behavior is sluggish.

Formerly thought to be ancestral to Amphibia, the lungfishes have surrendered this honor to the lobefins because they lack, and the latter possess, the necessary combination of characters. These will be dealt with in the chapter on the skeleton and in the final chapter.

It was not until 1938 that anyone knew that crossopterygians survived to modern times. (The erroneous use of this name for Polypterini has

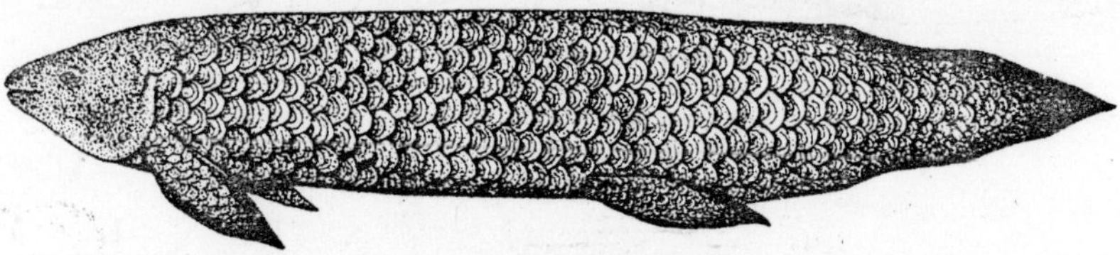

Fig. 23. *Epiceratodus forsteri,* the Australian lungfish. (Redrawn from Jordan, *Fishes,* Appleton.)

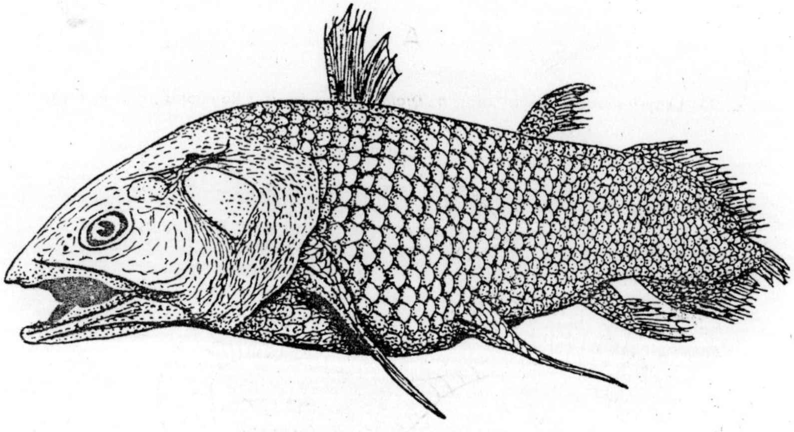

Fig. 24. The living crossopterygian, *Latimeria.*

been noted.) The group was known only from fossils, mostly Devonian, the very latest being Cretaceous, when presumably their line died out. Yet a living lobefin (Fig. 24), named *Latimeria,* was caught by fishermen trawling in fairly deep water off the east coast of South Africa, and its picture promptly appeared in magazines in this country and Europe, witnessing to the amazement of biologists that such a thing could happen. Little, unfortunately, was saved but the skin and part of the skull, and it remains for some expedition to haul up more of them in order that the soft parts and physiology of this remarkable relict may be studied.

The early Devonian lobefins therefore still give us most of what is known of the forerunners of the land animals. They were evidently lung breathers in part. Their paired fins show a jointed series of bones, smaller

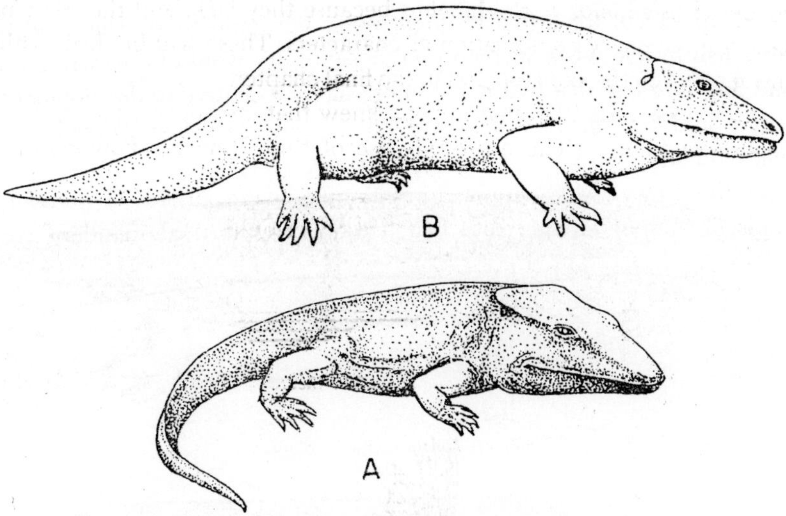

Fig. 25. Labyrinthodont amphibians, A, *Diplovertebron* (after Raymond), and B, *Eryops*.

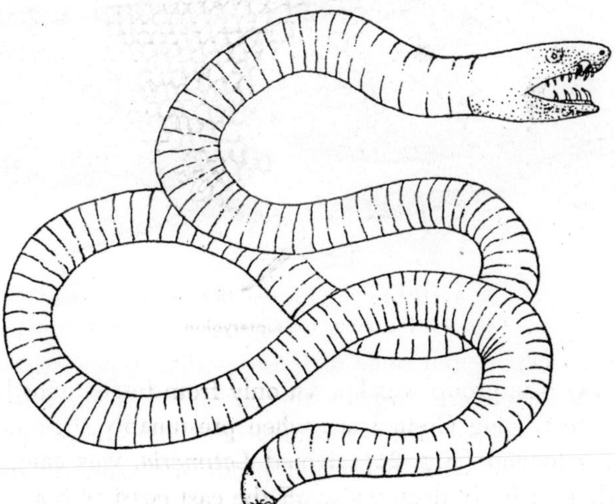

Fig. 26. A caecilian, *Caecilia pachynema*. (Based in part on Parker and Haswell, after Boulenger.)

and more numerous towards the tip, which provide the ground plan for the skeleton of the legs and feet of Amphibia. Romer suggests that the transition took place, paradoxically, as an adjustment making it possible to remain in the water. After all, the lobefins were fishes and could not have survived if thrown on land all at once. The early kinds of lobefins lived in fresh water, and if, as in Australia today, the rivers seasonally dried up except for a few stagnant pools here and there, then one way for a fish to live when its own pool became dry would be to wriggle out and flop into one that had not. A more likely incentive to development of legs would be difficult to imagine, but of course we may never have direct evidence.

In the Lower Mississippian period there were already a number of kinds of ***Amphibia,*** the class which today contains salamanders, frogs,

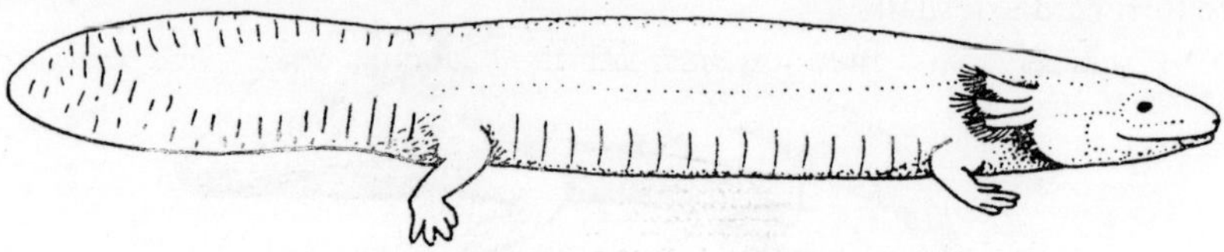

Fig. 27. *Necturus,* the mud puppy.

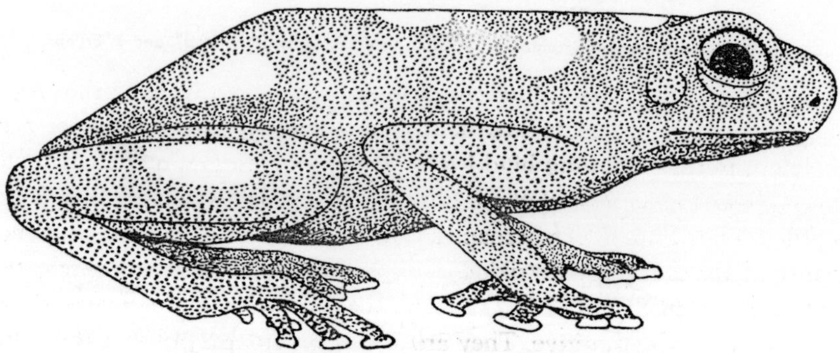

Fig. 28. A tree frog from Panama, *Dendrobates auratus.*

toads, and caecilians. But none of these modern types existed then, nor before the Mesozoic era. Instead, the earliest were large-headed, long-tailed, and short-legged labyrinthodonts (Fig. 25), direct descendants of the lobefinned fishes. They flourished, producing many different families and genera, up to the Triassic, by which time they had grown very large, with extraordinarily flat and heavy heads, and then they became extinct. From some of the early (not the late) labyrinthodonts there arose four separate lines leading to (1) Caecilia (Fig. 26), an order of burrowing wormlike tropical Amphibia, (2) Caudata (Fig. 27), the salamanders, newts, and mud puppies, (3) Salientia (Figs. 28, 29), the frogs and toads, and (4) the first ***Reptilia,*** to which we shall come in a moment.

We must say a bit more about the living kinds of Amphibia, for they are common and familiar, yet give us a false idea of vertebrate evolution if we assume that they are intermediate between fishes and reptiles. Breathing by gills during an aquatic larval stage and by lungs when adult, some of them indeed recapitulate hurriedly the change from fish to tetrapod ways of life. They lack scales, except for vestiges in the *Caecilia*, but certain labyrinthodonts had them. Eggs of Amphibia are laid either in water or out, they are surrounded by a jelly capsule filled with water, or sometimes by a tough membrane, and with few exceptions they are fertilized externally.

We scarcely need mention how far the Caecilia have gone from the

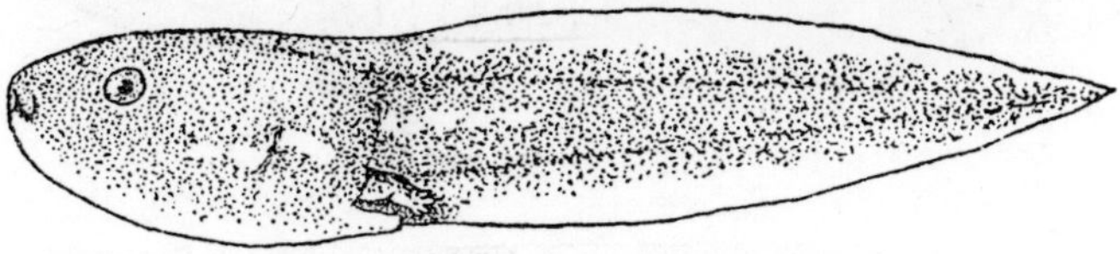

Fig. 29. Tadpole, *Gastrophryne carolinensis.* (After Noble.)

ancestral stock in losing all traces of limbs, nor how advanced the frogs and toads are likewise in loss of the tail, in many peculiarities of the skeleton, in their jumping hind legs, and the invention of a limbless tadpole; obviously neither of these orders has any direct relationship with reptiles. The Caudata are perhaps less conspicuously specialized, but one glance at the skull shows striking reduction from that of labyrinthodonts, and more detailed study makes it clear that scarcely any feature of their anatomy is really primitive. They are just as far from reptiles as the Salientia, but in different ways.

Reptilia now living reflect only a feeble glimmer of the glory that was theirs in Mesozoic days, when some twenty orders dominated land, sea, and sky. We shall look first at the living, then at a few of the major extinct groups. Reptiles as a whole wear either scales or plates or both. Their skins are dry since they lack the mucous glands of Amphibia; they either lay large eggs on land or give birth to active young (viviparity); they possess, like birds and mammals, the amnion, which is a membranous sac surrounding the developing embryo. The aquatic habits of many reptiles are secondary, representing a return from the land rather than a persistence of Amphibian ways. In size, form, and behavior the reptiles are the most diverse of all vertebrate classes except perhaps the mammals.

Lizards and snakes (Figs. 30, 31) make an order, Squamata, containing the majority of living reptiles. A lizard (suborder Lacertilia) is typi-

Fig. 30. Tegu lizard. (New York Zoological Society photo.)

Fig. 31. Southern copperhead, *Agkistrodon mokasen*. (American Museum of Natural History photo.)

cally coated with small scales, has four agile legs, eyelids, ear openings, a double-jointed articulation of its jaws (by way of the movable quadrate bone), and small teeth set on the edges of its jaw bones, not in sockets. Only two species, belonging to the genus *Heloderma* (the beaded

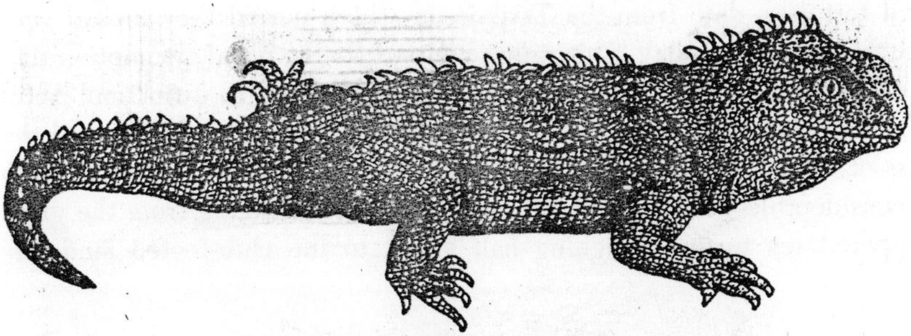

Fig. 32. *Sphenodon punctatum.* (After Gadow.)

lizards or Gila monsters) are poisonous. A few are viviparous. A few more, like the "glass snake," have no legs and may be mistaken for snakes. The great sea-lizards (mosasaurs) of the Cretaceous period were the largest and most completely aquatic Lacertilia.

Snakes (suborder Ophidia) have large ventral plates (except some primitive burrowing snakes), no eyelids, no ear openings, and an exaggerated freedom of the quadrate bone, so that the mouth can be stretched immensely. In the boas and the above-mentioned burrowing snakes are found vestiges of hind limbs. The viviparous habit is common among snakes, and a substantial number are poisonous.

Fig. 33. A sea turtle, *Caretta caretta,* just hatched.

Probably the lowest common denominator among living reptiles is the New Zealand *Sphenodon* (Fig. 32) which looks like a lizard but belongs to the more primitive order Rhynchocephalia. Its principal outward peculiarity is a small parietal eye, with a lens and retina, in the top of its head. This organ occurs as a vestige in a number of kinds of vertebrates. *Sphenodon,* to the sorrow of biologists, is on the verge of extinction. Other Rhynchocephalia are known as fossils as far back as the Triassic.

Since we are looking only at a few terminal twigs of the reptilian family tree, the connections will remain obscure until we climb down the branches towards the trunk. Two other twigs, however, call our attention first, the orders Crocodilia and Chelonia. The latter, turtles (Fig. 33) and tortoises, date from the Triassic, in which period they looked very much as they do today. This rather contradicts the usual assumption that a slow-going, heavily armored animal is on the way to extinction. Actually the wide casket of flattened bones overlain by horny plates, and the loss of teeth (another sign of "degeneration"), have not interfered with a considerable versatility of structure and habit, ranging from the great flippered sea turtles, weighing half a ton, to the club-footed land tor-

Fig. 34. American alligator, *Alligator mississippiensis.* (New York Zoological Society photo.)

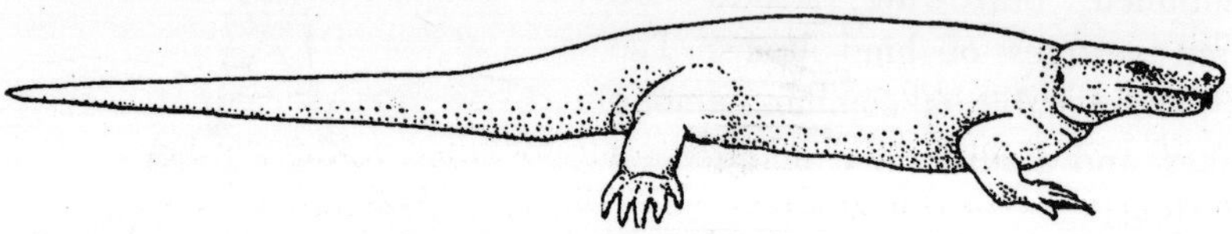

Fig. 35. A cotylosaur, *Limnoscelis.*

toises and the little domed box-turtles. All kinds lay eggs, for which even the most completely aquatic must crawl out of water and dig in the ground.

Crocodilia (Fig. 34) number only a few species, all large, all much alike in form and habits. They are aquatic or amphibious, predatory, and equipped with teeth in sockets. The skin is covered with plates, largest on trunk and tail. A bony palate separates the mouth from the nasal passages which run far back, so that the animal can breathe by putting the tip of its snout out of water, yet its mouth may be open below the surface.

The root of the reptilian tree was among the Carboniferous labyrinthodonts. At first no clear distinction can be made between these Amphibia and the reptilian order Cotylosauria (Fig. 35), the "stem" group. Cotylosaurs were bony headed, short legged, but presumably terrestrial all their lives. From them apparently several other orders arose before the Mesozoic era began. Rhynchocephalia, already mentioned, were one branch, and as we shall see in connection with the skull, the Squamata are very closely related to these.

Turtles were another by-product of cotylosaur expansion. Still another line, containing the strange pelycosaurs and mammal-like reptiles (Therapsida), led directly to mammals during the Triassic period. We shall

Fig. 36. A thecodont, *Saltoposuchus*.

deal with the Therapsida elsewhere, for this transformation is of great interest.

The only other branch from the prolific cotylosaur stem that we need deal with now is that which produced the so-called ruling reptiles of the Mesozoic; this term is a gracious compliment to their numbers and size, not to their political ability. Several orders came out of this branch, with a common ancestry apparently in the Thecodontia (Fig. 36). This name refers to the teeth occupying sockets in the jaws (unlike those of lizards, for instance); the thecodonts looked a little like crocodiles, but often travelled on enlarged hind legs. Crocodilia, one of their descendant orders, long ago gave up the bipedal habit, but the dinosaurs often did not.

Dinosaurs are placed in two orders which may have started separately from thecodonts back in the early Triassic. The Saurischia were those dinosaurs which kept a three-way pelvis, like most reptiles; some of them, the suborder Theropoda (Fig. 37), remained bipedal and carnivorous, culminating much later in the fearful *Tyrannosaurus*, largest of all terrestrial flesh eaters. Others, the Sauropoda, resumed the four-legged position and became herbivorous. In the late Jurassic their line produced the gigantic *Brontosaurus* (Fig. 38), *Brachiosaurus*, and *Diplodocus*, which bore about the same relationship to the theropods that a bison has to a wolf. Whether animals so ponderous could actually carry their weight

Fig. 37. A theropod dinosaur, *Gorgosaurus*.

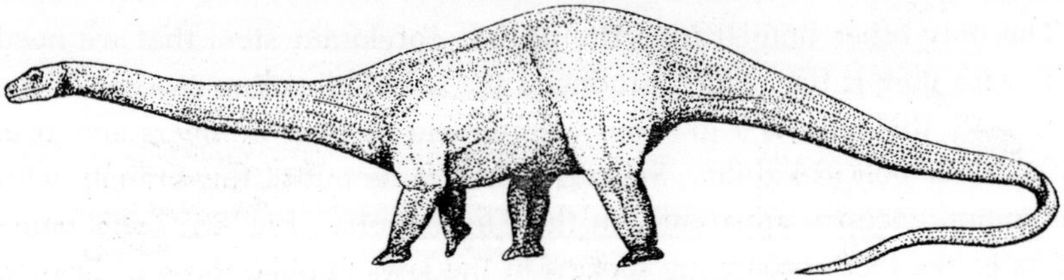

Fig. 38. A sauropod dinosaur, *Brontosaurus*.

without the buoyancy of surrounding water has been disputed, but the imprints of sauropod feet on what must have been exposed mud suggest that they could. The larger ones may well have been amphibious, like the hippopotamus among mammals.

Meanwhile another order of dinosaurs appeared and branched prolifically during the Mesozoic, the Ornithischia. As the name implies the pelvis was like that of birds. Again the bipedal habit came first, as in the duckbill dinosaurs (Fig. 39), but was superseded in the weird *Stegosaurus* (Fig. 40), Ankylosauria, and Ceratopsia (Fig. 41) by an obviously clumsy quadruped posture. The entire order seems to have been herbivorous. No dinosaurs, indeed no ruling reptiles except the Crocodilia,

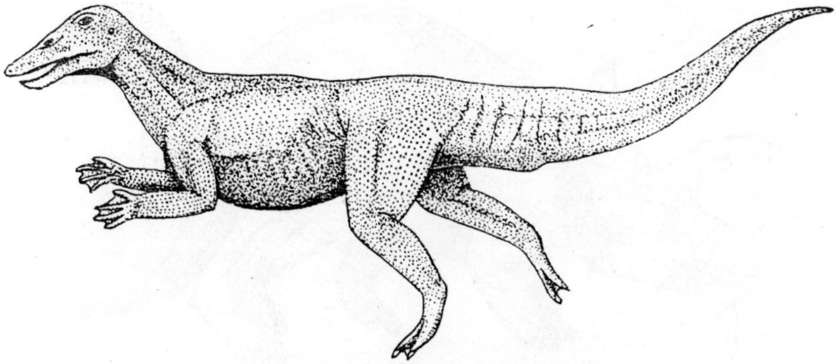

Fig. 39. A duckbilled dinosaur, *Trachodon.*

Fig. 40. *Stegosaurus.*

survived beyond the end of the Cretaceous. Extinction seems to have overtaken these and six or eight other Mesozoic orders that we have not discussed, at about the same time.

Before leaving the reptiles we must speak of two groups that everyone has heard about, the ichthyosaurs (Fig. 42) and the pterosaurs (Fig. 43). The first of these became porpoiselike in habits, size, and form. The feet made flippers, the nostrils migrated to the top of the head (as in whales), and the ichthyosaurs are known to have caught cephalopods. But the tail was unlike that of whales, being vertical instead of horizontal, and with the tail vertebrae curved down abruptly into its lower lobe. Skeletons of embryos found in place in adult bodies show that these marine reptiles

Fig. 41. A ceratopsian dinosaur, *Triceratops.*

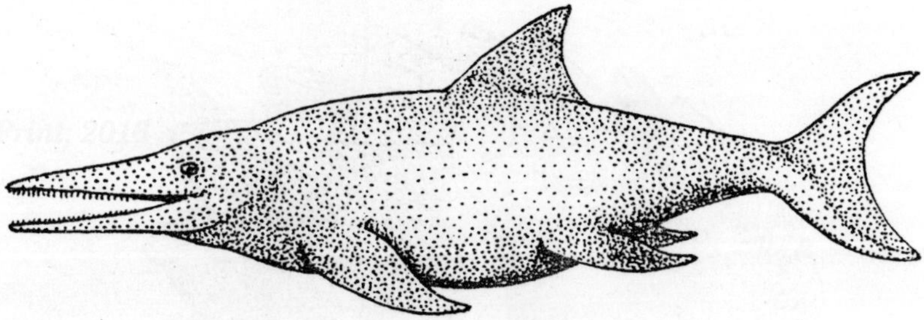

Fig. 42. *Ichthyosaurus,* a whalelike marine reptile.

were viviparous; obviously they could neither drag themselves out on land to lay eggs nor lay viable eggs in the water.

Perhaps the strangest of all reptiles that ever lived were the Pterosauria, familiarly called "pterodactyls." They belong somewhere among the offshoots of the thecodonts, hence were ruling reptiles. They were the only reptiles that ever accomplished real flight. Presumably they soared and skimmed much like an albatross or a pelican, and either dove for fish or snatched dead ones from the water, like a gull. But the wings of pterosaurs did not resemble those of birds. The last finger, greatly lengthened, supported a leathery wing of skin. Some pterosaurs were little fellows, but a few, with wings outstretched, exceeded twenty-five feet. Their skeletons were remarkably light and compact. *Pteranodon* seems to have thrown even its teeth overboard, but it grew a long bladelike crest on back of its skull, which may have acted as a vane to keep the long beak to windward.

To label a bird (class *Aves*) a "glorified reptile" emphasizes a number of points of similarity between the two, but it tends to obscure in our

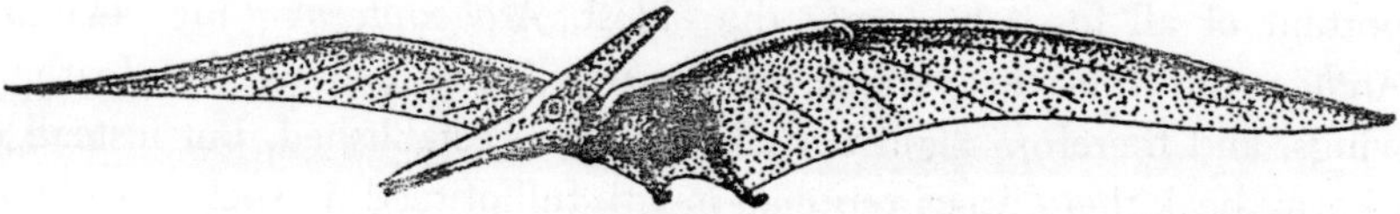

Fig. 43. One of the largest flying reptiles, *Pteranodon*. (Redrawn from Dunbar.)

Fig. 44. The earliest known bird, *Archeopteryx*.

minds the fact that there were almost as many radical changes involved in coming to modern birds as in coming to modern mammals. Some of these changes were physiological: a constant temperature well above 100° F; highly specialized sense organs and brain; remarkable neuromuscular coördination associated with flying and balancing. Outwardly the most obvious characters of a bird are its feathers (modified scales) and the feather-bearing wings, in which the supporting part is an abbreviated fore limb carrying only vestiges of fingers. Birds have no connection with Pterosauria.

Bird skeletons are delicate and seldom found fossilized. Therefore extinct birds add relatively little to our knowledge of the class. Most important of all fossil birds are the oldest, *Archeopteryx* (Fig. 44) and *Archeornis*, from the Jurassic of Bavaria. These tell us that feathers, wings, and therefore flight had already been established, but instead of a horny beak there was a reptilian mouth full of teeth in sockets (affinity with archosaurs), and the tail, unlike that of any later bird, was a long column of vertebrae with feathers projecting from both sides. By this tail we may set off the subclass ***Archeornithes***, for all other known birds have shortened it to, at most, a stub, the pygostyle, from which feathers extend back; the latter subclass, ***Neornithes***, numbers more than 20,000 species. Teeth, by the way, still persisted in the Cretaceous sea birds, *Hesperornis* and *Ichthyornis*, although these were modern as to tail.

Fig. 45. Hesperornis, a toothed Cretaceous bird with vestigial wings.

The Neornithes contain three very unequal divisions: ***Odontognathae*** (the Cretaceous toothed birds, Fig. 45); ***Palaeognathae*** (the ostrichlike flightless birds, Fig. 46, and the South American tinamous, which do not fly well; the validity of this division has been questioned); ***Neognathae***, all the rest. Curiously enough, the pattern of bones in the roof of the mouth gives a better clue to broad relationships than any of the bigger, more superficial differences with which we are familiar. Once the modern bird type appeared, say back in the Eocene or earlier, it was possible to evolve from it innumerable adaptive groups by means of relatively minor changes in shape, proportions of parts, size, and behavior (Figs. 47, 48, 49). Thus long-legged waders, web-footed swimmers, terrestrial running birds, seed eaters, insect eaters, scavengers, birds of prey, and so on, arose ***divergently***, and their characters superficially were duplicated again and again. For instance, swimming and diving birds include: (1) penguins, (2) grebes and loons, (3) ducks, (4) petrels, (5) gulls, terns, and skimmers, etc., yet probably each of these groups attained its present habits independently of the others. Similar ***parallelism*** occurs in mammals, and indeed in any great class of animals.

Fig. 46. South African Ostrich. (New York Zoological Society photo.)

Mammalia are warm-blooded animals with hair, which feed their young after birth by milk produced in paired mammary glands. They usually feed the young before birth by means of a placenta, which acts as intermediary between the embryo and the blood supply of the mother's uterus; but the very primitive monotremes (duckbill and echidna) of Australia lay eggs, as the first mammals must have done. Numerous anatomical details distinguish mammals from reptiles: the four-chambered heart, the single (left) aorta, the three tiny sound-transmitting bones of the ear, the fusion of a number of reptilian skull bones to make the complex mammalian temporal bone, the loss of several lower-jaw bones leaving just a dentary on either side—these mark the class that came, with birds, to share the world released from ruling reptiles at the

Fig. 47. A Shoebill stork, *Balaeniceps rex.* (New York Zoological Society photo.)

close of the Mesozoic era. Like birds, the mammals originated long before that. They came from certain primitive reptiles of the Triassic period, the Therapsida, or "mammal-like reptiles," and the stages of transition are now fairly well known.

We need not deal with these except to say that therapsid reptiles walked with their legs beneath, as mammals do, that their teeth were differentiated into incisors, canines, and molars, and that they showed the beginning of the mammalian hard palate. Of three extinct and very rare subclasses of Mesozoic mammals we shall mention none but the ***Pantotheria*** (***Trituberculata***), whose molar teeth bore three cusps each, in a

Fig. 48. Black-bellied plover. (American Museum of Natural History photo.)

pattern suggesting the source of the molar cusps of modern mammals. The trituberculates probably ate insects.

The ***Monotremata*** (Fig. 50) are different enough from all others to warrant a subclass, ***Prototheria,*** containing only the duckbill (*Ornithorhynchus*) and spiny anteaters (*Tachyglossus* and *Zaglossus*). Their mammary glands lack nipples; their digestive and reproductive systems have a single opening (hence the name monotreme, "one hole"); the mouth is equipped with a beak, while the teeth are degenerate; the shoulder girdle contains bones not known otherwise in mammals; and, worst of all, no early fossil monotremes are available to tell just how the living ones accumulated their awkward combination of characters.

The ***Marsupialia*** or pouched mammals may be placed with the placentals in the subclass ***Theria.*** They comprise an infraclass (***Metatheria***)

Fig. 49. Bluejay. (American Museum of Natural History photo.)

of relicts, limited, except for American opossums (Fig. 51), to the Australian and Papuan region. Long isolation from the other continents permitted them to evolve a large number of diverse types, superficially parallel to many orders of placental mammals. For instance, besides the familiar kangaroos and wallabies there are flesh eaters (Tasmanian wolf), digging types (wombat, and marsupial mole), several arboreal groups (Fig. 52) including the flying phalangers, and even a huge grazing marsupial (*Diprotodon*, of the Pleistocene period) analogous to the heavy ruminants. The young of marsupials are extremely small and undeveloped when born (those of a six-foot kangaroo may be only an inch long), but they crawl unaided into the pouch and attach there to the nipples, where they remain through a sort of second embryonic phase. In view of this demonstrated capacity to evolve into many adaptive types, it seems

Fig. 50. *Echidna* or spiny anteater, a monotreme, with its egg. (American Museum of Natural History photo.)

Fig. 51. Opossum and young. (American Museum of Natural History photo.)

that the main encumbrance which held the marsupials back, and allowed them to be beaten or by-passed by the rising placentals, was their second-rate brains. No marsupial pretends to be intelligent; it is somewhat better off than a reptile, but has less ability to profit by experience than the average placental.

To the ***Placentalia*** (infraclass ***Eutheria***), then, goes nearly all credit for the rise of intelligence during the Cenozoic era. Increasing size of the

Fig. 52. Koala, an arboreal marsupial. (American Museum of Natural History photo.)

forebrain and complexity of the cortex can be traced independently in several lines of advanced mammals, for example the horses, elephants, carnivores, and especially the monkey-ape-man order, Primates.

Our purpose now will be served best by showing the relationship between form and ways of life in some of the important placental orders. The Insectivora (shrews, moles) probably are nearest to the base of the subclass, although it takes a form like the tree shrews, or elephant shrews (Fig. 53), to demonstrate this.[1] Feet and teeth are the best indicators of mammal specialization; the tree shrews have plain, clawed, five-toed

[1] According to Simpson, 1945, tree shrews are Primates.

feet, while their teeth retain the ancestral number, 44 (3 incisors, 1 canine, 4 premolars, 3 molars, on each side of each jaw).

Bats (order Chiroptera) are close to insectivores, but have modified the forelegs into wings; each wing is a thin sheet of skin spread between the delicate, elongated fingers (except the thumb), and of course reach-

Fig. 53. Elephant shrew, *Elephantulus rufescens.* (New York Zoological Society photo.)

ing the side of the body, hind leg, and tail. Most bats are small and catch insects in flight, but in the Malay-Australian region there are much larger fruit-eating bats, the "flying foxes" (Fig. 54).

Unquestionably the Primates also arose from very early arboreal insectivores. The fossil record of Primates is poor, however, and we shall simply review the kinds now living: (1) Suborder Prosimii. A lemur (Fig. 55) looks like a cross between a squirrel and a monkey. To see lemurs in the wild go to Madagascar; a very few occur in Africa and India. Lemurs live in trees, feeding mainly on insects. They show the beginning of flattened nails, in contrast to pointed claws, a tendency to bring the eyes forward on the face so that vision is partly binocular, and a certain amount of brain enlargement. But, compared with monkeys and apes, they are not intelligent. The tarsier or specter (Fig. 56) is a harmless little arboreal animal with enormous eyes, living in the Malay jungle. It is active at night, and its fingers and toes bear expanded tips

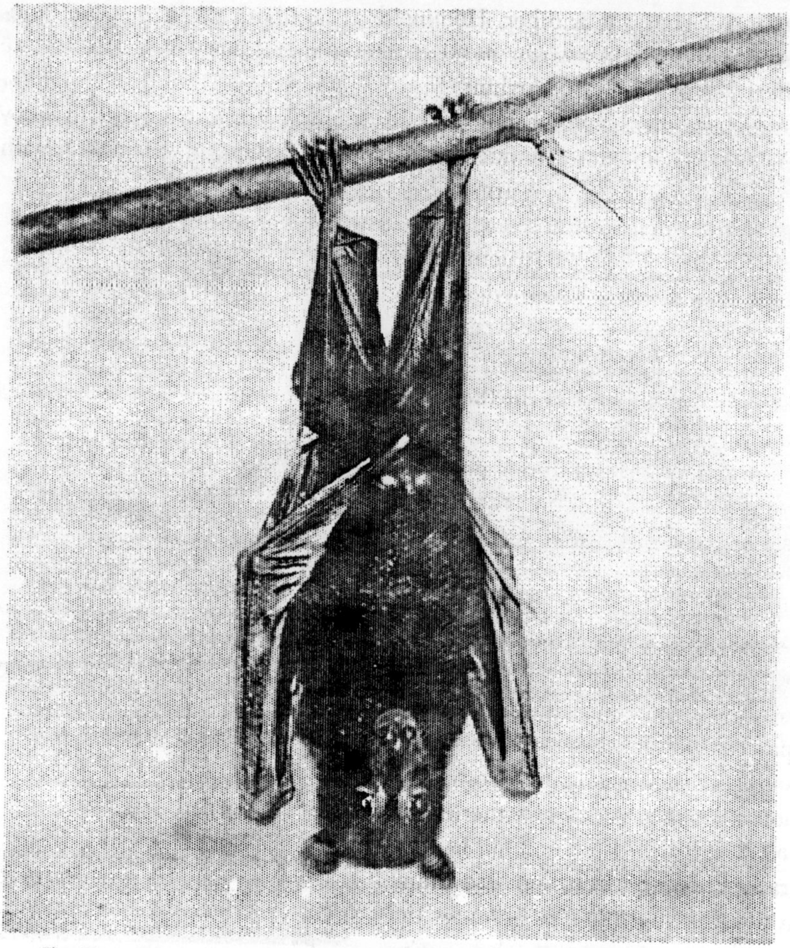

Fig. 54. Indian fruit bat, *Pteropus giganteus.* (New York Zoological Society photo.)

like those of a tree frog, to help in climbing. (2) Suborder Anthropoidea. Here we must use another category (superfamily) since the monkeys of the New World are distinct from the Old World monkeys, apes, and man. New World monkeys, with a broad nasal septum and (often) a prehensile tail, are the superfamily Ceboidea (= Platyrrhini). Of these the marmosets, Hapalidae, are most primitive, having bushy, squirrellike tails, and claws on every digit except the hallux. Cebidae are the spider monkeys (Fig. 57), howlers, capuchins, and a few others, most of them with prehensile tails. Old World monkeys, including the short-tailed terrestrial baboons (Fig. 58), are in the superfamily Cercopithecoidea; apes and man belong in another, Hominoidea, and both of these superfamilies, agreeing in the narrow nasal septum and lack of prehensile tail, are collectively known as Catarrhini.

Apes, like men, have no tail, and are generally larger than any monkey (Fig. 59). Their arms are much longer than their legs, in correlation with the habit of brachiating or arm-swinging from branch to branch. They approach quite closely in structure and intelligence to man. The modern apes are: gorilla (largest) and chimpanzee, in Africa; gibbon (smallest, several species) and orangutan, in the Indo-malayan region.

Fig. 55. Ring-tailed lemur, *Lemur catta*. (New York Zoological Society photo.)

The human stock most probably arose one or two million years ago in Asia or Africa. We shall speak of the early types of man in our last chapter. Although man is, both physically and mentally, a giant among primates (Fig. 60), most of his peculiarities are distinctly foreshadowed in the apes. Among anatomists there is little doubt of our arboreal ancestry, remote though it was. And there is no doubt that the ability of the hand to grasp objects by opposing the thumb to the fingers (as most Primates do) contributed more than any other single factor to developing the brain, for Primates constantly explore their environment by handling things, turning them over, associating touch with sight, and, sometimes, remembering the association.

We have intentionally included the Primates, with man, in the first part of the placental mammals because in many respects the order is quite primitive. Teeth, for example, and limbs, are obviously less modified than those of the majority of living orders.

Not very distant from ancestral insectivores were the ancestral Carnivora, the flesh eaters, represented today by a great many mammals: (1) Fissipedia, the cats (Fig. 61), dogs, weasels, bears, and all other ter-

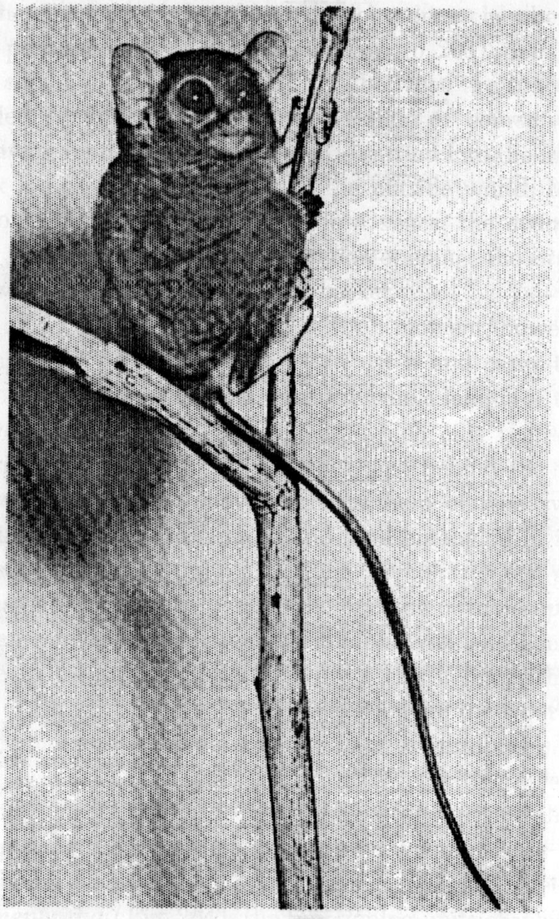

Fig. 56. Tarsier. (New York Zoological Society photo.)

restrial carnivores; (2) Pinnipedia, the seals (Fig. 62), sea lions and walruses; both of these suborders arose from the extinct (3) Creodonta, which were the small-brained carnivores (Fig. 63) of the early Tertiary. Compared with the latter, most living Fissipedia possess larger brains, specialized cutting teeth, greater speed and agility, and a tendency to raise the heel off the ground so as to run on the toes (*digitigrade* locomotion).

Whales and porpoises (order Cetacea) have gone so far in adjustment to marine life that it is very difficult to determine their early relationships, but there is some suggestion of a creodont connection. At any rate,

Fig. 57. Spider monkey, *Ateles ater*. (New York Zoological Society photo.)

Fig. 58. Hamadryas baboon, *Papio hamadryas*. (New York Zoological Society photo.)

Fig. 59. Chimpanzee, *Pan troglodytes.* (New York Zoological Society photo.)

their structural peculiarities are evident: loss of hind legs; conversion of forelegs into flippers; development of horizontal flukes on the tail; migration of nostrils (in toothed whales these are combined into a single opening) to the top of the head, so that a whale can "blow" merely by rolling the back just above the surface; loss of differentiation among the teeth and vertebrae so that the whole series of each becomes essentially alike. The toothed whales (Odontoceti, Fig. 64) are the majority of Cetacea, ranging from small porpoises to the seventy-foot sperm whale, which dives for the giant squid to a depth of half a mile or more. But the greatest of all whales are the Mysticeti (Fig. 65), the whalebone whales, and one of these, the sulphur-bottom, is probably the largest animal that ever lived. It may reach a length of 100 feet, a weight of 150 tons. Yet the food upon which these leviathans attain their bulk is mostly minute swimming Crustacea, strained out of tons of water by dangling fringes of horny "whalebone," taking the place of teeth.

The edentates or "toothless ones" of the order Xenarthra deserve notice for certain very peculiar features. Anteaters (Fig. 66), for instance, are truly edentate, feeding only on ants and termites which they mop up with a slender sticky tongue. But armadillos (Fig. 113), the animals with a flexible armor of bony plates in the skin, and sloths, which climb inverted in trees and have the wrong number of neck vertebrae, are not toothless. Xenarthra occur only in South and Central America, with the exception of some giant ground sloths that went as far north as Canada a few thousand years ago, glyptodonts which came into the southwestern states, and armadillos, which reach Louisiana. The order seems to be a product of the long isolation of South America from the rest of the world during the first half of the Cenozoic era.

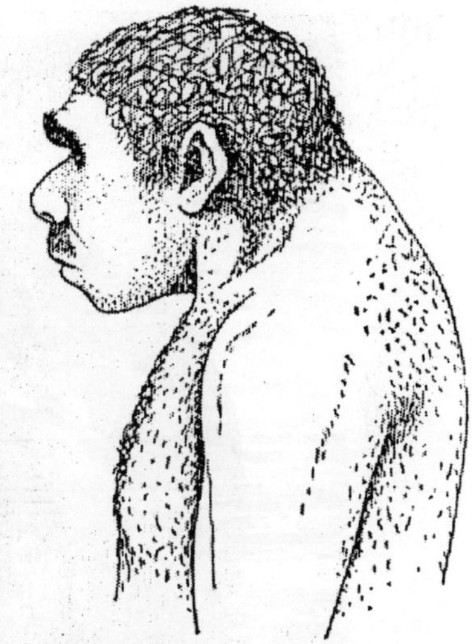

Fig. 60. Restoration of Neanderthal man. (Drawn from bust in Chicago Natural History Museum.)

Much more familiar are the Rodentia, most numerous order of mammals, and the Lagomorpha. The first contains all those mammals that gnaw by means of two upper and two lower incisor teeth: the squirrels, woodchucks, mice, rats, beaver, guinea pigs, porcupines, and their innumerable relatives. There is little information from fossils as to their connection with other mammals. The lagomorphs (rabbits, hares, and pikas) agree

Fig. 61. Leopard. (American Museum of Natural History photo.)

Fig. 62. Fur seal, female. (American Museum of Natural History photo.)

Fig. 63. A creodont, *Sinopa*.

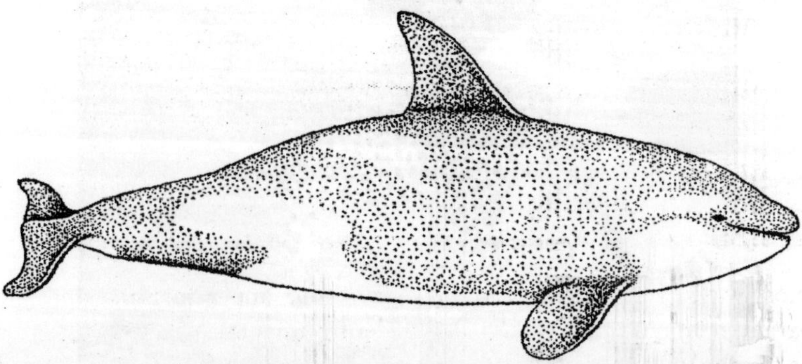

Fig. 64. Killer whale, *Orca gladiator*, one of the Odontoceti. (Redrawn from Parker and Haswell, after True.)

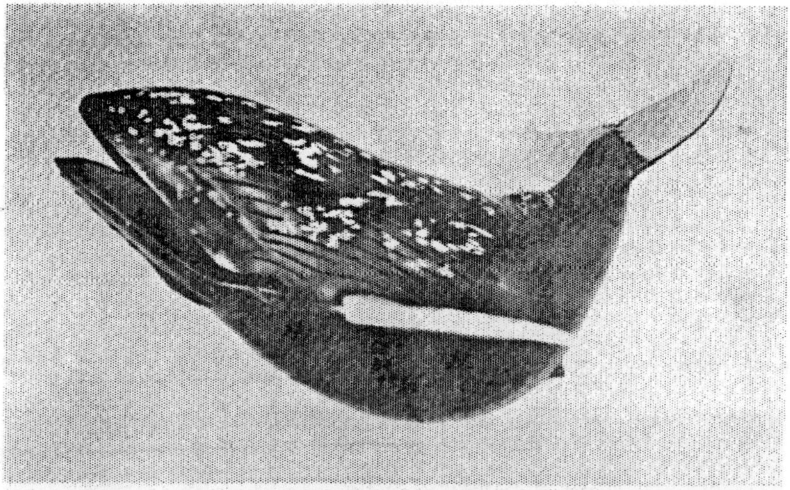

Fig. 65. Model of a humpback whale, *Megaptera nodosa*, one of the Mysticeti. (American Museum of Natural History photo.)

Fig. 66. Great anteater, *Myrmecophaga tridactyla*. (New York Zoological Society photo.)

in most ways with rodents, but they have four upper incisors, and the apparent relationship may actually be quite remote.

As to the odd-toed ungulates (order Perissodactyla), the even-toed ungulates (Artiodactyla), the elephants (Proboscidea), conies (Hyracoidea), and the sea cows (Sirenia), they all have something in com-

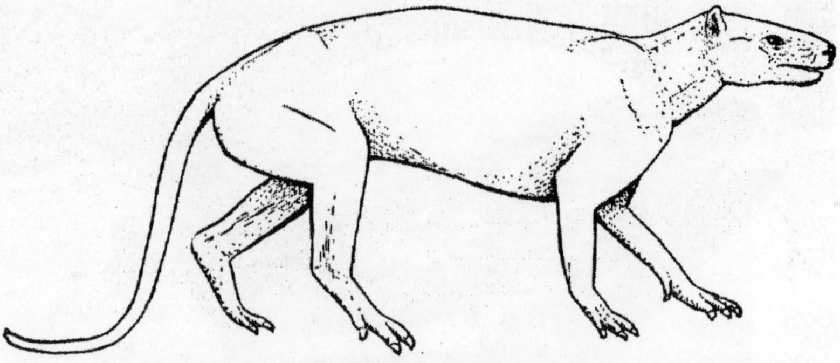

Fig. 67. *Phenacodus*, a condylarth, somewhat like the ancestor of modern ungulate mammals.

Fig. 68. African rhinoceros, *Diceros bicornis*. (New York Zoological Society photo.)

mon, mainly the presence of hoofs in the terrestrial kinds, but their relationships are not close; they have been separate since the Eocene at least. Then, and before, there existed a poorly defined ancestral stock, the Condylarthra (Fig. 67), from which radiated eight or ten different lines of hoofed, herbivorous animals. Some of these failed and died out millions of years ago; others survived through remarkable sequences of progressive evolution.

Perissodactyla today are rhinoceroses (Fig. 68) and tapirs (Fig. 69), with three toes on each foot (four on front feet in tapirs), and horses (Fig. 70), with one toe. Progress in this order has consisted of increasing

Fig. 69. Malay tapir, *Tapirus indicus.* (New York Zoological Society photo.)

Fig. 70. Grant's zebra. (American Museum of Natural History photo.)

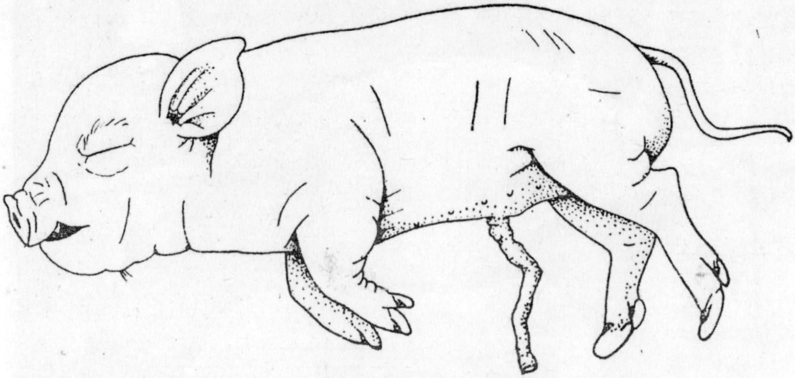

Fig. 71. Fetal pig.

Fig. 72. Newfoundland caribou, *Rangifer caribou*. (New York Zoological Society photo.)

size (the earliest horse could not have weighed over twenty-five pounds), reducing the inner and outer toes of each foot by gradual stages, and slowly improving the form of the molar teeth, to serve as grinders.

Artiodactyla are the two- or four-toed ungulates; the same rules of progress apply broadly to them, but with a great deal more leeway al-

Fig. 73. Manatee or sea cow. (American Museum of Natural History photo.)

Fig. 74. African elephant, *Loxodonta africana*. (New York Zoological Society photo.)

lowed for modification of other features. That is, a pig, deer, hippopotamus, giraffe, bison, cow, camel, and goat are all artiodactyls, and an immense variety of others are living or have lived (Figs. 71, 72).

Sea cows or manatees (Sirenia, Fig. 73) interest us here because they are another example of complete aquatic adaptation, in which the front legs became flippers, the hind legs disappeared, and the tail broadened to a round, horizontal paddle. These huge and harmless animals, like the mermaid legends to which they may have given rise, are on the way out.

Elephants and their extinct relatives, the mammoths and mastodons, not only show an extraordinary combination of characters, but also a wonderful array of diverging evolutionary lines that began with Eocene ancestors no larger than sheep. Some features of an elephant (Fig. 74), such as his pillarlike limbs, short neck, long and agile trunk, and thick hide (hence pachyderm), are probably correlated with his bulk; but the molar teeth, tusks, and his intelligence are equally remarkable.

A number of minor groups of mammals have gone unmentioned here, but we may have said enough to suggest how complex is the history of the class, and how the position of man is that of a twig somewhere near the top of an infinitely branching tree. The importance of that twig, of course, is not in its location on the tree, but in the kind of fruit it bears.

CHAPTER II

Development

GENERAL

The organism is no less an individual in the egg or the embryonic stages than when development is finished. At no point can anatomy be distinguished from embryology. With many animals the developmental stages are practically the whole life-span. To mature means to achieve all the structures and functions normally used in living and reproducing.

Although the parts of an animal may look different, or may be unfinished, the study of anatomy in an early stage is just as valid as in a later one, for immature or embryonic organs often throw a flood of light on puzzling features of the adult. Sometimes ancestral traits appear vaguely and then vanish as development continues. In any case embryology shows the organism undergoing an amazingly swift and delicate transformation of parts. New tissues, new organs, new systems appear and arrange themselves as if modeled by invisible hands. Certain structures arise, working perfectly for a time, and then may disappear or transform as the way of life changes. This business of development goes most rapidly in the beginning, when a few hours may reveal tremendous advances, but it slows down gradually until at maturity the only significant changes are those of normal physiology, of healing wounds, or of wearing out.

We used the word "development" in the first chapter to express the evolutionary transformation of some structure in the history of a group of animals. We now use it in another, stricter, sense, for ***the changes during the life of an individual from the beginning to maturity.*** Development is not growth. Growth means increase of mass. An organism can

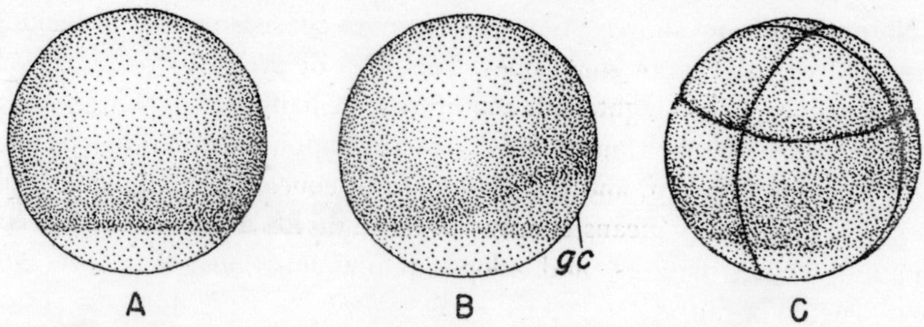

Fig. 75. Egg of frog, *Rana pipiens.* A, unfertilized; B, fertilized (zygote); C, 8-cell stage of cleavage. gc, gray crescent.

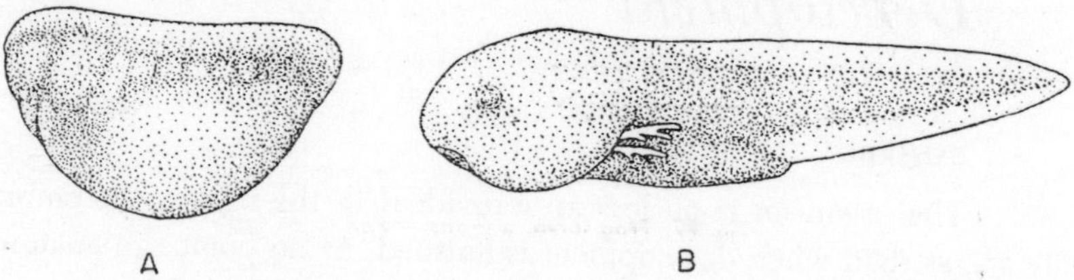

Fig. 76. Frog embryos. A, as somites appear; B, at time of hatching.

grow without developing; more often it develops without growing; but as a rule they go together. So development means a change of form, of materials, the construction of new parts not previously present, or sometimes even loss of parts. It is the making of cells, tissues, organs, and systems from the undifferentiated substance of the egg. It is the theme of embryology.

Every kind of chordate, including those that give birth to fully formed young, goes through a developmental sequence of which the first stage is an *egg* (Fig. 75). From this single cell, by transformation of materials in it or brought to it, there follow the elaboration of parts and refinement of functions which presently make it self-sufficient, independent. Up to this time (before birth, or before hatching) it is an *embryo* (Fig. 76). Of course much development is yet to come, sometimes gradually, as with the infancy, childhood, and adolescence of man, or sometimes with abrupt transformations which go with a radical change in the ways of life, as when a tadpole in a few days becomes a frog. ***Postembryonic*** development may therefore include a stage called the ***larva*** (Fig. 77), which may have little resemblance to the adult. Most amphibians and fishes have larvae, but reptiles, birds, and mammals do not. The change from larva to adult is ***metamorphosis.*** Occasionally, as with some fishes, there is more than one larval stage.

No one will now suppose that development consists merely of making large individuals out of small ones. This idea of *preformation* was current, however, in the eighteenth century (see Chapter XII). Later, when microscopes were used more widely, preformation was abandoned as a theory of development, and the more natural concept of *epigenesis* took its place. Epigenesis means simply that an individual develops by accumulating new structures, and adopting new functions, that were not there when it began.

Is there any general rule that animals follow in this epigenetic development, any "law" that makes it possible to understand embryology as a whole? The broadest generalization perhaps is von Baer's law, that different kinds of animals resemble one another more closely in earlier than in later stages of development. There are exceptions, but the statement is mainly true. A close corollary, subject to still more exceptions, is Haeckel's famous "Biogenetic Law," that animals tend to repeat, very hastily of course, during their own development, the changes undergone in evolution by their ancestors from the earliest times. This obviously would explain the close resemblance between different types in early embryonic stages, and it could (and sometimes does) shed a good deal of light on what happened in the evolution of special characters. The trouble is that larvae, embryos, and even eggs may become so much modified in adaptation to circumstances which they meet (but which the adult does not and never did meet) that they refuse to remain photostatic copies of ancient history. Another difficulty in making Haeckel's rule work is that embryos are capable of taking short cuts, wholly spoiling the effect of continuity which we would look for.

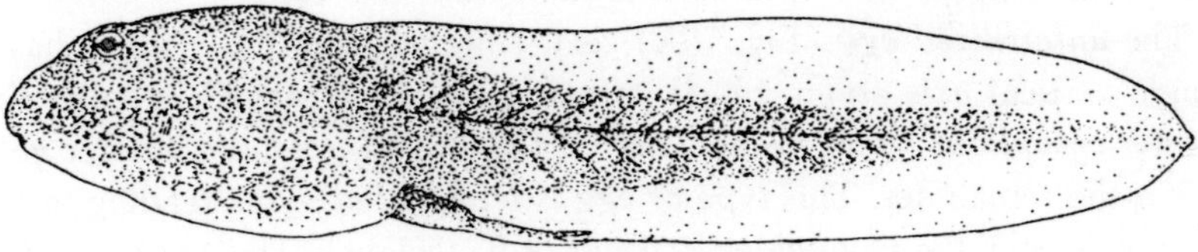

Fig. 77. Frog larva, a tadpole, *Rana*.

To the comparative anatomist an understanding of embryology is essential for the precious, if intermittent, light it sheds on evolution, for solving puzzling relationships of adult structures, for correct interpretation of many highly complex organs as well as some useless vestiges, and for realization of how heredity works, through epigenesis, to renew again and again the marvel of a finished organism.

DEVELOPMENT OF A FROG

We shall use the frog to illustrate the sequence of structural changes from egg to adult because it is a familiar animal, easy to observe, and shows some important phenomena such as metamorphosis. On the other hand, frogs are not primitive or generalized vertebrates. In their development, as well as anatomy, they are a peculiar side branch of the vertebrate line, and thus do not adequately represent vertebrates as a whole. The commonest laboratory frog in this country is *Rana pipiens;* many others agree with it in most of the details that follow.

The ***unfertilized egg*** (Fig. 75A), as it emerges from the oviduct, has a single vertical ***axis*** about which its materials are arranged radially, the lighter ***cytoplasm*** (with ***pronucleus***) at the ***animal pole,*** the heavier ***yolk*** filling the remainder. This type of egg is ***telolecithal*** (yolk tending to occupy one pole, the ***vegetal,*** on account of its weight). The egg has ***radial symmetry.***

Immediately following fertilization, that is, the entry of the sperm through the surface of the egg, there is a gradual concentration of cytoplasm in a crescent area lying a little below the equator on one side (Fig. 75B). This ***gray crescent*** obviously creates a new axis at right angles to the first, for now it is possible to say the gray crescent is ***posterior,*** and the opposite face ***anterior.*** By the same token the egg now has a right and left side and a dorsal and ventral aspect, even though no structures have appeared yet. It has, therefore, ***bilateral symmetry*** as soon as the gray crescent appears.

Cleavage (Fig. 75C) is the division of the fertilized egg (zygote) into the first embryonic cells, the ***blastomeres.*** Cleavage in the frog is ***holoblastic,*** that is, the cleavage planes pass entirely through the egg including the yolk, which would not be true in an egg with larger yolk than this. Early cleavages begin with grooves that sink into the surface of the animal pole, the ***first*** plane across that surface through the pole, the ***second*** also through the pole but at right angles to the first (now there are four equal blastomeres). The ***third*** cleavage plane is ***horizontal,*** that is, a circular groove around the egg somewhat above the equator. Now there are eight blastomeres, the upper four being smaller than the lower. The fourth and succeeding cleavages lose their regularity of position, but maintain a rhythmic time sequence. Presently cleavage comes in planes parallel to the surface so that new blastomeres lie under the surface in several layers. And always cleavage is most rapid towards the animal pole, being slowed by the yolk below, with the result that the upper blastomeres become rapidly smaller and smaller, while the lower are still fairly large. The rounding-up of egg substance into cells tends to create intercellular space, so that shortly the ***blastula*** is formed (Fig. 78), a stage in which there is a discrete cavity lying within and somewhat above the center of the ball of cells. During this time no growth, of course, has occurred.

Recalling now the ***gray crescent,*** we find that it assumes a fundamental importance, for along its lower edge appears a small horizontal

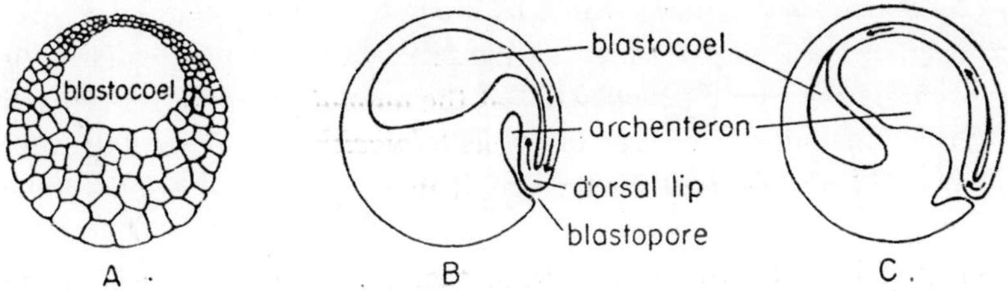

Fig. 78. Diagrams of A, blastula; B, C, movements taking place in formation of the gastrula of frog.

groove, below the equator. This thrusts deeper into the blastula, its lips remaining closed yet gradually stretching farther and farther horizontally. This slit is the ***blastopore*** (Fig. 78B, C). It marks the posterior aspect of the future embryo. The cells of its ***dorsal lip*** move slowly inward and fold back and up in a layer lying against the inner surface of those that still cover the upper half of the egg. At the same time as this involution is going on there is an overthrust (***epiboly***) or pushing down of the dorsal lip towards the vegetal pole, like the dropping of a curtain over the yolk cells. The blastopore continues to stretch sideways until it completely encircles the egg, now having moved nearly to the bottom of the yolk. At this time we see with the naked eye a round white spot beneath, which is the nearly enclosed yolk, and say that the egg (or embryo) is in the yolk-plug stage. The process described above is ***gastrulation;*** during it, the embryo is a ***gastrula.*** The most important achievement of gastrulation is to make an inner layer of cells, containing a new cavity, the ***archenteron*** (primitive gut), whose temporary opening is the blastopore. That portion of the invaginated cells which forms the roof of the archenteron is the ***chordamesoderm,*** while that of the floor and sides is ***endoderm.*** Cells remaining on the outer surface are ***ectoderm.*** Strictly speaking, the gastrula has ***two germ layers,*** for chordamesoderm and endoderm are not at first distinguishable.

Many years ago Spemann showed that the cells of the dorsal lip, coming in as the chordamesoderm to lie against the ectoderm on its inner side, seem to act in some way as a controlling influence on the future development of that dorsal ectoderm, for it now begins the process leading to development of the central nervous system. Further, it will ***not*** do so

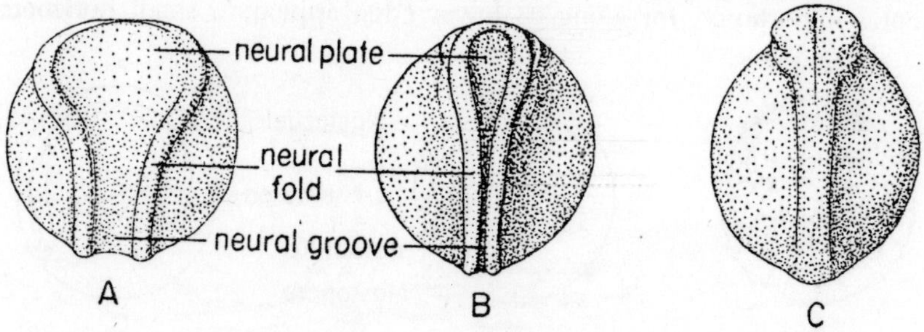

Fig. 79. Steps in formation of the neural tube, frog.

if the chordamesoderm has been cut out or prevented from occupying its normal place. And finally, chordamesoderm (or dorsal lip) transplanted to a spot elsewhere beneath the ectoderm of an embryo in this stage will induce that particular area to develop a prospective central nervous system, even though such may already have been induced in the regular position. So Spemann called the dorsal lip an ***organizer,*** and both it and other more recently discovered organizers have been the subjects of innumerable experimental studies. We shall mention one or two more instances later, in the frog.

Development of the central nervous system (Fig. 79) starts by the appearance of a pair of low longitudinal ridges in the dorsal ectoderm, which meet in a semicircle anteriorly; these ***neural folds*** heighten and come closer together, assuming the shape of a tennis racket, with a deep ***neural groove*** down the handle. Presently this groove closes, making a tube lined with that part of the ectoderm which lay between the neural folds, as is best seen in a cross section. The ***neural tube***, however, closes more slowly anteriorly, leaving a much larger cavity there, where the spread part of the tennis racket lay. Finally we have a closed, tubular, central nervous system with a bulblike enlargement anteriorly, which, of course, is the prospective brain. ***Neurulation*** is this process, and the phase during which the folds and groove are patent is the ***neurula.*** At this stage the embryo begins to lengthen. It becomes oval and somewhat flattened on the sides, but its bulk is no greater than that of the single egg cell. The yolk is being consumed by tissue formation.

Internal changes are taking place in the roof of the archenteron (chordamesoderm) which we must sketch briefly (Fig. 80). Along the median line a rodlike organ forms, the ***notochord;*** its position is parallel to and just beneath the neural tube. Mesoderm cells lying to its right and left

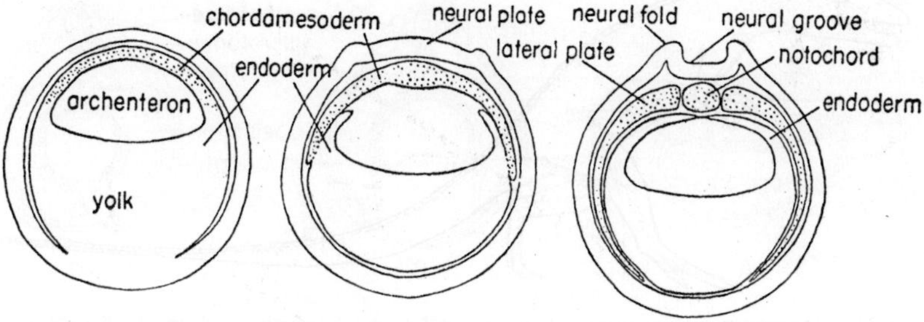

Fig. 80. Cross sections of frog embryos showing formation of mesoderm (stippled).

(***lateral plates***) divide in such a way as to produce two layers with a narrow space between. At the same time the endoderm of the side walls grows dorsad and mediad beneath mesoderm and notochord, making a complete endodermal roof for the archenteron. We can now say the embryo has three germ layers, for mesoderm is sandwiched between ectoderm and endoderm. The last will produce the mucosa or inner lining (only) of the digestive tract, and most of the liver and pancreas. The ectoderm will produce epidermis, lining of mouth and anus, central nervous system, and certain parts of the paired sense organs (eyes, ears, nasal cavities). The mesoderm will produce virtually all other structures in the body.

The lateral plates, extending laterad and downward, between endo- and ectoderm, approach the midventral line as the last yolk is being used up, and thus they envelop the archenteron. The space between the inner (***splanchnic***) and outer (***parietal***) layer of mesoderm is the ***coelom*** (Fig. 81).

The parietal mesoderm develops a shelflike fold inward, partially and then completely separating the dorsal part (***epimere***) from the ventral (***hypomere***); henceforth the latter contains the *peritoneal* cavity, which widens, remaining undivided. The coelom of the epimere, however, narrows on account of a great thickening of its inner wall, and becomes divided into a lengthwise series of cavities. These boxlike units are the ***somites.*** Their inner, thickened walls become the *myotomes* or prospective body muscle segments, while the outer layer is called the ***dermatome*** because it enters primarily into the formation of dermis (the deep layer of the skin). The embryo is now, in part, a segmented animal. The somites show externally as a series of slight swellings along the dorsolateral surfaces. The two-layered wall between each somite and the next is

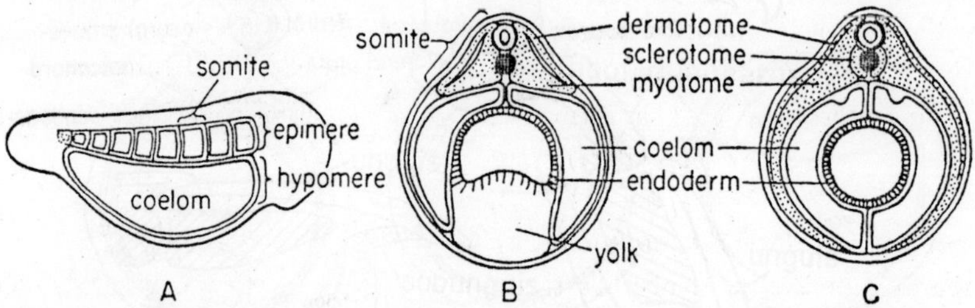

Fig. 81. Diagrams to show development of somites and myotomes in frog. A, lateral view, as transparency; B, C, cross sections.

a myoseptum (later a white sheet of connective tissue separating one muscle segment from the next). Now the myotomes begin to grow laterally and down, crowding between the ectoderm and the parietal peritoneum, until they finally reach the midventral line. Here they meet but do not fuse with the corresponding myotomes of the opposite side. The double layer of mesoderm so formed along the midventral line becomes a white line of connective tissue, the ***linea alba,*** separating right and left portions of the body muscles. No coelom remains between myotome and dermatome.

Returning for a moment to the time at which the lateral plate mesoderm (with prospective peritoneal cavity) approaches the midventral line, we find that each splanchnic layer (next to the endoderm) thickens and fuses with the one opposite, then forms a groove in the midventral line. Within the groove are prospective ***cardiac endothelial cells,*** the forerunners of the heart and associated blood vessels. These cells form into a pair of tubes, becoming a single tube at the level where the heart is to be, while around them the thickened wall of mesoderm closes like a sac and makes the ***pericardium.*** No blood forms and no heartbeat can be seen until a system of arteries, veins, and capillaries has grown sufficiently to provide a circuitous course for the blood to travel. But this is not long delayed. The ectoderm of the outer surface of the future throat grows out as three pairs of ***external gills,*** a short time after the embryo has hatched and become a larva. Within these gills run capillaries derived from ***branchial arteries,*** which in turn are branches of the ***ventral aorta*** running forward from the heart.

The story of development of the excretory and reproductive organs is best deferred to the chapter dealing with their structure.

We left the central nervous system at the end of the neurula stage. Subsequently the brain becomes swollen at three levels (Fig. 82), anteriorly the ***forebrain*** (***prosencephalon***), then ***midbrain*** (***mesencephalon***), then ***hindbrain*** (***rhombencephalon***). The neural tube cavity continues through these divisions. Later the forebrain shows a new differentiation into (anteriorly) a ***telencephalon,*** with right and left ***cerebral hemispheres,*** and a smaller ***diencephalon.*** The midbrain develops dorsal swellings, the ***optic lobes,*** while the hindbrain divides into a ***metencephalon*** (the dorsal part of this becomes the ***cerebellum***) and ***myelencephalon;*** this, with the remainder of the metencephalon, makes the ***medulla.*** Cranial nerves and other features will be treated in the chapter on the nervous system.

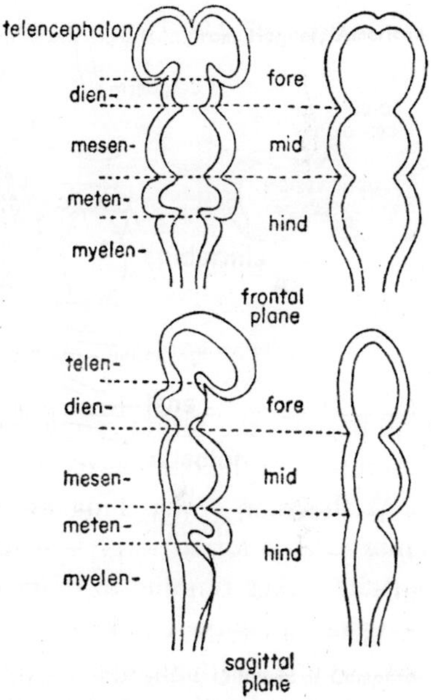

Fig. 82. Diagrams of development of the parts of a vertebrate brain. On right, earlier 3-lobed stage; on left, later 5-lobed stage.

We should look, however, at the very interesting development of the eye, for its early stages are related directly to the development of the brain, and it is an excellent example of how an organizer works (Fig. 83). The diencephalon very early produces two hollow lateral outgrowths, the ***optic vesicles,*** which grow toward the adjacent ectoderm and presently meet it. As they do so their swollen ends turn inward again, thus making two-layered ***optic cups.*** At the point where each optic vesicle reached the ectoderm the latter also sinks in, thickens, and becomes wholly transparent; the ***lens*** of the eye now lies just within the rim of the optic cup. Outside new ectoderm crosses over the lens, becomes transparent likewise, and so makes the ***cornea.***

The important point is this: it has been shown that the ectoderm will *not* make lens and cornea if the optic vesicle is prevented from reaching it, and also that a transplanted optic vesicle *will* induce formation of a lens if placed beneath the ectoderm elsewhere on the body of an embryo at this stage. Hence the optic vesicle is another example of an ***organizer.***

It is sufficient here to add that the invaginated lining of the optic cup

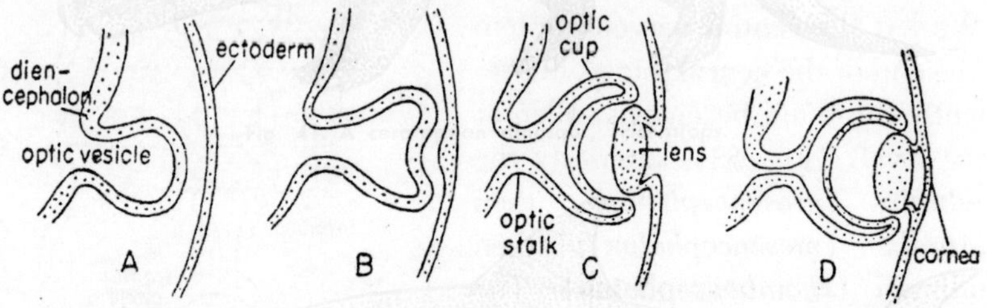

Fig. 83. Stages in the early development of a vertebrate eye, diagrammatic.

will become the ***retina,*** while the stalk leading to the diencephalon is the forerunner of the ***optic nerve.***

At the time of hatching (emergence from the jelly surrounding the embryo) the ***larval*** stage is reached. By now the body muscles and segmental nerves running to them are sufficiently complete to permit swimming movements. The body is moderately streamlined, with a flattened tail. Mouth and anus open a little later, for the larva has not quite consumed its yolk; the ectoderm invaginates to meet the anterior and posterior ends of the archenteron, and when the breakthrough occurs, the mouth and anus are lined with ectoderm. A temporary sucker or adhesive organ is found at hatching just beneath the head, by which the larva clings to objects. It has, of course, no legs.

Development of the legs in frogs, unlike most tetrapods, is delayed until nearly the end of a prolonged larval stage—in *Rana pipiens,* about a month from hatching. The hind legs are apparent first because the forelegs are concealed by the operculum, a sac of skin which grew back over the external gills shortly after the latter appeared. Each leg forms as a minute ***limb bud,*** a pimple of ectoderm enclosing undifferentiated mesodermal cells. After a period of growth we find in the bud the rudiments of cartilages and muscles, formed in place, and presently the digits project from the tip. As soon as joints have developed and branches of appropriate spinal nerves grow into the limb muscles, the frog can begin to kick.

Other parts of the skeleton also are cartilaginous during the larval stage, bone not appearing before ***metamorphosis.*** This astonishing change is much more than a mere development of legs and lungs, and loss of tail; it is a rebuilding of most of the organs of the whole body: skull, vertebrae, mouth, tongue, pharynx, skin, eyes, muscles, circulation, intestine, excretory and reproductive systems, endocrine organs, all undergo pro-

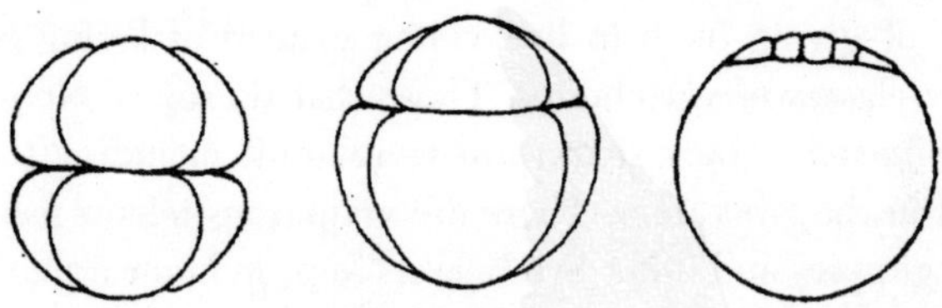

Fig. 84. Three types of cleavage of vertebrate eggs.

found and rapid readjustments. In this respect the frog is a highly exceptional animal.

DEVELOPMENT OF OTHER CHORDATES

General

We are now ready to see how development takes place under special conditions met in other groups than Amphibia. Some of these we shall describe briefly, to give the general pattern of changes for reference in dealing later with anatomy of special systems.

Small eggs with little yolk usually divide all the way through; cleavage of this sort is ***holoblastic*** (Fig. 84). Such are the eggs of protochordates, of lampreys, of a few fishes (e.g., lungfishes), of Amphibia (as we saw in the frog), and finally, of the marsupial and placental mammals. If such eggs (e.g., Amphioxus, mammals) have their yolk rather evenly distributed, because there is scarcely any of it, they are ***isolecithal.*** A ***telolecithal*** egg has enough yolk to make a distinct contrast between the heavier, vegetal pole (yolk) and the lighter, animal pole (cytoplasm), where cleavage goes most rapidly. Many groups of vertebrates have telolecithal eggs with so much yolk as to prevent cleavage entirely, except at the animal pole; this kind of cleavage is ***meroblastic.*** It results in a pad or island of small cells, the ***blastoderm,*** which gradually spreads down around the undivided yolk, enclosing it in a ***yolk sac.*** The animals with meroblastic cleavage are not necessarily related to one another, for the condition has come about several times independently: hagfishes, Elasmobranchii, Actinopteri (at least most of them), reptiles, birds, and the monotreme mammals. It follows that the minute holoblastic eggs of higher mammals have been reduced from the meroblastic type because, in feeding the embryo within the uterus, there was no need to retain a big yolk.

The habit of giving birth to live young instead of laying eggs is met among many classes of vertebrates. Those that do so are ***viviparous;*** egg layers are ***oviparous.*** Many sharks are viviparous, including the common dogfish used in the laboratory. There are viviparous teleost fishes, such as the familiar guppies and other live bearers kept in home aquaria. At least one salamander (*Salamandra atra,* of Europe) is viviparous. Snakes with the same habit include rattlesnakes, garter snakes, ringnecks, and others, while some of the "horned toads," among lizards, do likewise. Finally, of course, all mammals except monotremes are viviparous. Nobody has yet reported a viviparous bird, unless the Phoenix of ancient mythology meets the specifications.

Development of *Branchiostoma* (Amphioxus)

We shall follow the earliest stages in *Branchiostoma,* the egg of which is minute and has practically no yolk, hence is isolecithal. Cleavage is therefore holoblastic. The result of cleavage is a blastula (Fig. 85) with a relatively larger blastocoel than that of the frog, and the lower cells are slightly larger than the upper, which shows that the egg was not absolutely isolecithal. Gastrulation comes by broad invagination of the lower hemisphere of the blastula, accompanied by epiboly and involution, to make a two-layered cup with no remnant of the blastocoel. The opening of the cup is, of course, the blastopore; its lining is endoderm, its outer surface ectoderm. The blastopore then closes as the neural tube forms, but remains connected with the latter temporarily by the ***neurenteric canal.***

The neural tube forms by sinking down of the medullary plate until it is closed over by more lateral ectoderm, after which it rounds into a tube. The notochord develops from the roof of the archenteron in a roughly similar way, but of course without remaining tubular.

Coelom formation is quite different from that in the frog, and the difference is important because of its bearing on the origin of body segmentation in chordates. Dorsolaterally along the sides of the archenteron

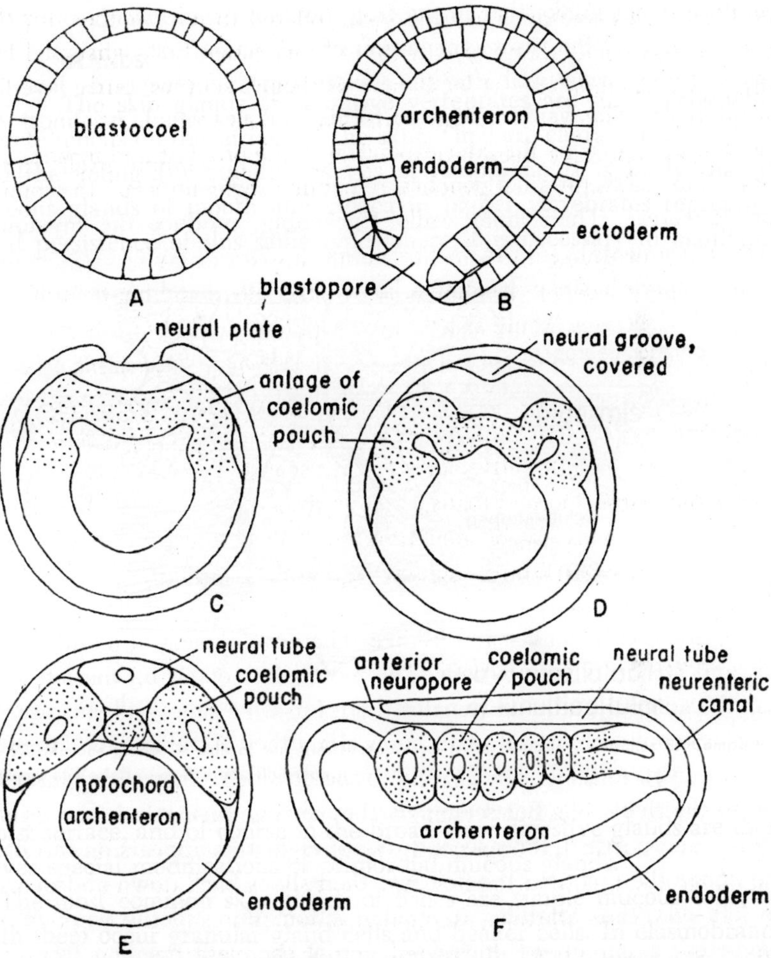

Fig. 85. Early development of *Branchiostoma* (Amphioxus). A, blastula, vertical section; B, gastrula, longitudinal section; C, D, appearance of coelomic pouches and neural tube; E, cross section after notochord forms; F, lateral view, as transparency.

there arises a paired series of pouches, commencing anteriorly and adding on behind; while the later ones are appearing, the early, anterior ones are closing and thus separating their cavities from that of the archenteron. These are *coelomic pouches;* their walls are mesoderm. As the pouches grow they extend ventrad between the endoderm and ectoderm, just as we saw with the equivalent lateral plate mesoderm in the frog, and presently the ventral part of each pouch unites with those before and behind, and on the opposite side, making a continuous coelom. The end result therefore is not unlike that in the frog, but *Branchiostoma* (more primitively) accomplishes its segmentation of mesoderm first, while the frog delays segmentation until after the coelom appears between the layers of lateral plate. The ventral coelom, formed as described, is almost obliterated anteriorly by the later expansion of the pharynx and atrium. The dorsal parts of the original pouches, remaining segmented, are the somites (or epimeres). Their inner walls, thickening, produce the myotomes (muscle segments).

The embryo has now become a larva (Fig. 86), hatching from its egg

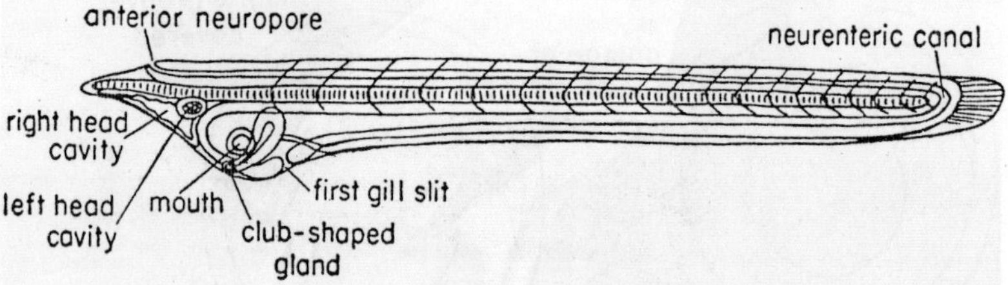

Fig. 86. Larva of Amphioxus. (After Parker and Haswell.)

membrane, and is slender and fishlike. The development of mouth, pharynx, and atrium is very curious. The mouth first opens on the left side, near the anterior end, and remains there for some time. Gill slits appear as a single row of fourteen small pores along the ventral surface (there is no atrium yet, so they open directly to the outside). They slide gradually over to the right side. But seemingly they change their minds about both number and position, for a second row of eight slits appears on the right side ***above*** the first, and the first row then slides back down and over to the left side, thus attaining its final position. But still the work is not completed; a bar grows downward across each gill opening from the dorsal side, dividing each slit into two; this is the secondary gill bar. Thus all the way along the finished pharynx secondary bars alternate with primary ones, which were there first. Additional slits continue to be added behind the original ones, increasing them to a hundred or more pairs.

Before this is completed, two metapleural folds arise on the outer wall of the body, running lengthwise. They lie laterally or dorsally from the gill slits, presently grow down over these, and forward from behind, so enclosing a space, the atrium. This remains open only at its posterior end, the atriopore, so that water from the gill slits can escape there.

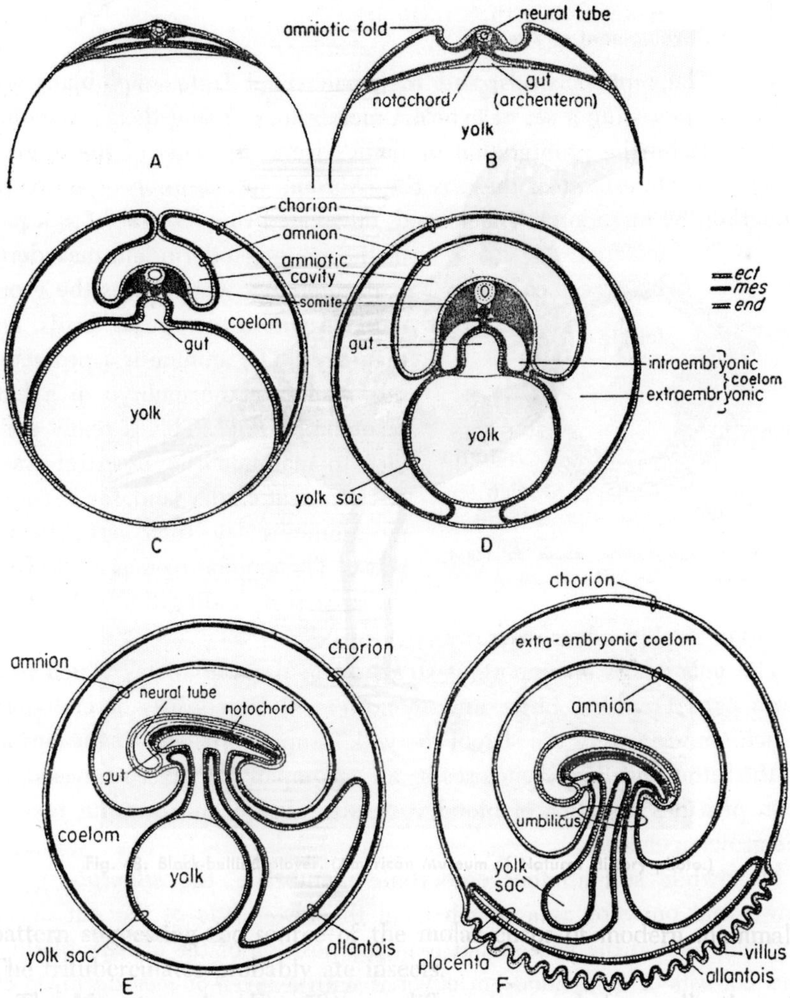

Fig. 87. Development of germ layers and extraembryonic membranes in amniote embryos. A, B, C, D, cross sections through large-yolked type (reptiles, birds); E, sagittal section of the same; F, sagittal section through embryo of a placental mammal.

By the time this has happened, and the mouth has moved down to the ventral side, making the animal virtually symmetrical, we can say it is no longer a larva. As we shall emphasize in the last chapter, *Branchiostoma* cannot be considered strictly ancestral to vertebrates, for much of its anatomy is open to the suspicion of being specially modified from an earlier primitive type.

Development of Amniota

The reptiles, birds, and mammals differ from amphibians and fishes in possessing a set of external membranes during their embryonic life, made by the proliferation of tissues from the body of the embryo (Fig. 87). Outermost of these is the ***chorion,*** an extension of ectoderm underlain by mesoderm. The second, developed simultaneously as a part of the same ectoderm and mesoderm, is the ***amnion,*** which gives the group name ***Amniota*** to reptiles, birds, and mammals. The amnion is a protective sac, enclosing the embryo in a bath of ***amniotic fluid.*** This is really a device to maintain the ancestral water environment on dry land, for of course no amniotes lay their eggs in the water. The amnion rises as a fold from the area surrounding the body of the embryo, until its edges meet over the top.

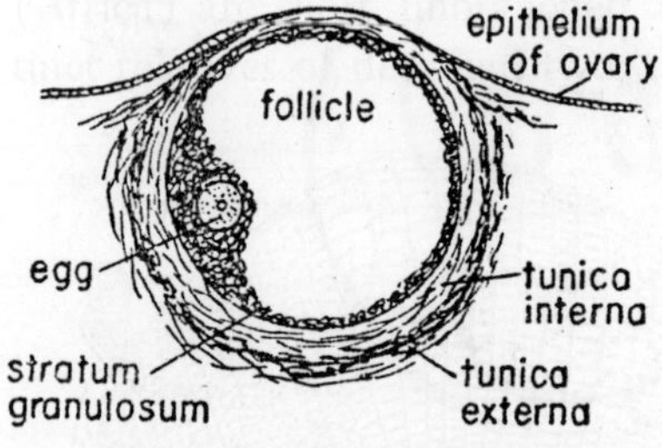

Fig. 88. Human follicle about to erupt. (After Arey.)

The ***yolk sac*** is an important structure in those amniotes which have large eggs (reptiles, birds, and monotremes). It consists of endoderm which, on account of the size of the yolk, simply spreads over the surface of the latter and finally encloses it; an accompanying layer of mesoderm soon provides a system of blood vessels to absorb and transmit food to the growing embryo.

Meanwhile still another embryonic membrane, the ***allantois,*** grows down and outward as a pouch from the lower side of the gut of the embryo, behind the yolk sac. It is in the position of the urinary bladder, and consists of gut endoderm covered with a layer of mesoderm. It expands into the space beneath the growing amniotic fold (extra-embryonic coelom), and serves, with the help of blood vessels, as an organ of respiration during embryonic life.

Among mammals, the marsupials and placentals have become viviparous and at the same time substituted a uterine food supply for that of the yolk. Consequently the yolk sac, prominent in only a few mammals, is vestigial in the embryos of most. The egg is extremely small (Fig. 88); cleavage is holoblastic; and the membranes external to the embryo are altered to carry on a metabolic exchange with the blood of the surrounding uterus. To see how this is done we shall trace the early stages of a mammal embryo to its establishment as a ***fetus*** (dependent upon the mother's blood supply) in the wall of the uterus.

The egg is fertilized in its passage down the Fallopian tube, and cleavage therefore begins considerably before its arrival in the uterus proper (Fig. 89). Instead of a blastula, the result of cleavage is a ***blastocyst,*** that is, a spherical ball of cells containing a cavity and also a group of inner cells; this ***inner cell mass*** is to be responsible for the formation

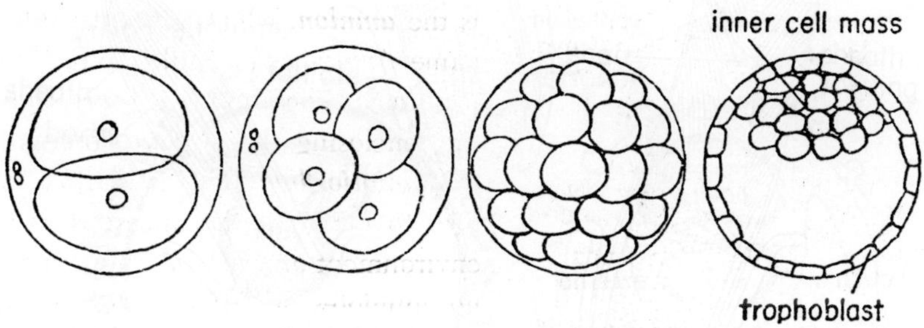

Fig. 89. Cleavage of a mammal egg.

of the embryo, while the outer layer, called ***trophoblast,*** remains in contact with the surrounding tissues of the uterus and, becoming the chorion, plays an important part in making the placenta.

The blastocyst sinks into the mucosal lining of the uterus about ten days (human) after fertilization of the egg, thus becoming entirely surrounded by uterine tissue. Development now takes so many short cuts that it seems quite unlike that of primitive vertebrates, or even of other Amniota. The ***amnion cavity*** (in man) appears as a space in the inner cell mass (Fig. 90A). The cells lining that space are ectodermal, and the body of the embryo presently appears on the floor of the cavity.

Gastrulation, if we can call it that, comes not by any invagination but by separation of a layer of endoderm cells from the lower part of the inner cell mass, and its spreading to occupy the inside of the blastocyst. The sac so formed corresponds to both archenteron and yolk sac, for if a yolk had been present it would have been enclosed in the same layer of endoderm. (This is further demonstrated a little later by the appearance of typical yolk-sac blood vessels over the surface of the pouch of endoderm.) The endoderm with its cavity fills most of the available space within the trophoblast, below the amnion cavity.

But presently, from cells pushing in from the ectoderm (***primitive streak***) there arises mesoderm, which in turn crowds between the endoderm and surrounding trophoblast. In this mesoderm there opens the *coelom,* which soon becomes very large, and the mesoderm now makes a thin lining against (1) the inside of the trophoblast, (2) the ectoderm of the amniotic cavity, thus completing the amnion, (3) the prospective embryo itself, and (4) the yolk sac. Naturally this causes the structures already formed to float almost free in the big extra-embryonic coelom, except at one place, the *body stalk* (Fig. 90B), by which the embryo

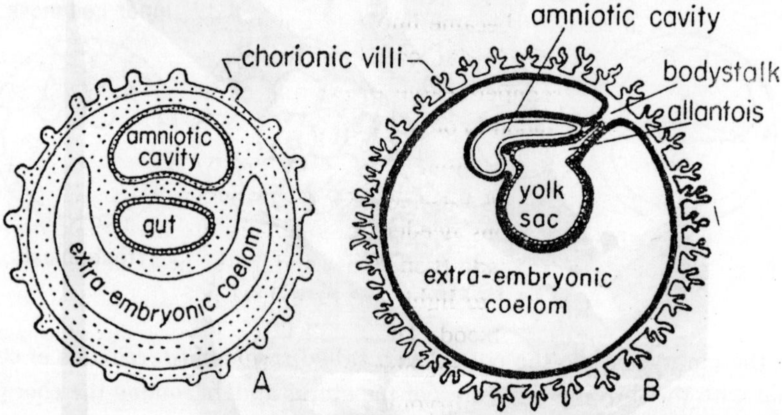

Fig. 90. Development of chorionic villi and body stalk of mammal embryo.

keeps continuity with its enveloping case. We can now speak of the latter as the *chorion* (trophoblast plus a lining of mesoderm), and the extra-embryonic coelom is also a *chorionic cavity.* But the amnion, during later growth, presses out against the chorion and practically eliminates its cavity.

The *allantois,* of little consequence in human embryos but a rather important organ in those of most mammals, grows out from the hindgut into the mesoderm of the body stalk. The yolk sac dwindles to a vestige. We shall not give details of the establishment of the organ systems, for these are not basically different from the corresponding stages of lower vertebrates. But for an adequate picture it is necessary to deal with the placenta and umbilical cord.

Two umbilical arteries and an umbilical vein grow out along the body stalk, and are enclosed in mesoderm and ectoderm by the spread of the amnion outward. Thus the cord is formed, carrying the embryo's own blood supply out to the chorion. In the meantime the latter has been growing branched villi out into the tissue of the uterus around it (this does not happen in marsupials). The villi form all over the chorion, but later disappear except in an area where they face the thickness of the uterus wall directly, and there they become elaborately developed. Capillaries take the embryo's blood into these villi and back again into the umbilical vein. There is no actual connection with the mother's blood system. The thick pad of chorion with villi and adjacent tissues of the uterus makes the ***placenta.***

The chorion, during early stages, has been eroding away the mucosa of the uterus in which it became implanted, until the villi lie almost free in sinuses filled with maternal blood, but always the exchange of materials is through the semipermeable membrane covering these villi, and *not* by the passage of blood to or from them.

So the fetus floats in its amniotic sac, attached by a vascular cord to the placenta, until time for birth (***parturition***), when it has developed the structures and functions needed for open-air life. Several different adjustments need to be made then rather suddenly: breathing air, feeding, excreting, responding to light, sound, and touch, and a number of changes in circulation of blood. These will be mentioned in later chapters. The placenta, cord, and membranes are shed as the afterbirth, having served their purpose as embryonic adaptations.

CHAPTER III

Integument

GENERAL NATURE OF INTEGUMENT

Integument is the covering of animals,—the skin and associated organs and structures. Meeting the environment directly, it repels a part of what it meets, helps maintain constant conditions inside the body, reacts in highly specialized ways to certain influences in the environment, and so (speaking loosely) makes the animal aware of important matters outside. We might then include sense organs as a part of the integument, but it is customary to treat them in relation to the nervous system because they translate stimuli into nervous impulses. Teeth are integumentary structures, often studied as part of the digestive system, or in connection with the skull, but it expresses more clearly their morphological meaning to put them here.

The unadorned skin is a complex sheet of tissues many cells in thickness (Fig. 91). The outer part, ***epidermis,*** develops from ectoderm in the embryo, but the inner layer, ***dermis*** (or ***corium***), arises from mesoderm. The epidermis is an essentially uniform epithelial tissue, modified here and there as it contributes to scales, hair, feathers, sweat glands, and the like, but in no case is it a mixture of diverse tissues. The human epidermis grows continually, new cells being added by division of those in its ***germinative*** or ***Malpighian*** layer, which is the one nearest the underlying dermis. As successive strata of cells appear they are pushed towards the outside by those beneath them, and go through a sequence of changes of shape and appearance, recognized by the names ***granulosum, lucidum,*** and ***corneum.*** In the last of these the cells are thin, flat, horny, and dead, but instead of accumulating they drop off in small flakes, and their place

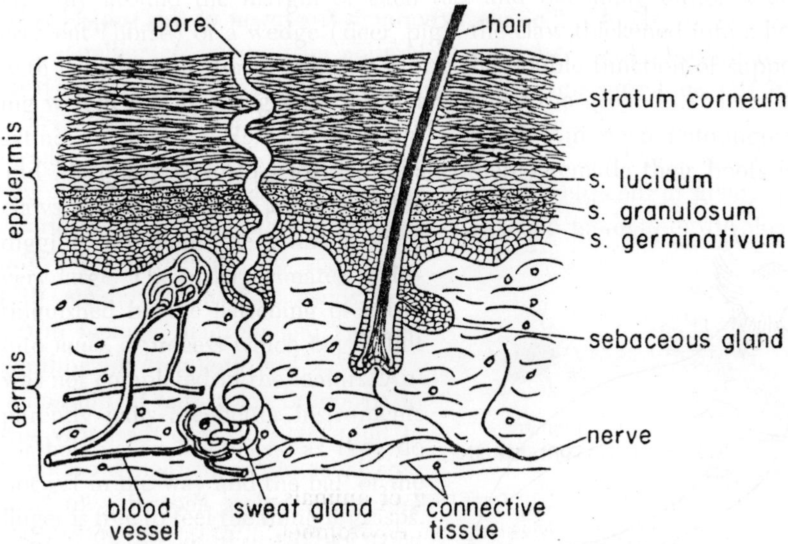

Fig. 91. Diagram of section through human skin.

is taken by others coming from below. So it is with mammals in general. The other vertebrates seldom show more than two distinct strata, germinativum and corneum. Reptiles and Amphibia shed their stratum corneum as a whole, periodically. This is called ***molting.*** The same term applies to the seasonal change of feathers in a bird, or of fur in many mammals. Fishes do not usually have a horny epidermal layer because their skins are protected by scales, mucous glands, or both.

The ***dermis*** is complicated, with a network of connective tissue into which penetrate small blood vessels and nerves. Smooth muscles are found here, and small end organs for the senses of touch and temperature. The dermis hurts when a hornet stings us; the dermis bleeds when we scratch ourselves. Leather, made of tanned skins, gets its toughness from the fibrous connective tissue of the dermis.

COLOR

Color occurs either in the skin itself or in scales, plates, beaks, feathers, hair, and other products of the skin. Usually it is not evenly distributed over the body but forms a pattern characteristic of a species. In a very great number of cases, but not all, this color and pattern have some adaptive relationship with the surroundings, such as to make the animal difficult to see (tree frog), or, less often, to make it exceedingly conspicuous (skunk). The most thorough treatment of this subject is by Hugh B. Cott (1940: *Adaptive Coloration in Animals*, Macmillan).

In the epidermis and structures derived from it, color results from pigment (either diffused or in granules), from diffraction (physical color, iridescence), or from a combination of the two. Human skin is colored by pigment in the epidermal cells, as is that of many mammals. Hair gets its color from pigment deposited during its growth, sometimes uniformly, sometimes in alternating light and dark zones, as in many rodents. Bird feathers receive their pigment in the same way, but often with the addition of structural color; that is, diffraction of light takes place in the horny surface layer of the feather, breaking it into many colors according to their wave lengths, as in the throats of male hummingbirds. The beaks of birds and the shells of turtles often contain a variety of pigments.

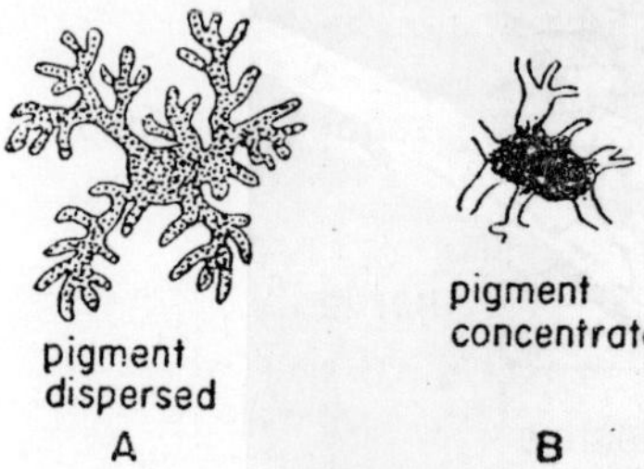

Fig. 92. A melanophore in frog skin.

In contrast with these epidermal colors, those of the dermis usually are produced by pigment within specialized cells, the ***chromatophores.*** This is true of the lower vertebrates, whose epidermis is transparent enough to allow color to show through. Thus when a toad pulls off his stratum corneum, this is dull and colorless, but the body appears fresher and brighter than before. Chromatophores branch irregularly, and as a rule the distribution of pigment in them changes slowly or not at all, but there are many fishes, frogs, and lizards capable of rapid adjustment to match a change in their surroundings. Pigment flows out from the center of a cell along all its branches, thus causing intensification of that color, or it withdraws to the center, allowing a different color to show from adjoining cells (Fig. 92). ***Neurohumors*** (hormones released from nerve endings) cause these changes, which are reflexes initiated by visual stimulation, temperature, humidity, and other factors; they are also induced (not as reflexes) by a hormone from the intermediate lobe of the pituitary gland. The kinds of chromatophores are: ***melanophores,*** with brown or black melanin pigment, ***xanthophores,*** with red or yellow pigment, and ***guanophores,*** which are white owing to reflection of light by guanin crystals. Chromatophores often occur in deep connective tissue, as the peritoneum or the myosepta; they are sometimes able to migrate from one part to another.

GLANDS

The skin glands of primitive vertebrates are single cells which secrete mucus on the surface of the skin. They are characteristic of fishes and the young (or aquatic) stages of Amphibia. It is likely that the mucous glands of mouth and cloaca in higher vertebrates represent a local persistence of this same type where it is necessary to maintain a

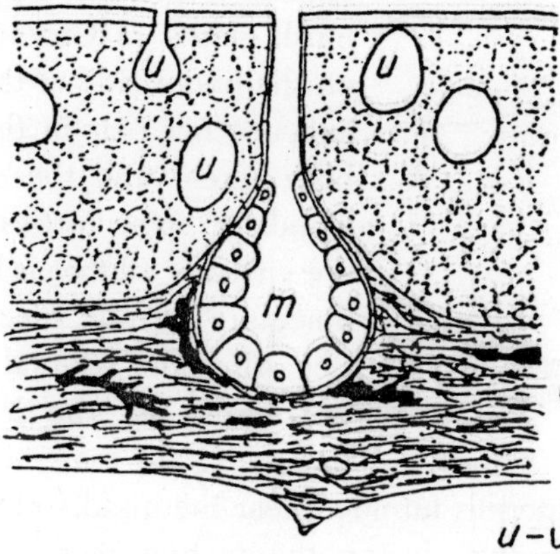

Fig. 93. Mucous glands in skin of lungfish, *Protopterus*. (After Kingsley, *Outlines of Comparative Anatomy of Vertebrates*, 3rd ed., Blakiston.)

moist surface, and of course in the broad view digestive glands are themselves special modifications of primordial mucous glands.

The most common skin glands of fishes are simple mucous cells, but with them occur granular gland cells and beaker cells. In elasmobranchs enlarged cells arise from the Malpighian layer, move to the surface, and pour out their secretion. Some teleosts have developed multicellular poison glands, as in catfishes at the base of the pectoral spines (*Noturus, Schilbeodes*), or in some scorpaenids on the dorsal spines. The African lungfish, *Protopterus,* has both uni- and multicellular mucous glands, the latter pushing in from the surface like sweat glands (Fig. 93).

The egg capsule of some embryonic frogs and salamanders is thought to be digested by a secretion from unicellular glands. At metamorphosis to the terrestrial (adult) stage the mucous glands in Amphibia become multicellular and alveolar (branched, with pockets). One type of these, the granular glands, pour out a thick, frequently poisonous secretion when violently stimulated; the ordinary mucus is also poisonous in a few kinds of frogs (*Dendrobates,* etc.). It is common for either granular or mucous glands to produce an odor. In tree-climbing frogs and salamanders the toes may have adhesive swollen tips.

Reptiles, being terrestrial primarily, have almost given up any skin glands whatever, for it is necessary to conserve water in the body. Therefore reptiles are not slimy; their skin is dry. Even those which returned secondarily to aquatic life (most turtles, crocodiles, etc.) have not regained a mucous secretion. The musk glands of Crocodilia (a pair on the under side of the lower jaw and another pair in the cloaca, in both sexes) and the cloacal stink glands of some turtles and snakes are virtually all that is left of integumentary glands among reptiles. Regarding the femoral pores (Fig. 94) found on the legs of most lizards, Smith (1946, p. 13) says, "The pores are really integumental glands formed of the epidermis infolded near the middle of a scale. The glands are particularly active during the mating season, secreting a corneous substance that may project fingerlike a considerable distance from the glands. . . . Apparently the organs have the sole function of stimulation of the female in courtship and mating activities. Therefore in females they are vestigial."

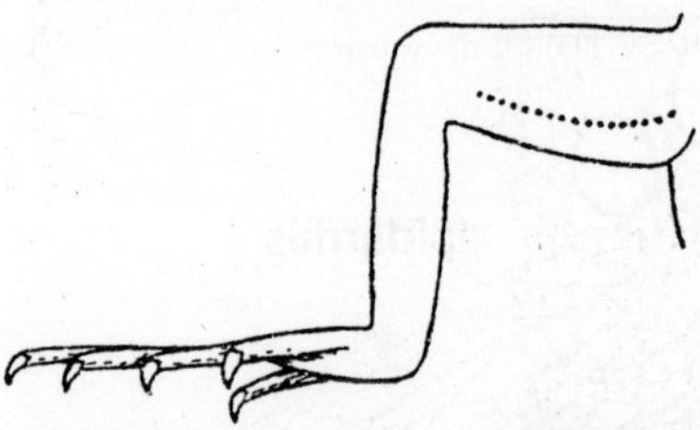

Fig. 94. Femoral pores on ventral side of thigh of a lizard, *Crotaphytus collaris.*

So likewise in birds the skin glands are few and of little consequence. The only ones commonly met are the uropygial glands just above the tail in ducks, geese, chickens, pigeons, and some other rather primitive types. They provide oil which is used in preening the feathers.

Mammals have either ***tubular*** or ***alveolar*** (***acinous***) (Fig. 95) skin glands. Wax glands in the ear canal are tubular. Sweat glands (Fig. 91) are also tubular, much coiled at the inner end, and not branched; evaporation of water through perspiration is an important device to regulate body temperature, but the loss of urea and carbon dioxide in sweat is so slight as to have no value in excretion. The distribution of sweat glands on the body is limited, especially in hairy mammals; in man they are most abundant on the palms and soles. Sirenia and Cetacea, being completely aquatic, have none, nor do the scaly pangolin, the spiny anteater (*Tachyglossus*), the golden mole (*Chrysochloris,* of South Africa), or the European mole, *Talpa.*

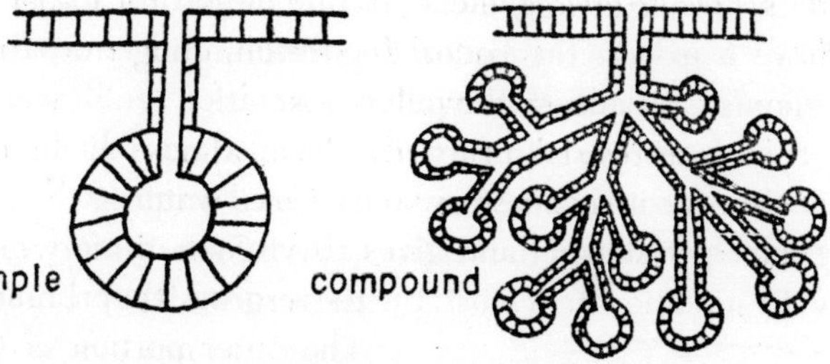

Fig. 95. Simple and compound alveolar glands, diagrammatic.

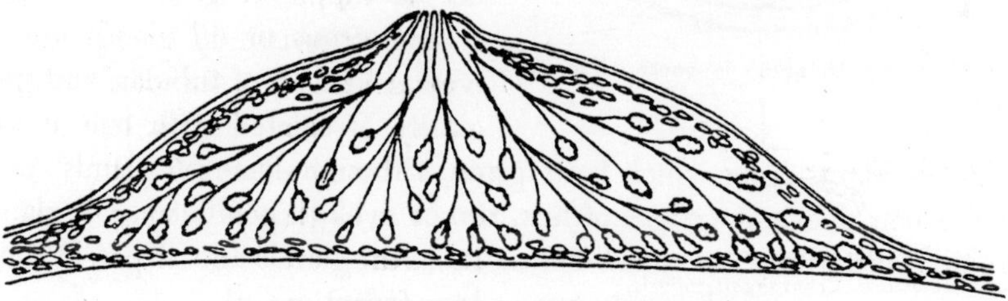

Fig. 96. Section through human mammary gland.

The third important type of tubular skin gland is the ***mammary*** or milk gland (Fig. 96), present in all mammals but with its functioning limited almost entirely to the female after birth of young. In monotremes the nipples are lacking, but the glands consist of two clusters of follicles whose ducts open on the ventral side of the body near the middle; they are said to function in both sexes. A pair of mammary pouches develops at the breeding season, and they deepen to form a pocket for the young in *Tachyglossus*. Marsupials, excepting some of the opossums, have such a pouch or marsupium in which the young are carried after a very brief period of gestation in the uterus. The pouch contains nipples, varying in number according to the usual size of the litter in different species. The maximum number is probably thirteen (six pairs and one extra) in one species of opossum. Newly born marsupials, extremely small and unfinished, creep among the hair of the mother until they enter the pouch, then take hold of a nipple by swallowing it, and remain attached through what amounts to a second embryonic phase of development, before they are ready to escape from the pouch (Fig. 97).

Among placental mammals the number and location of mammary glands is correlated with the size of litter and with accessibility to the young. For example, carnivores, rats, and pigs, which lie on their sides when feeding their offspring, have nipples in two series along the abdomen (the primitive arrangement), while elephants, sea cows, bats, and primates have a pair in the pectoral position, and in whales and ruminants the glands are beneath the pelvis. Lactation (milk secretion) in the great majority of mammals is seasonal, because reproduction, on which it depends, is also seasonal. The mammary gland originates as an epidermal thickening which pushes in, and gives rise to one or more milk ducts. As these grow they branch inwardly. The outer portion of the skin surrounding the ducts then protrudes as the nipple (Fig. 98).

Fig. 97. Young kangaroo in pouch, diagrammatic.

Sebaceous or ***oil*** glands are alveolar instead of tubular, and generally associated with hair in the follicles or pits from which hairs grow, although some oil glands occur where hair does not (edges of lips, rim of eyelids, genitalia). Oil glands are present in those mammals that have sweat glands and absent in those that do not; this rule has only one or two exceptions.

As a special modification of skin glands, many deep-sea fishes have light-producing organs called ***photophores.*** The property of luminescence is common among invertebrates from Protozoa to insects, and especially in those swarms of marine animals that live beyond the reach of sunlight in the sea, or that come to the surface only at night. Among fishes we meet photophores in several unrelated groups, nearly all in the deep sea;

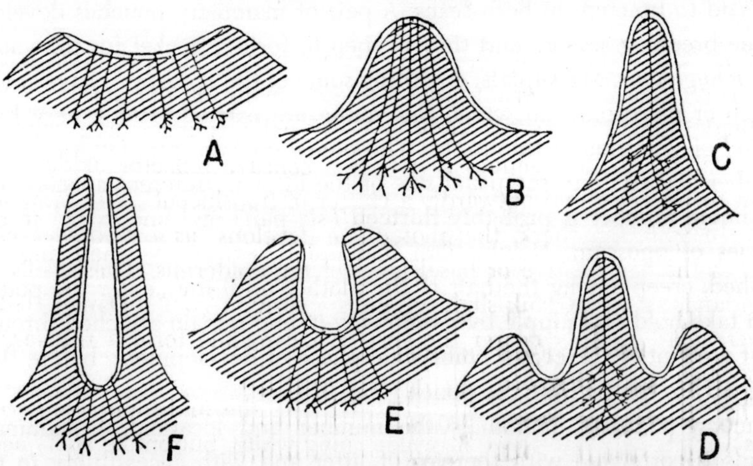

Fig. 98. Types of nipples. (From Kingsley, *Outlines of Comparative Anatomy of Vertebrates*, 3rd. ed., Blakiston.) A, *Echidna;* B, human; C, *Didelphys,* before lactation; D, *Didelphys,* at lactation; E, cow, embryonic; F, cow, adult.

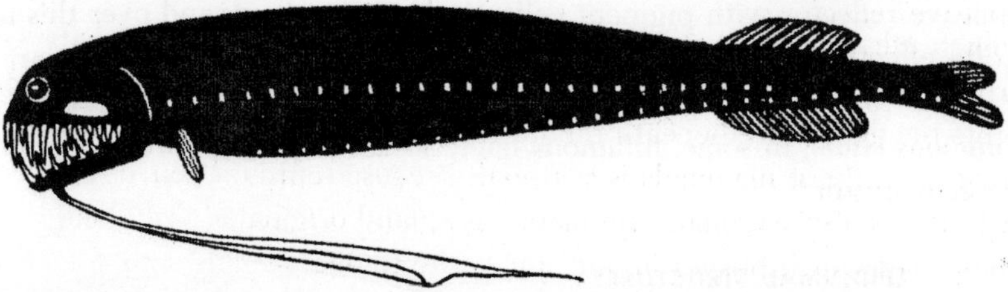

Fig. 99. A deep-sea fish, showing light organs (photophores). (Redrawn from Hegner, American Museum of Natural History photo.)

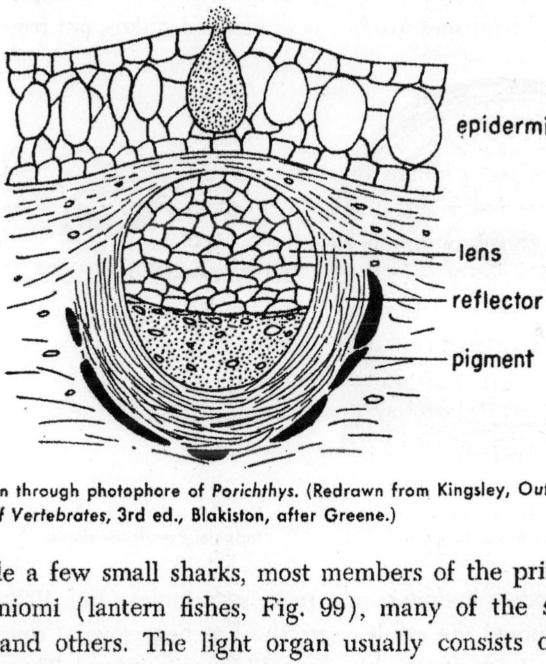

Fig. 100. Section through photophore of *Porichthys*. (Redrawn from Kingsley, *Outlines of Comparative Anatomy of Vertebrates*, 3rd ed., Blakiston, after Greene.)

they include a few small sharks, most members of the primitive teleost suborder Iniomi (lantern fishes, Fig. 99), many of the suborder Stomatioidei, and others. The light organ usually consists of a modified gland producing a luminous secretion, a concave reflector of dark pigment behind this, and outwardly a cluster of translucent cells acting as a lens. In *Spinax*, a shark, the photophore develops "as a modification of cells in the germinative or basal layer of the epidermis. These cells enlarge and, as a cup, sink slightly into the corium. . . ." (Daniel, 1934). Other enlarged cells nearer the surface meanwhile form a transparent lens which emits the light.

There is, however, one shallow-water fish, the midshipman (*Porichthys*), related to toadfishes, in which shiny white, buttonlike spots occur in rows on the skin and have been shown to produce light under strong stimulation. Here again, each photophore (Fig. 100) is made of a deep, concave reflector with pigment cells, enclosing a gland, and over this is a quite effective lens. In most light-producing fishes the luminous part is thought to be derived from skin glands which originally secreted a luminous slime; in some, luminous bacteria in the photophore accomplish the same result.

EPIDERMAL STRUCTURES

Horny Teeth, Beaks

The teeth found in the oral funnel and on the tongue of lampreys (Fig. 101) and hagfishes are horny epidermal spikes, not true teeth. They

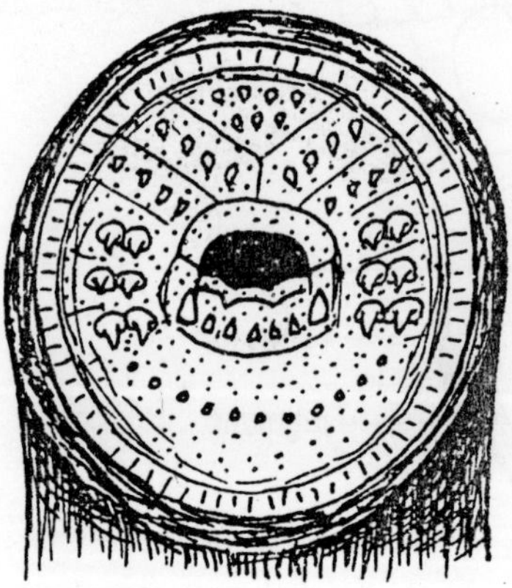

Fig. 101. Buccal funnel and horny teeth of lamprey, *Entosphenus lamottenii*.

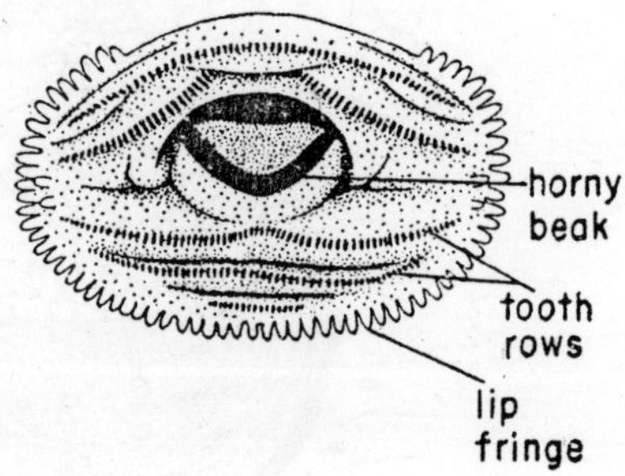

Fig. 102. Horny beak and tooth rows of tadpole, *Pseudacris clarkii*.

can be removed as hollow cones. In most tadpoles (Fig. 102), before the jaws have assumed the adult form, the lips bear several rows of minute horny teeth on folds of the skin, for nibbling plant food. They are shed at metamorphosis and have nothing to do with true teeth.

Beaks are horny epidermal sheaths enclosing the bones of the upper and lower jaws (Fig. 103). Usually beaks do not keep company with teeth (turtles and modern birds, for example), but sometimes, as in ceratopsian dinosaurs, the two seem to have gone together. In monotreme mammals a beak has almost but not quite displaced teeth, which appear only temporarily in the duckbill. It is among birds that we see the greatest diversity of beaks, from the simple, straight cone of a flycatcher to the stout, short, seed-cracking type found in sparrows or grosbeaks, the hooked predatory weapon of hawks and owls, the delicate tube of nectar-drinking hummingbirds, the powerful chisel of an ivory-billed woodpecker, or such monstrosities as the banana beak of a toucan or the spatula-and-pouch combination of a pelican.

Claws, Hoofs, Nails, Spurs

Claws are horny points produced on the ends of the digits by the epidermis, not only to protect the tip from abrasion but for many

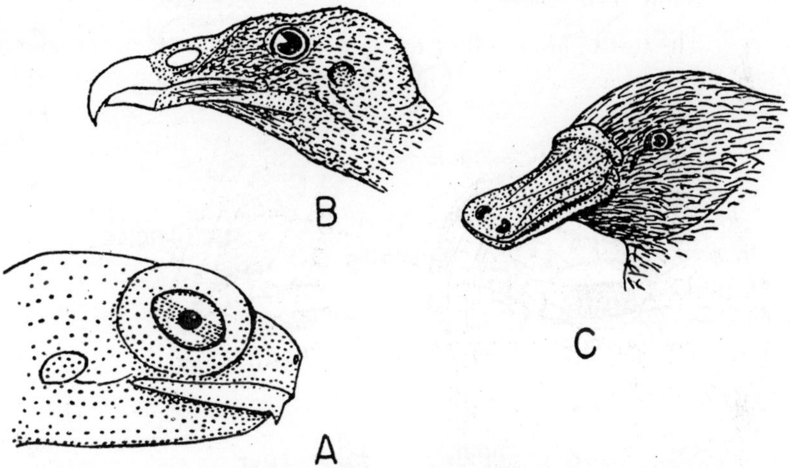

Fig. 103. Types of beaks. A, painted turtle, young; B, buzzard (after Chapman); C, Duckbill (after Parker and Haswell).

special functions which the digit alone could not perform. They appear first, sporadically, in Amphibia, where they may be found in a few mountain-brook salamanders (*Onychodactylus*) and certain primitive frogs (Pipidae). But among reptiles claws are standard equipment, being useful for clinging to rough surfaces, digging holes, or merely for running. Typically a claw grows continuously from the germinative layer of the epidermis, and by the curvature and convexity given to it as it forms, the distal part continues out and down beyond the tip of the toe. A zone of softer material, the ***subunguis,*** lies beneath the claw where it leaves the toe. (See Fig. 104.)

Birds use plain straight claws for scratching on the ground (chickens), slender curved ones for grasping twigs (almost any songbird), sharp hooks for seizing prey (falcons, eagles), and a variety of others, including what amount to hoofs on the two toes of the ostrich, for heavy running. Wings, which are modified fore legs, have lost most external

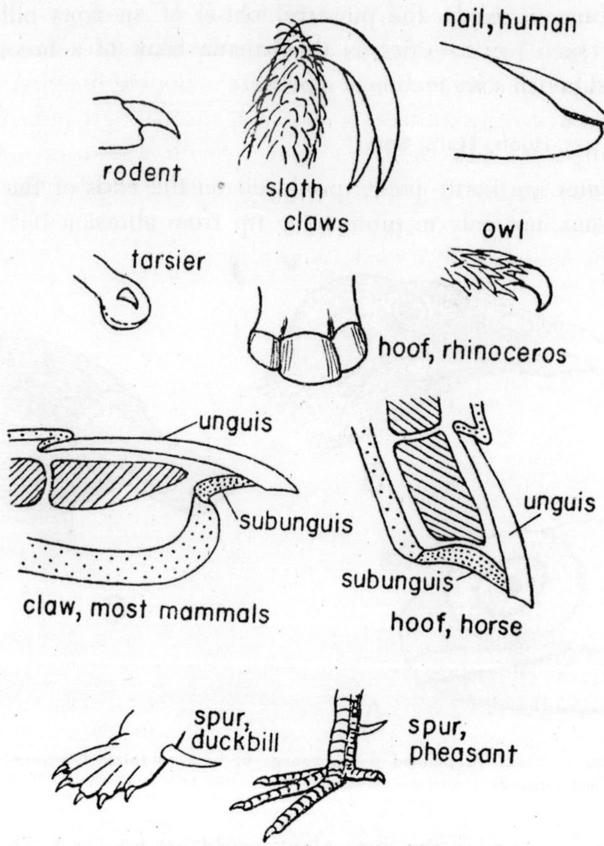

Fig. 104. Claws, hoofs, nails, and spurs.

evidence of digits, but the South American hoatzin has two movable clawed fingers on the wing in its nestling stage; these aid in climbing. A claw (first digit) is present in the screamers, and from one to three claws in the ostrichlike birds.

In mammals the history of claws is of great interest. Upon a heritage of simple, sharp claws, persisting through the majority of mammals (marsupials, insectivores, rodents, carnivores, etc.), a few groups made fundamental innovations. The various "ungulate" orders, independently of one another, developed these claws into ***hoofs,*** on which a heavy grazing or running animal could bear its weight. Generally this happened after the digitigrade (toe-walking) habit was established, for otherwise a claw would not be in the right position to support weight; enlarging laterally around the margin of each toe, and becoming either a deep crescent (horse) or a wedge (deer, pig), the claw thickened into a hoof, with its subunguis for a pad, to perform the double function of supporting weight and resisting wear and tear against the ground. At least two extinct ungulates (*Homalodotherium*, a South American notoungulate, and *Moropus*, a North American chalicothere) remade their hoofs into claws, apparently as an adaptation for digging, although the animals were very large. The order Primates is distinguished by the flattening of claws into *nails*, a process which began but was not completed in the lemurs. Apparently nails are better instruments for holding small objects, as they do not get in the way and the ball of the finger is free to feel the thing it grasps.

Spurs are another matter. Roosters, but not hens, develop a horny spur on the back of each leg, used for fighting. It is not a modified claw. Males of the tropical jacanas and screamers (Fig. 105) bear a spur on the front of each wing. The male duckbill (monotreme) has a spur on the inner side of each hind foot, hollow and provided with a poison gland.

Fig. 105. Crested screamer, showing wing spurs. (After Evans.)

Epidermal Scales

Scales make the exoskeleton of most reptiles, and they occur in particular parts of the body in birds and many mammals. They are rigid pieces of the stratum corneum, and it is only the presence of a narrow zone of thin epidermis between one scale and the next that makes a scaly armor flexible (Fig. 106). In snakes and lizards the stratum corneum is shed as a whole, periodically, because otherwise the animal would be unable to grow beyond the limits set by its exoskeleton. There is no morphological relationship between the epidermal reptilian scales and the dermal scales of fishes.

Frequently dermal bone develops in the skin immediately below each scale, as in the alligator lizards (*Gerrhonotus*) and "glass snakes" (*Ophi-*

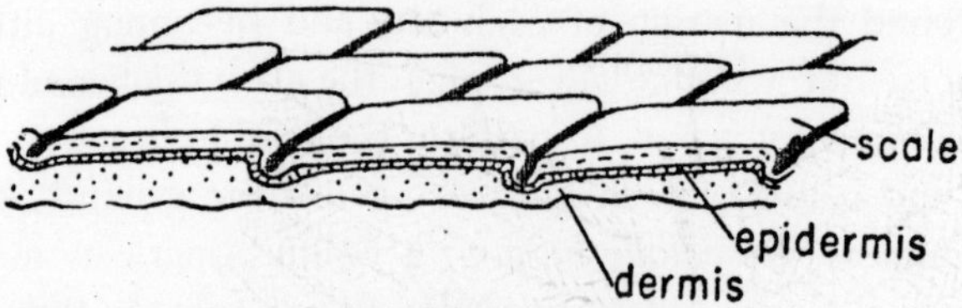

Fig. 106. Reptilian type of scales.

saurus). The plates of the carapace (above) and plastron (below) of turtles are horny and correspond in their development to scales, but beneath them is a layer of bony plates whose edges do not match those of the horny plates. In Crocodilia the skin of back and tail is also provided with bony plates (Fig. 34).

Feathers

The earliest known bird, the Jurassic fossil *Archeopteryx*, had feathers like those of modern birds. From the study of development of individual feathers we infer that they are derived from the horny scales of reptiles, but a feather is infinitely more complicated, and adapted to the two functions of flight and temperature control. By specially developed flight feathers (***remiges***) on the margins of the wings the flying surface is made great enough to be effective. Layers of air, a poor conductor of heat, between the overlapping feathers prevent too rapid radiation of heat from the body. The "tail" of all modern birds consists of feathers (***rectrices***) extending out from the small movable stump of the true tail, the ***uropygium.*** Feathers also contribute to a streamlining of the body, without which flight, considering a bird's shape, would be awkward or impossible; such are the ***contour*** feathers. The arrangement of feathers on the skin is in certain distinct zones, the ***pterylae*** or ***feather***

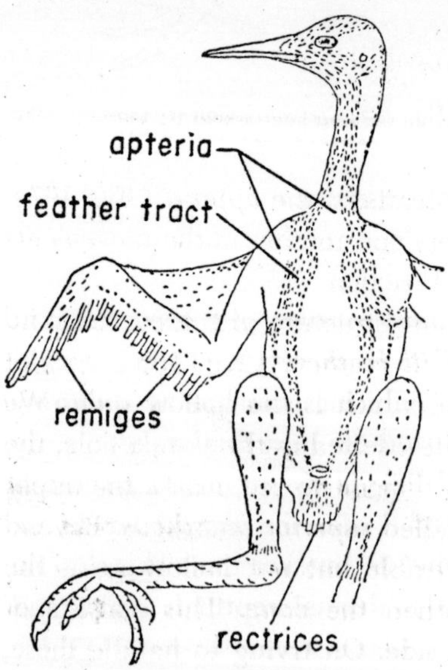

Fig. 107. Feather tracts of roadrunner, *Geococcyx*, ventral side. (After Kingsley, *Outlines of Comparative Anatomy of Vertebrates*, 3rd ed., Blakiston.)

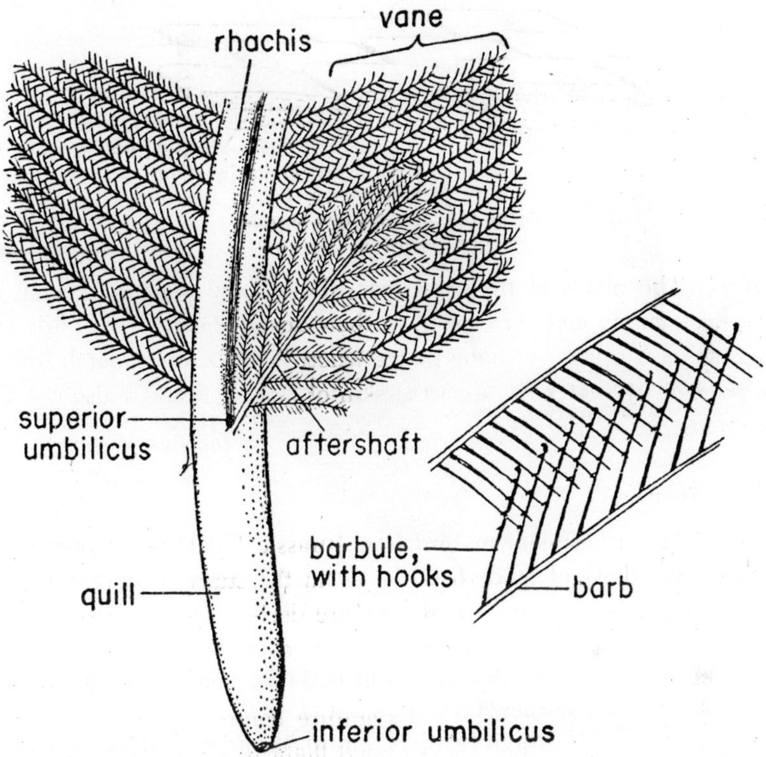

Fig. 108. Diagram of quill feather, with detail showing interlocking barbules.

tracts, between which lie areas without feathers, the ***apteria*** (Fig. 107). These allow groups of feathers to slip past one another in the movements of wings, legs, or neck, with a minimum of friction.

We should take a moment to look at the structure of the contour and flight feathers, known collectively as ***quill feathers*** (Fig. 108). Any of these has a median ***shaft,*** at the base of which is the hollow ***quill.*** We find at the lower tip of the quill, where it inserted in the skin, a hole, the ***inferior umbilicus.*** A similar hole, often difficult to see, marks the upper end of the hollow (quill) part, and is called ***superior umbilicus.*** Beyond this the central shaft continues as a flexible but not hollow axis (the ***rhachis***) for the wide part of the feather, the ***vane.*** This consists of numerous fine ***barbs*** emerging on each side. On trying to handle these, we find that they cling to one another with considerable tenacity, and a lens shows that each barb has smaller ***barbules*** branching from it; the barbules carry hooks which interlock with barbules of the next series.

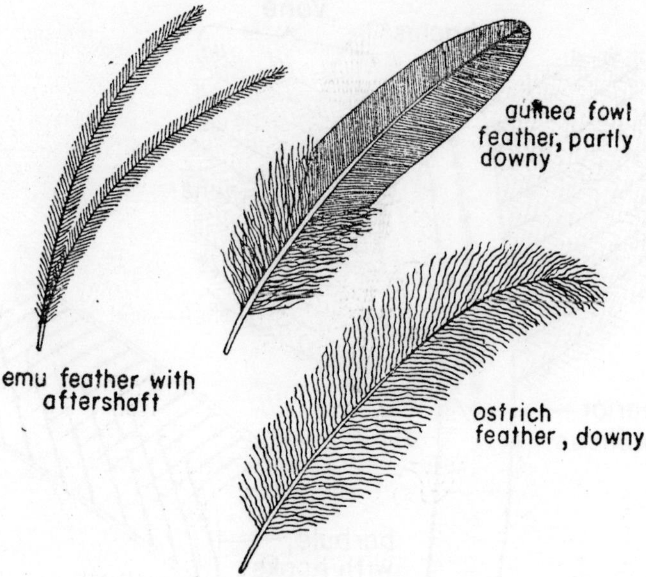

Fig. 109. Feather types.

Thus the vane, without being a solid surface, is so tightly knit that it accomplishes the same effect. (Compare Fig. 109.)

Besides the familiar quill feathers, a bird has ***down*** feathers, consisting mainly of fine soft barbs arising directly from a short quill, with no shaft and no vane. The barbs are not linked together but form a loose fluff. ***Nestling down*** is the feathering of newly hatched young birds, while ***powder down*** persists in adults as an extra cushion or padding under the contour feathers. There are also ***filoplumes*** or ***hair feathers***, in which few barbs or none remain, but only the shaft, and such a feather often performs the function of a hair or bristle.

Birds regularly go through a ***molt,*** or change of plumage, according to age or season. Starting with a downy ***nestling*** plumage, quill feathers gradually replace the down, so that the bird can fly, but this ***juvenal*** plumage is not that of the adult. ***Nuptial*** plumage, put on as the breeding season approaches, will in most birds show a marked contrast between the sexes, the male being far more conspicuous, but the pattern of the female remaining similar to that of the immature bird. ***Postnuptial*** plumage is seen in many birds in the fall and winter after breeding, and they may repeat the nuptial pattern as many years as they live to reproduce. Not all these statements would be correct for all birds.

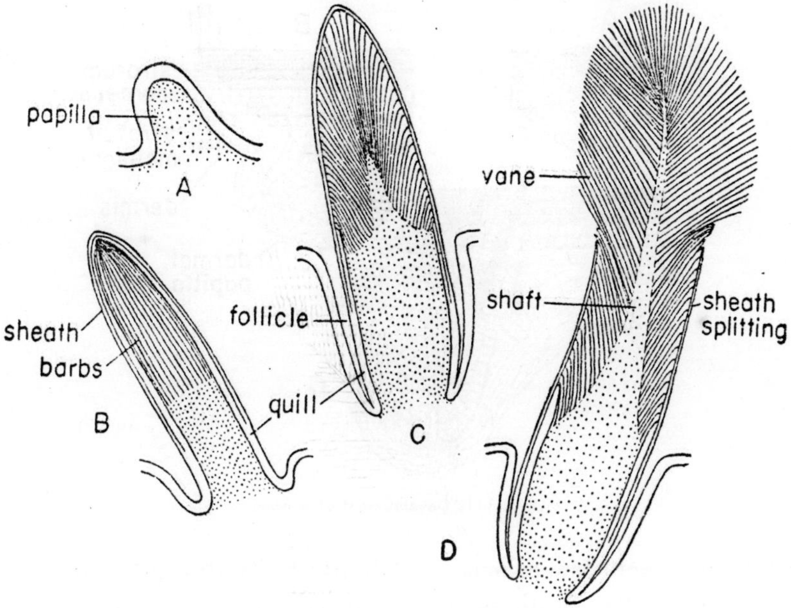

Fig. 110. Development of a feather.

We can better understand the relationships between the types of feathers by watching their development (Fig. 110). The earliest stage in a feather resembles the corresponding stage in a reptilian scale; that is, a cap of ectoderm overlies a dermal papilla containing a blood supply. The *papilla,* with its cap, lengthens into a cylindrical cone, and the pit surrounding its base deepens into a ***follicle,*** so that any suggestion of a scale is promptly lost. Now the basal portion of the prospective feather grows as a cylinder of ectoderm, producing a layer of horn (keratin) around itself and enclosing a ***pulp cavity*** within; this region is to be the ***quill.*** The surface layer of keratin splits away from the deeper, making a ***sheath.*** The deeper layer, while still covered by the sheath, begins to produce distally a series of hairlike ridges, at first still clinging together in a bundle; these are the future ***barbs.*** If the feather is to be a ***down*** feather, little more development is necessary except the breaking of the sheath and release of the barbs, which split apart distally but remain attached to the top of the hollow quill. The pulp cavity of the latter dries up. But if a ***quill*** feather develops, its ***shaft,*** beyond the quill, arises by the exaggerated growth of one ridge, among the prospective barbs, while still in the sheath; others then are drawn out, as it were, along both sides

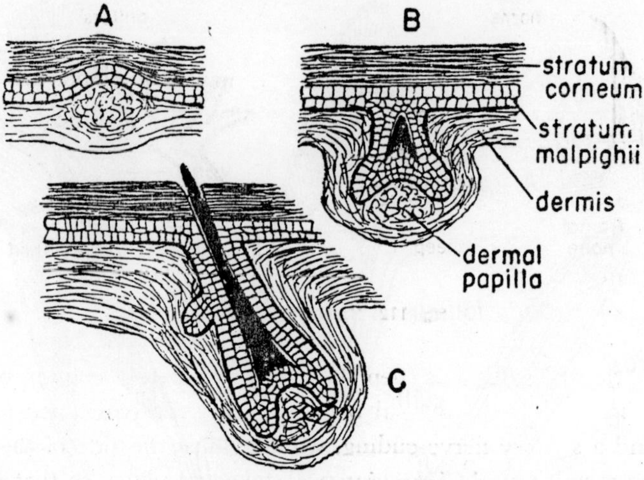

Fig. 111. Development of a hair.

of this shaft, until few or none are left attached to the quill directly. At the splitting of the sheath, the barbs separate only slightly, spread flat to make the ***vane***, and the pulp cavity of the quill dries and makes it a hollow tube, open at its two ends, the ***umbilici.***

Hair

Hair is limited to mammals, in which it performs at least two of the functions served by feathers in birds, namely insulation against temperature change, and coloration. Overlapping hair, especially thick fur or wool, carries a blanket of trapped air, slow to transmit heat. Hair pigments are the basis for protective coloration, or, as with skunks, for display. Hair also occasionally becomes a defensive weapon, as in the porcupine (rodent), European hedgehog (insectivore) and spiny anteater (monotreme).

Essentially, then, hair is a specialized outgrowth of the stratum corneum (Fig. 91), possibly originating from reptilian scales, although there is disagreement about this. A single hair consists of a dense rod of cornified dead cells, in which one can distinguish a ***medulla*** (middle), surrounded by a ***cortex,*** and sometimes a ***cuticle.*** This hair is produced rapidly from a cluster of germinative cells in the bottom of a pit, the ***hair follicle.*** In order to meet the special needs of a growing hair (speaking unscientifically) the epidermal germinative layer sinks into the dermis (Fig. 111), but at the bottom of the pit it makes a cap over a

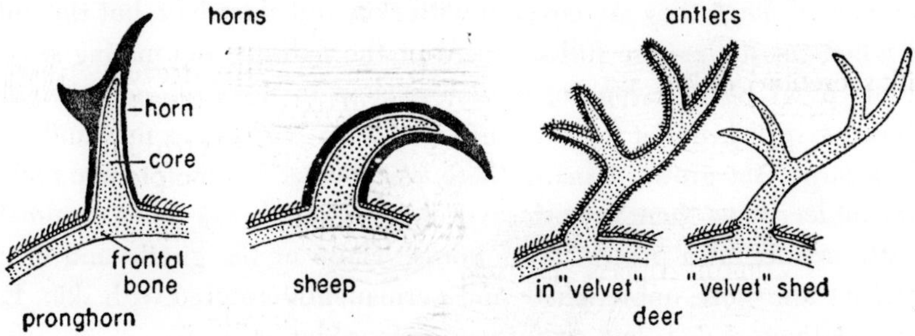

Fig. 112. Horns and antlers.

papilla of dermal cells, and then starts to proliferate a column of horny material. Meanwhile the dermal papilla usually receives a special blood supply and a sensory nerve ending, while against the side of the follicle a minute smooth muscle (***arrector pili***) takes its place, so that the hair can be moved. This is helpful in allowing air to enter the spaces among hairs, or for display of anger or fright.

The hair of most mammals is of ***definitive*** growth; that is, it reaches a certain length and then drops out. This happens at least twice a year in many mammals of temperate or polar climate, often with a change of color and density. The pubic and axillary hair of man is definitive, but that of the head is not. During late embryonic stages (human) there is a nearly complete coat of fine hair, the ***lanugo,*** on the head and body, but it is usually shed before birth.

There are many local specializations of hair, as the short screening hairs of nostrils and ear openings, to discourage entry of insects and dirt; the eyebrows and eyelashes, which help keep rain and sweat out of the eyes; the whiskers or vibrissae of a cat, which are delicate organs of touch; the tail switch used by nearly all grazing animals to drive away annoying insects; the broad winter fringe on the feet of snowshoe hares; the sex-limited mane of the lion.

SPECIAL ARMOR

Horns and Antlers

Under this heading we place, for convenience, structures of altogether different origins and quite unrelated composition. ***Antlers*** (Figs. 72, 112) of deer, elk, and moose are of bone, formed beneath the "velvet," connected with the frontal bone, and naturally cannot continue to grow except while they are supplied with blood vessels; therefore during the "velvet" stage they are covered with skin and short hair, but this rubs off when the antlers are full-sized. After the fighting and mating season antlers break off at a zone of weakness close to the skull, and then the following spring, under the influence of the sex hormone, a new and usually a larger set grows. Usually these weapons are limited to the males, but reindeer have them in both sexes. A great number of other mammals, mostly extinct, also possess bony horns. Those of the giraffe and okapi (Africa) are short, unbranched, and permanently covered with skin. Extinct relatives of deer had two, three or four "horns."

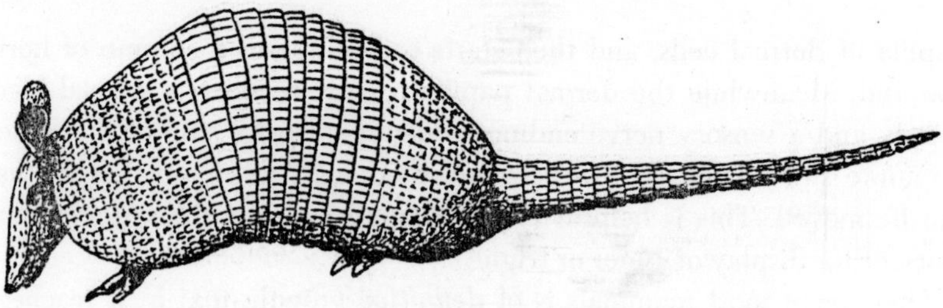

Fig. 113. Nine-banded armadillo, showing bony plates in skin.

The "hollow-horned ruminants" are those in which an outer epidermal shell grows as a cap over an inner core of bone, as in cattle, sheep, goats and antelope (Fig. 112). In all these the horn is permanent and unbranched, continuing to grow from the germinative layer between it and the bony core. Again, types with two, three, and four horns are known. But the pronghorn and Saiga antelope (not true antelopes) shed their branched horns annually, although they do not lose the core. Sexual differences in size and shape of horns are the rule, notably in sheep and cattle.

The extinct uintatheres and titanotheres had bony horns, the former as many as six. Curiously enough, there have been horned gophers and horned armadillos. The rhinoceroses belong in a different category, for their nose horns are integumental entirely, consisting of closely fused horny fibers; whether these fibers are modified hairs is a disputed question.

Scales and Plates in Mammals

Reptile-like scales are not uncommon in mammals, mingled with hairs, as on the tails of rats and their relatives, or the feet of kangaroos. But a more extreme sort of armor develops in the scaly anteater or pangolin (*Manis*, of the Malay region), encasing it, like a pine cone, in large overlapping plates. On the other hand, armadillos (Fig. 113) and the extinct glyptodonts have a mosaic of bony plates or knobs embedded in the dermis; these have nothing to do with hairs or with the stratum corneum.

DERMAL ARMOR

Scales, Plates, and Denticles

A dermal armor characterized the oldest known groups of fossil fishes, including ostracoderms, placoderms, paleoniscids, and Choanichthyes. Only in some secondarily modified descendants of these is there reduction to a naked or unarmored skin. The armor we are considering here is ***dermal***, in contrast to epidermal, for it develops in the dermis of the skin and has, so far as known, no epidermal components, even though the scales may push through the epidermis and be exposed to the outside.

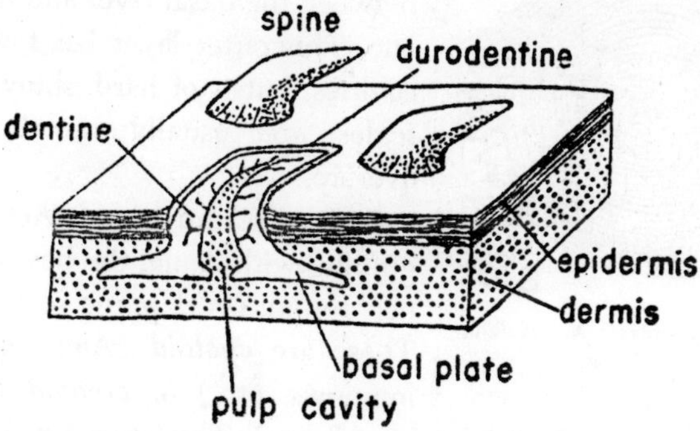

Fig. 114. Placoid scales or dermal denticles of shark, diagrammatic.

Sharks and skates (but not chimaeras) have an incomplete covering of fine broad-based *denticles* (Fig. 114) whose points project up and posteriorly through the skin, and whose general form is like a tooth. These are commonly called ***placoid scales.*** The supposed cap of enamel covering their tips is a hard form of dentine, and is produced, like the deeper dentine, by the dermis. ***Dentine*** is much like bone, and is provided in these scales as in teeth with a ***pulp cavity*** and ***dentine canals.*** A relationship clearly exists between teeth and dermal scales, although we may not be sure in which early group the transition took place.

Among Choanichthyes (lungfishes and lobefins) the ancestral type of scale is a *cosmoid* scale, a flat and rather elaborate structure. The deep

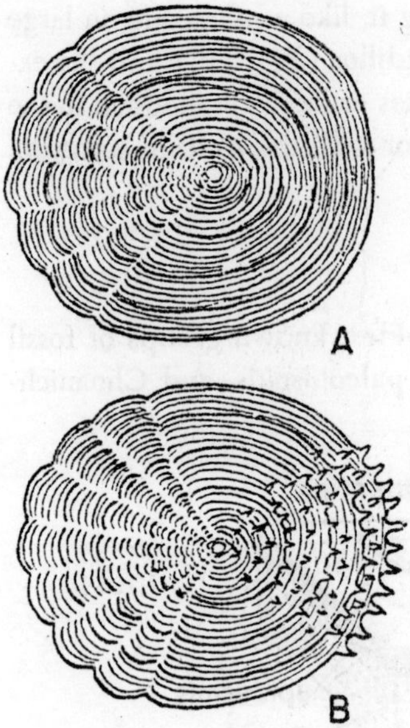

Fig. 115. Fish scales. A, cycloid; B, ctenoid.

layer is of bone. Over this is a middle layer of porous bone through which capillaries run. Outwardly there is a cover of cosmine (dentine, with many pulp cavities and a hard surface). These scales project throught the skin and overlap one another.

Ganoid scales, found in acanthodians and the primitive ray-finned fishes (Actinopteri) have little or no spongy bone between the basal layer and the superficial one. This outer layer has few or no pulp cavities, but is of hard, shiny ganoin. The scales are usually rhomboid, without overlap.

But more advanced Actinopteri, beginning with *Amia*, the bowfin, have thinner, round, overlapping scales (Fig. 115). These are ***cycloid*** (*Amia*, herring, carp, minnows, etc.) or ***ctenoid*** (perch, bass, flounder, and most advanced teleost fishes). The latter differ from cycloid in having toothed outer margins. During the life of a fish the scale grows around its edges, and since its rate of increment differs seasonally, concentric rings of growth show on the scale, by which it is possible to tell the age of the fish.

Among many families of fishes the scales are reduced, concealed or entirely lost, as in lampreys and hagfishes, chimaeras, most eels, catfishes, the ocean sunfish, etc. Sometimes the body becomes encased in a secondary armor of bone, as in trunkfishes (Fig. 116), or in spiked bony plates, as the porcupine fishes. Bony plates of many shapes and sizes occur in

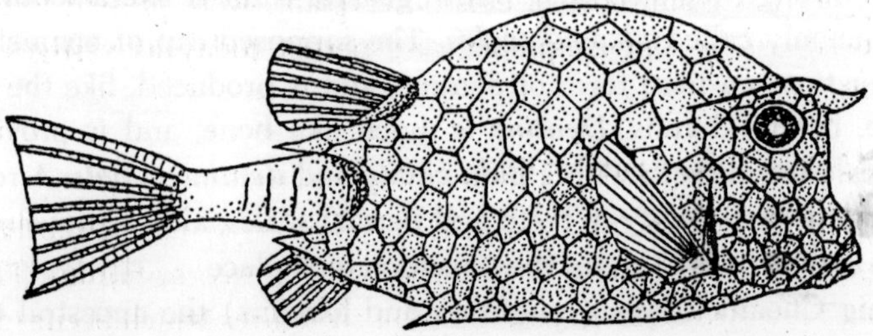

Fig. 116. Cowfish, showing plates of bone (not scales) in skin.

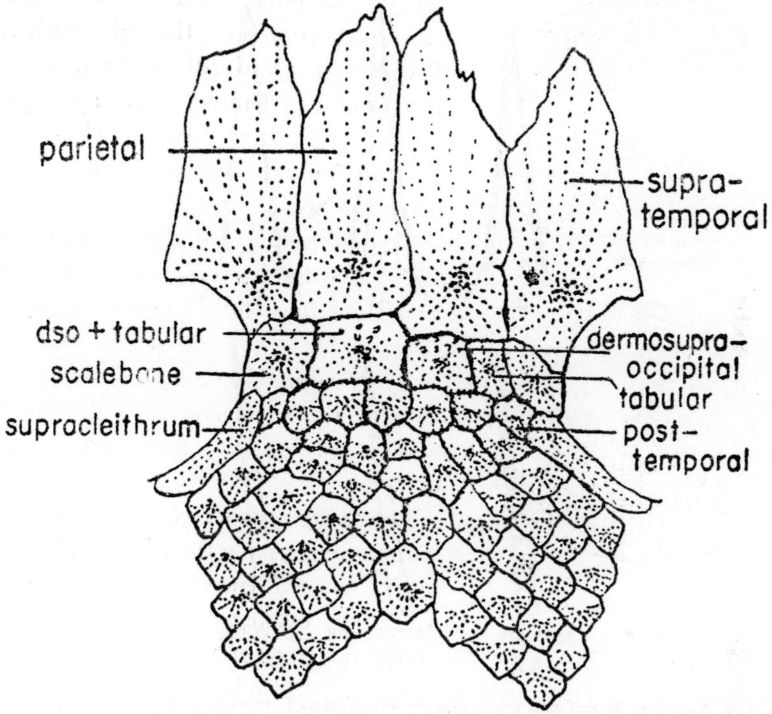

Fig. 117. Occipital region of gar, *Lepisosteus*, showing transition from ganoid scales to skull bones.

sturgeons, the extinct placoderms, and ostracoderms. A few show small denticles, surprisingly similar to placoid scales, as in some South American catfishes.

We need only add here that there is a direct, close relationship between dermal scales, as found in primitive fishes, and dermal bone of the skull and superficial parts of the pectoral girdle (Fig. 117). This will be discussed more fully in the chapter on the skeleton. Bony scales, presumably derived from those of crossopterygians, were present in the skin of some labyrinthodont Amphibia, and persist today as vestiges in the caecilians.

Dermal Spines and Fin Rays

Fin rays are the rods supporting the fins of fishes; they may be cartilaginous (lampreys and hagfishes), bony, horny, or fibrous. Those of most fishes are formed from two different structures which occur together or separately in various groups. These structures are: (1) paired rods

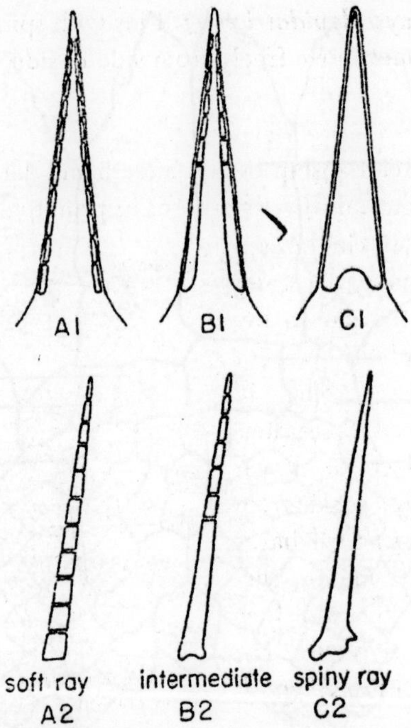

Fig. 118. Diagrams of soft and spiny rays in Actinopteri. 1, transverse sections; 2, lateral views.

called *ceratotrichia,* which are either (cf. Eaton, 1945) horny, fibrous, or bony, and in the latter case jointed; (2) *scale rows,* which tend during development to align themselves with the paired rods, and often to fuse with them or replace them entirely.

Elasmobranchii have fin rays composed of horny nonjointed ceratotrichia only. These may be supplemented by stiff spines at the anterior margins of dorsal and anal fins.

In bony fishes the paired rods, either horny or fibrous, are usually replaced by scale-rows forming jointed rays called *lepidotrichia* (Fig. 118). In lungfishes, lobefins, and some of the primitive Actinopteri the paired rods were often bony and jointed, with scale rows only partially fused with them.

Dermal spines, a common kind of armor in primitive fishes, seem to result from fusion of rows of bony scales. They are defensive weapons and secondarily may serve a fin as a cutwater. These primitive spines should not be confused with the *spiny rays* of many advanced teleosts (perch, sculpin, catfish, etc.), which arose by fusion of the separate units of soft or jointed rays (*lepidotrichia*). Fins with spiny rays may be lifted or lowered but cannot move freely from side to side, as the softer fins do.

TEETH

There is a relationship between teeth and dermal scales, as shown by their development and by their form in primitive fishes. When vertebrates began to use jaws in the capture of prey, presumably the scales at the margins of the mouth became modified to hold, cut, or tear, and an infinite variety of special uses followed thereafter. Teeth do not, in most lower vertebrates, occupy sockets in the jaw bones but rest their bases on the surface of bone or cartilage. Also it is the rule that such teeth, being lost easily, are replaced by new ones developing in series.

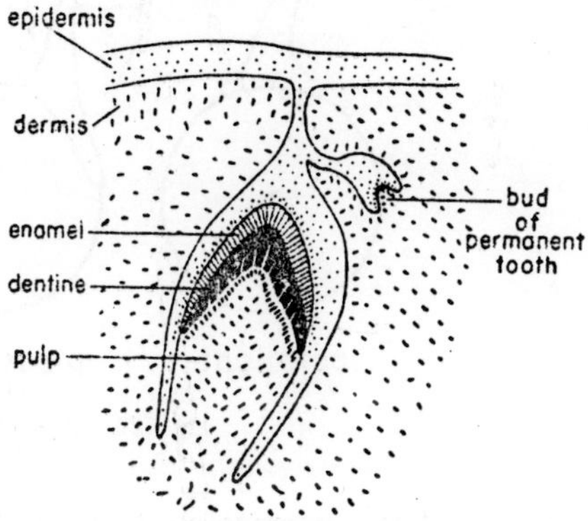

Fig. 119. Development of mammal teeth; a milk tooth preceding a permanent tooth.

In the development of a tooth (Fig. 119), the germinative layer of the epidermis sinks down into the dermis, and there spreads over a papilla of dermal cells. The inner, dentine layer of the tooth is then secreted by this ***dental papilla,*** while the enamel capping it is produced by the ***enamel organ*** from the epidermis. As the tooth enlarges, the dental papilla becomes the pulp, which contains ***odontoblasts*** (dentine-secreting cells), a blood vessel, and a nerve. As long as the tooth remains below the surface, ***ameloblasts*** (cells of the enamel organ) add to the enamel, but presently it erupts through the epidermis, and at the same time becomes attached to the bone or cartilage of the jaw. If this attachment is on the rim only, as in most fishes and amphibians, the dentition is ***acrodont*** (Fig. 120); if on the inner margin of the bone, as in most lizards, it is ***pleurodont;*** if in sockets, as with crocodiles, dinosaurs, toothed birds, and mammals, it is ***thecodont.*** In the latter case the tooth does not push down into the bone, but the bone grows up around the root of the tooth, and at the zone of contact ***cement*** is deposited by dermal cells, to anchor the roots against the periosteum of the jaw.

When teeth may be replaced several times, the succession is ***polyphyodont.*** Beneath the integument just behind or below the functioning tooth rows there is a series of developing buds, those most nearly complete being nearest the functional teeth, so that when one of the latter drops out the new tooth next behind replaces it; this occurs in many fishes, amphibians, and some reptiles (lizards and snakes). Mammals are generally ***diphyodont,*** having only two sets of teeth, the ***deciduous*** or "milk"

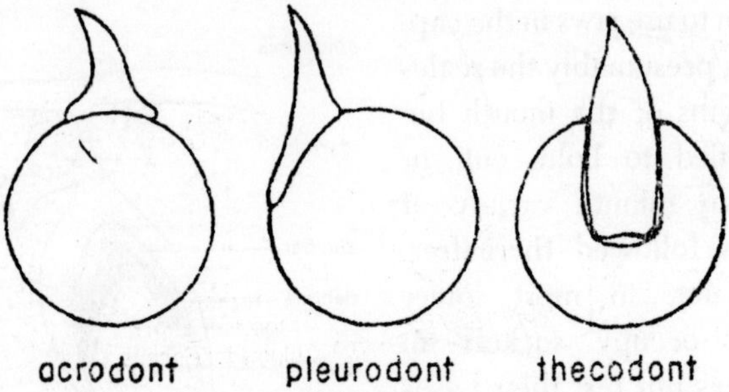

Fig. 120. Types of tooth attachment.

teeth and the ***permanent*** teeth (Fig. 121). A few are ***monophyodont*** (moles, some rodents, and the toothed whales).

The comparative anatomy of teeth is of great interest. Sharks, which are predatory, have sharp conical teeth, with from one to five prongs (Fig. 122). Rays have a flat pavement of grinders for crushing hard-shelled food. The chimaeras (Holocephali), like lungfishes (Dipnoi), have large crushing plates in the mouth, but not separate teeth. Most bony fishes have simple conical teeth in single or multiple rows, with additional patches on the palatines, vomers, tongue, and pharyngeal bones, but a number have become toothless, for example sturgeons, carp, minnows, suckers, pipefish, and seahorses. The lobefins (Crossopterygii) established a tooth pattern which was carried on into the Amphibia and many reptiles, viz., two rows in the upper jaw, the outermost of these on maxillary and premaxillary bones, the inner on vomers, palatines and ectopterygoids. In the lower jaw there was also a double row, becoming single in tetrapods. Toads are toothless, while frogs have only their uppers left. Turtles seem to have been toothless from the beginning. Many poisonous snakes have grooved or hollow fangs which transmit the poison from the ducts of the poison glands into the prey (Fig. 122).

The ruling reptiles and primitive extinct birds were thecodont; so also were mammal-like reptiles, and of course the mammals inherited this. In these two groups there comes a distinction between ***incisors*** (front nippers), ***canines*** (enlarged fangs), and ***molars*** (chewing or grinding). Such dentition is *heterodont* (Fig. 123), in contrast to the uniform or ***homodont*** dentition of most lower vertebrates. It is interesting that the toothed whales reverted to homodonty as they became marine (Fig. 124).

We should now look at the specializations of teeth among mammals,

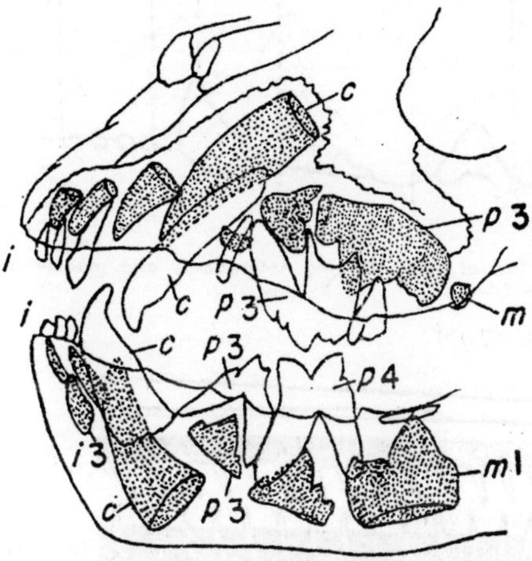

Fig. 121. Jaws of six-months lion, showing tooth replacement. c, canine; i, incisor; m, molar; p, premolar. (From Kingsley, after Weber.)

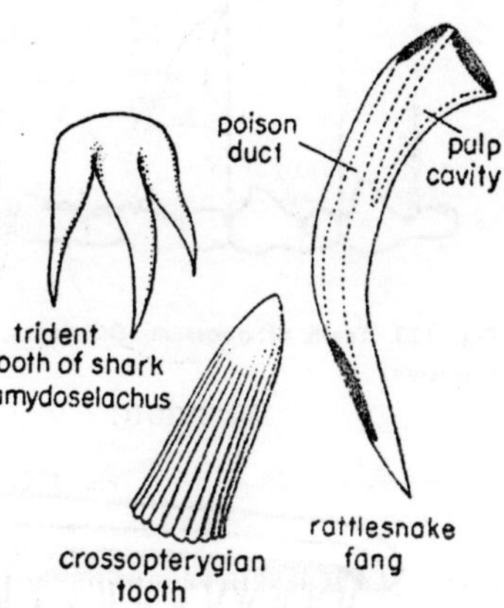

Fig. 122. Examples of functional modifications of teeth.

and in so doing pass over the practically toothless monotremes (only the duckbill has temporary rudiments of teeth). The primitive number in marsupials is about 50; their arrangement is expressed by a ***dental formula,*** which indicates the number of each kind of teeth on one half of each jaw. For the opossum (Fig. 123) this is: $\frac{5\text{–}1\text{–}3\text{–}4}{4\text{–}1\text{–}3\text{–}4}$, meaning 5 incisors (4 below), 1 canine, 3 premolars, 4 molars. Multiplying these numbers by 2 (for the two sides) gives 50. Most marsupials have reduced the number considerably. Placental mammals seem to have started with 44: $\frac{3\text{–}1\text{–}4\text{–}3}{3\text{–}1\text{–}4\text{–}3}$. This formula is found in tree shrews, early Primates, early ungulates and pigs, but again, it is reduced in the progressive types. The human formula is $\frac{2\text{–}1\text{–}2\text{–}3}{2\text{–}1\text{–}2\text{–}3} = 32$; cat $\frac{3\text{–}1\text{–}3\text{–}1}{3\text{–}1\text{–}2\text{–}1} = 30$; porcupine $\frac{1\text{–}0\text{–}1\text{–}3}{1\text{–}0\text{–}1\text{–}3} = 20$. It should be noted that the incisors, canines, and premolars in placental mammals (but not in marsupials) are deciduous, while the molars come only once, in the permanent set.

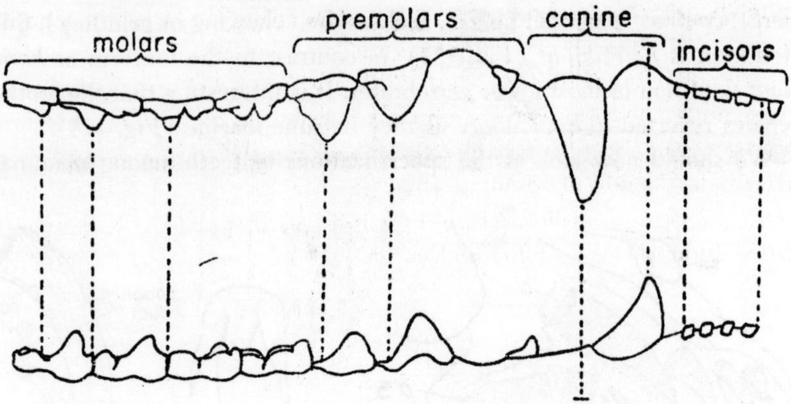

Fig. 123. Teeth of opossum, *Didelphys*, right lateral view, showing heterodonty and manner of occlusion.

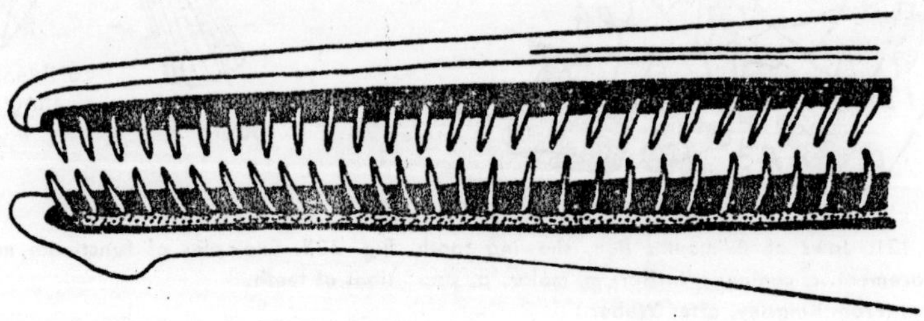

Fig. 124. Jaws of porpoise, *Lagenorhynchus*, showing homodont teeth. (After Parker and Haswell.)

It remains to say something of the form of molar (and premolar) teeth, for these present some of the most important characters indicating relationships in mammals; it has even been remarked that some investigators act as if a mammal were made of nothing but molar teeth. The incisors and canines, being simple in form, present no particular problem, but the premolars and molars (Fig. 125) usually have a complex, uneven surface of rounded cusps (***bunodont***), crescents (***selenodont***) or ridges (***lophodont***) of enamel. It is now thought that the primitive molar pattern in certain Mesozoic mammals was with three cusps, so arranged that those of the lower teeth alternated (approximately) with those of the uppers (Fig. 125C, D). Additional cusps later appeared on the posterior ends of the lower molars, and the further evolution of these teeth was correlated with specialized feeding habits. Herbivores, no matter of what order, generally develop a more and more elaborate selenodont or lophodont pattern, since this is most effective for grinding resistant vege-

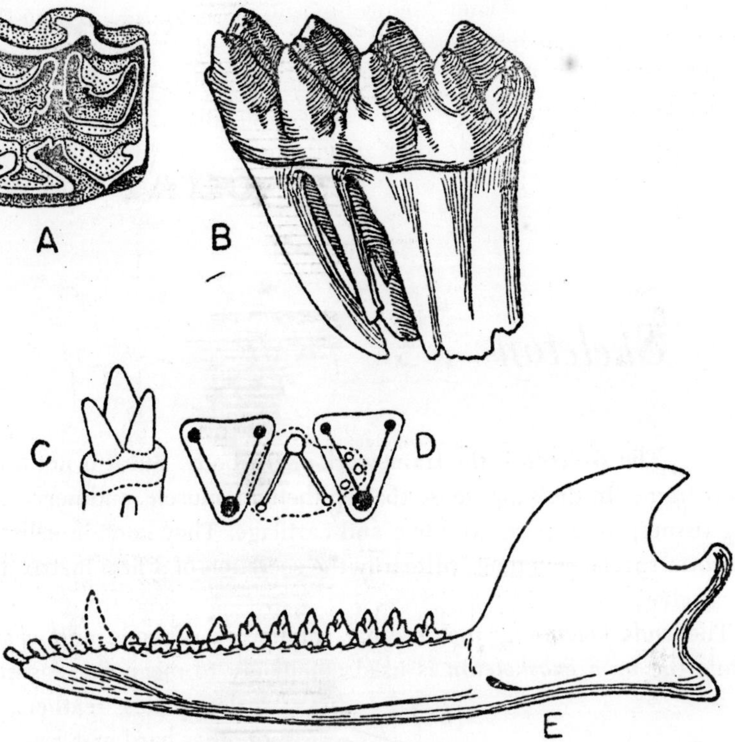

Fig. 125. Types of teeth. A, molar tooth of horse, showing selenodont pattern; B, tooth of mastodon, bunodont type (after Scott); C, tritubercular tooth of *Spalacotherium* (after Kingsley); D, occlusion of upper (dark) and lower (light) tritubercular molars (after Kingsley); E, lower jaw, inner side, of *Amphitherium prevostii*, a Jurassic trituberculate mammal (redrawn from *A Catalogue of the Mesozoic Mammalia in the Geological Department of the British Museum* by Simpson).

tation with the side-to-side motion of the lower jaw. Those of carnivores and primates usually remain bunodont or develop shearing edges, as in cats, for cutting meat; such shearing molars are *carnassials.* Rodents have no canines, but a space, the *diastema,* remains between the enlarged gnawing incisors and the premolars. Anteaters (Xenarthra) have no teeth at all.

CHAPTER IV

Skeleton

The skeleton is the framework of the body. But it is not a wholly inert frame. In the same sense that epithelium, muscle, and nerve are living tissues, so, in part, are bone and cartilage. They contain cells which perform special functions, primarily the secretion of a firm matrix around themselves.

The ***endoskeleton*** is the internal framework of bone and cartilage, while the term ***exoskeleton*** is used commonly to mean the covering of scales, plates, feathers, hairs, or other hard materials on the surfaces of most animals. This is an arbitrary distinction, for the scales of fishes are internal in origin, and one cannot, in many cases, draw the line between scales, plates, and the superficial bones of the skull, which are part of the "endoskeleton."

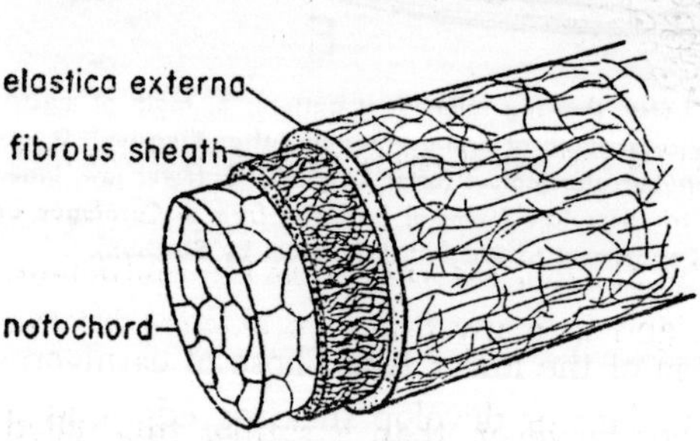

Fig. 126. Diagram of notochord and sheaths.

The earliest sign of a skeleton that appears in the development of man or any other chordate is a semi-stiff rod along the axis of the body, the ***notochord*** (Fig. 126). This is neither typical cartilage nor bone, for its cells secrete almost no matrix. Instead they grow to a large size, so distended by watery vacuoles that the protoplasm is crowded against the cell membrane, while the notochordal epithelium secretes a tight sheath around the notochord so that the turgor of the cells within keeps it rigid.

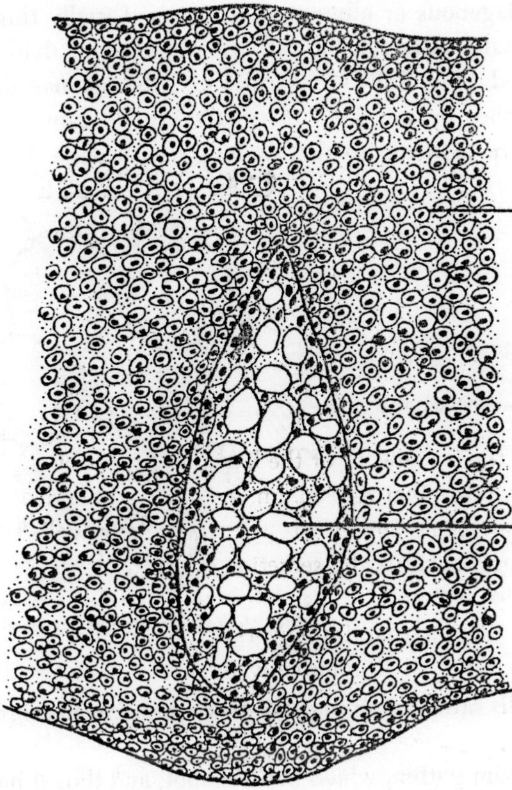

Fig. 127. Section through notochord in basal plate of skull of embryo snapping turtle, *Chelydra*, showing two types of cartilage.

This sheath, of elastic fibers, is the ***elastica externa.*** Next the epithelium secretes an inner ***fibrous sheath,*** and in many fishes an ***elastica interna*** too.

The notochord is no more rigid, however, than a rubber tube filled with water under pressure. It bends readily, but springs back straight. In an embryo, therefore, such a simple axial rod gives partial support to the body and serves as a central spring against which the body muscles can pull, first on one side and then on the other, in fishlike swimming motions.

Cartilage or gristle is soft enough to cut with a knife, and contains a large amount of water so that it shrinks when dry. But it is firm enough to provide support for other tissues, attachment for muscles, or protective enclosure for delicate part. Cells lie throughout cartilage in minute spaces called ***lacunae*** (Figs. 127, 128), singly or in pairs or fours, all secreting a protein (collagenous or albuminous) matrix. Usually this contains no hard calcium salts, although the cartilage skeletons of sharks and rays are partly calcified (Fig. 129). As a rule, too, the lacunae do not connect with one another by any kind of canals, nor do capillaries or nerve endings penetrate to them, so that as long as cartilage does not

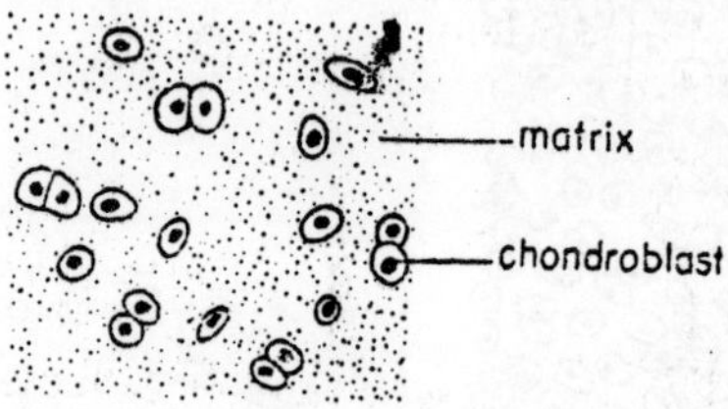

Fig. 128. Detail of hyaline cartilage.

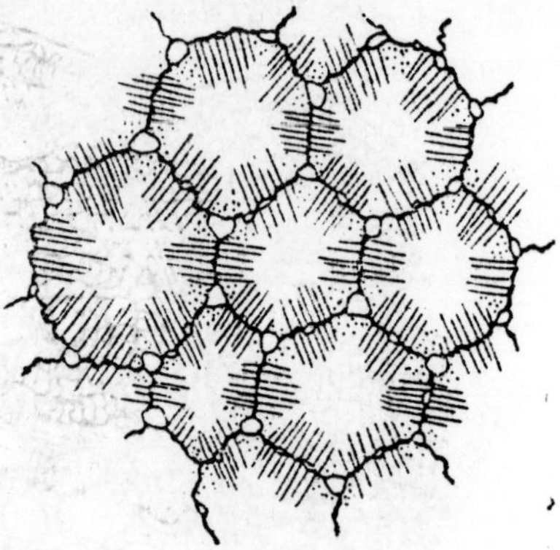

Fig. 129. Plates of calcified cartilage in surface of chondocranium of ray, much enlarged.

grow too thick these cells satisfy their physiological requirements through the water of the matrix.

Cartilage can grow from within, which bone cannot, and thus it has an advantage over bone in the early stages of development when the shape of the skeleton must change rapidly. The skeletons of embryos consist of cartilage, and bone is then added afterwards, either replacing the cartilage or appearing in new positions.

Precartilage is the name given to cartilage when its matrix first begins to form around young ***chondroblasts*** (cartilage cells). ***Hyaline*** cartilage has a firm, translucent matrix; it occurs in the head as a forerunner of the bony skull, in growing limbs and the vertebral column, in the series of rings which keep the trachea and bronchial tubes from collapsing, and in many other parts of the skeleton. When this matrix becomes filled with white connective tissue fibers, as in the tough intervertebral discs, it is ***fibrous*** cartilage. ***Elastic*** cartilage is occupied by fibers of yellow elastic tissue, where flexibility is necessary, as with the external ear and the epiglottis.

Although ***bone*** never comes as early in life as cartilage, it seems to be fully as old historically, for the most ancient known fishes were more bony than cartilaginous. But its matrix is much harder, has a higher specific gravity, and adjusts its shape less readily during developmental changes. Instead of protein, it is almost entirely lime. Therefore the bone cells cannot depend on the matrix for water, food, and gas exchange, but must all be reached by minute ***canaliculi,*** which in turn are supplied

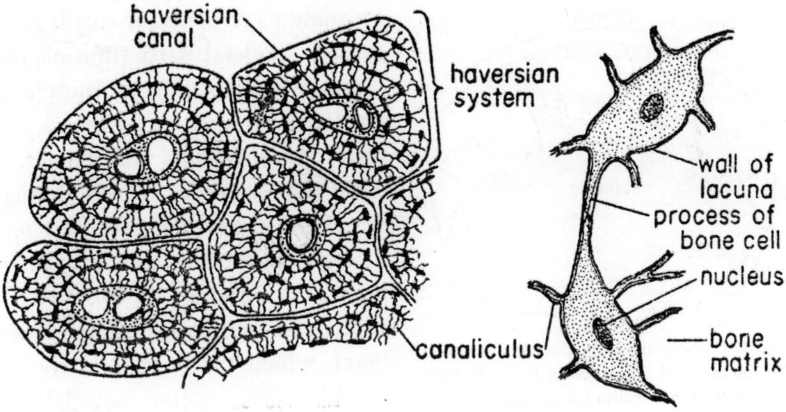

Fig. 130. Section through a group of Haversian systems in bone, and detail of two cells.

from capillaries. During development bone forms around microscopic blood vessels and nerves, and frequently encloses other kinds of tissue as well. The part of a bone which bears the most strain, which contains no visible cavities, is ***compact*** bone; other portions, as the deeper part of large bones, are full of small spaces called ***cancellae;*** this is ***cancellous*** bone.

Bone cells lie in regular order, grouped concentrically around capillaries (Fig. 130). The capillary occupies a ***Haversian canal.*** Radiating from this are countless extremely fine ***canaliculi,*** which pass through a narrow layer or ***lamella*** of bone and reach the lacunae occupied by the cells which secreted that lamella. By other canaliculi these cells connect with one another and with a further series beyond the next lamella, and so on for several such layers. But presently they overextend their communication lines, so to speak, and come under the influence of another Haversian canal, which has its own series of concentric lamellae and cells. Thus bone consists of innumerable ***Haversian systems,*** each built around a blood vessel.

Since the matrix is calcareous, the cells can never change their positions or secrete more bone than will fill up the spaces between them.

Therefore bone does not grow from within, but only by addition on the outside. It can, however, be absorbed on the surface, by ***osteoclasts*** or "bone-devouring" cells, and thus change its shape while developing.

Functions of the skeleton are largely based upon resistance to strain. Muscle attachment is perhaps the primary function, since movement and support of the body depend upon interrelations of bone (or cartilage) and muscle. Skeletal parts transmit movement, which begins as muscle contraction, by acting as levers. They often amplify this movement; the hand, for example, swings through a wider arc than the basal part of the forearm where the biceps muscle attaches. In order to do this the bones of the arm must resist the tendency to bend, which the weight of the hand would cause a softer material to do. The sound-transmitting bones of the middle ear are a special case of lever action, for they translate vibrations of the ear drum into vibrations of endolymph.

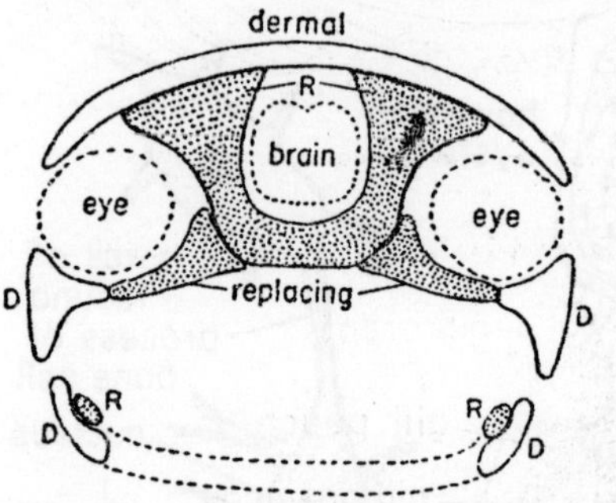

Fig. 131. Diagram of relationship between (D) dermal and (R) replacing bones in cross section of a generalized skull.

Sesamoid bones, forming as nodules in tendons, are a special example of resistance to strain, for they occur where tendons pass over a joint (kneecap or patella). The majority of bones replace, during development, cartilages which occupied corresponding positions in the younger stages (Fig. 131). Such are called ***replacing*** or ***cartilage*** bones. This does not imply that the cartilage becomes bone. Other bones, especially superficial parts of the skull and shoulder girdle, develop in membranes under the skin, or in the skin, without reference to any previous cartilage. These are ***dermal*** or ***membrane*** bones. In the detailed structure of these two categories of bone there is no distinction, but only in their topographic relation to cartilage.

We customarily divide the skeleton into an ***axial*** portion (the skull and vertebral column with their accessory parts), and an ***appendicular*** portion (limbs, with pectoral girdle and pelvic girdle). This distinction does not hold in all vertebrate animals, but is convenient.

The skull, as shown by its development, is essentially a ***cranium*** or braincase, with which are incorporated the nasal capsules (cartilage enclosing pits for the olfactory organs) and otic capsules (cartilage sur-

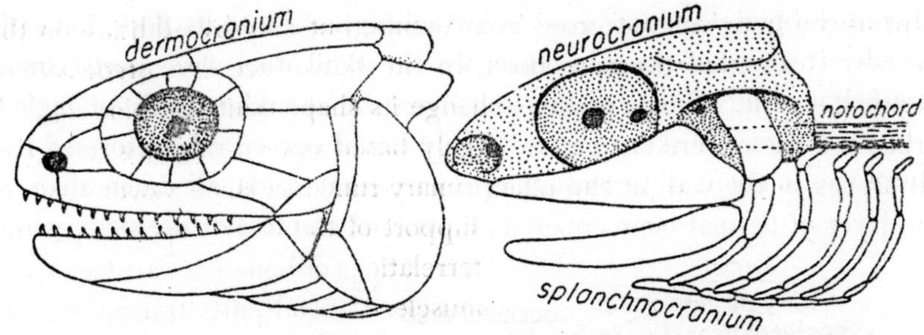

Fig. 132. Diagrams of dermocranium, neurocranium, and splanocranium in fishes.

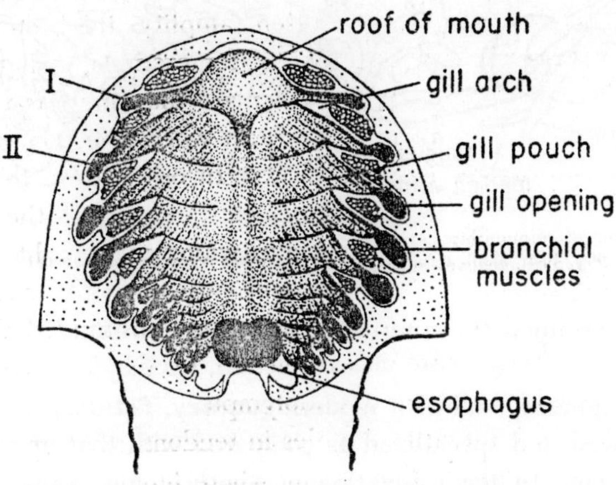

Fig. 133. Ventral aspect of skull of cephalaspid ostracoderm, showing oropharynx from below. I, mandibular cleft; II, hyoid cleft. (From Gregory, 1933, Transactions of the American Philosophical Society, vol. 23, after Stensiö.)

rounding the inner ears). It is not correct to include "optic capsules," which form the sclerotic coats of the eyeballs, for there is no evidence that they were ever a part of the skull. Over this ***neurocranium,*** preformed in cartilage, there is added in most vertebrates a ***dermocranium*** consisting of dermal bone (Fig. 132). The jaws, hyoid arch, and cartilages found in the throat are the ***splanchnocranium*** or ***visceral skeleton,*** originally developed in association with gill support.

COMPARATIVE ANATOMY OF THE SKULL

The Agnatha (Cyclostomes)

In its oldest stages, as in ostracoderm fishes, the skull formed a protective covering for the brain, sense organs, and most other soft parts of the head (Fig. 133). At the same time it was a strong base for attachment of the body muscles used in swimming, and into its floor, beneath the brain, extended the notochord. In the skull, too, were parts corresponding to the visceral arches and even the shoulder girdle of higher forms, but these were not jointed. Ten pairs of curved ridges formed the gill arches in the wall of the pharynx in cephalaspids, all much alike. A thin layer of dermal bone covered all external surfaces of the skull, while

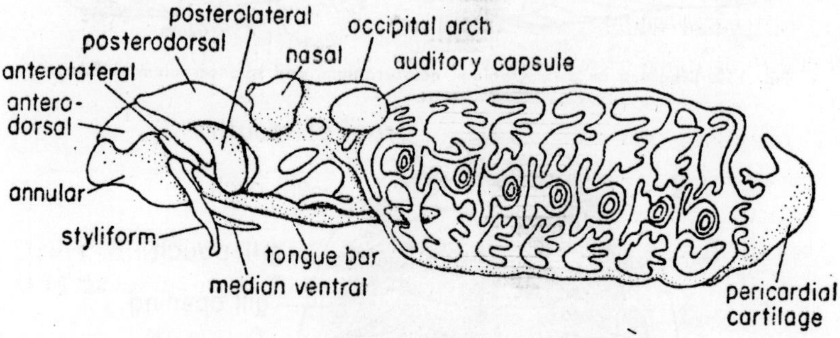

Fig. 134. Skull, adult lamprey. (Modified from various sources.)

replacing bone lined the internal cavities and surrounded the cartilage portions.

The cartilaginous skull of a modern lamprey, *Petromyzon* (Fig. 134), is both reduced and specialized as compared with that of cephalaspids. It gives no clue to the derivation of gnathostome skulls (those with jaws). But the ***larva*** of *Petromyzon* (Fig. 135) has perhaps the simplest skull of any vertebrate. A pair of diverging rods, the ***anterior parachordal*** cartilages, lie at the sides of the forward end of the notochord; two ***posterior parachordals*** come a little farther back, next to the ear capsules. Soon these capsules and the four parachordals meet and fuse, making an incomplete floor and side walls for the brain. A single nasal capsule appears just in front. Meanwhile small bars develop between successive gill pouches, which lie farther back than the brain; these ***visceral cartilages*** unite to make a jointless gill basket, fusing anteriorly with the parachordals.

Earliest Gnathostomes

Probably the greatest advance ever to take place in the evolution of the skull was the formation of jaws from a gill arch (Fig. 136). This may have accompanied a change in feeding habits from mud grubbing to capture of active prey. With this came a jointing of all the arches.

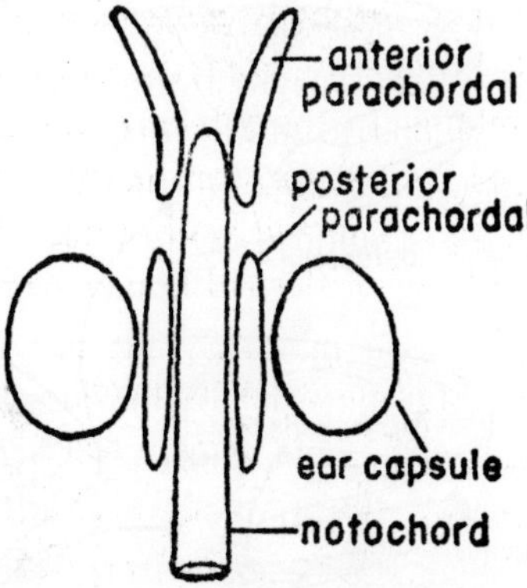

Fig. 135. Early larval skull of lamprey.

The arch at the rim of the mouth appears to have bent forward both above and below the mouth, forming on each side an upper and a lower jaw (***palatoquadrate*** and ***mandible***). This condition we meet in acanthodians. The jaws now bore teeth, yet had not forsaken their gills, a full set of which remained along their posterior margins. When each upper jaw, as in this case, articulates movably in two places with the cranium but receives no hyoid support, it is said to have ***autodiastylic*** suspension (De Beer and Moy-Thomas, 1935). The manner of jaw suspension is important in skull evolution, and the major types will be noted as we go along (Fig. 137).

Although little is known of the inner cranium of the oldest gnathostomes, they had a striking diversity of dermal armor. In antiarchs and arthrodires there were large headplates bearing lateral-line canals, and the anterior part of the trunk was armored like the head (Fig. 186). In acanthodians the head, body, and fins were covered with small diamond-shaped scales. In all the eyes were far forward, usually well in advance of the angle of the mouth, and the front part of the brain cavity was between the eye sockets; this type of cranium is ***platybasic,*** in contrast to the ***tropibasic*** type found later in teleost fishes, many reptiles, birds, and a few others, where the orbits practically meet each other in the middle of the head, and the brain lies behind them.

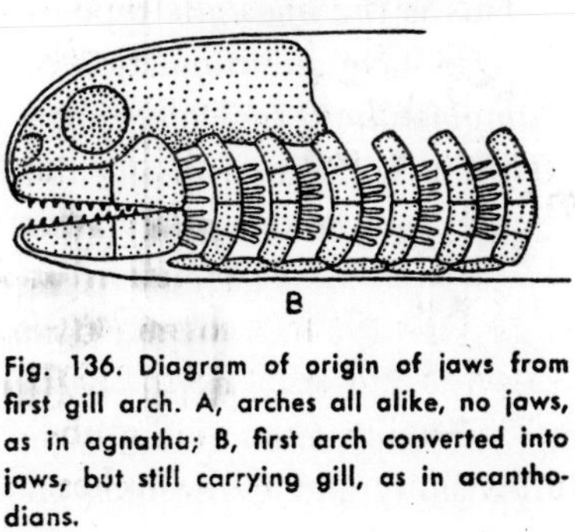

Fig. 136. Diagram of origin of jaws from first gill arch. A, arches all alike, no jaws, as in agnatha; B, first arch converted into jaws, but still carrying gill, as in acanthodians.

Elasmobranch Fishes

The ***chondrocranium*** of elasmobranch skulls is simply the cartilage braincase, persisting without the addition of bone (Fig. 138A). In lacking bone the elasmobranch skull is specialized (degenerate, if you

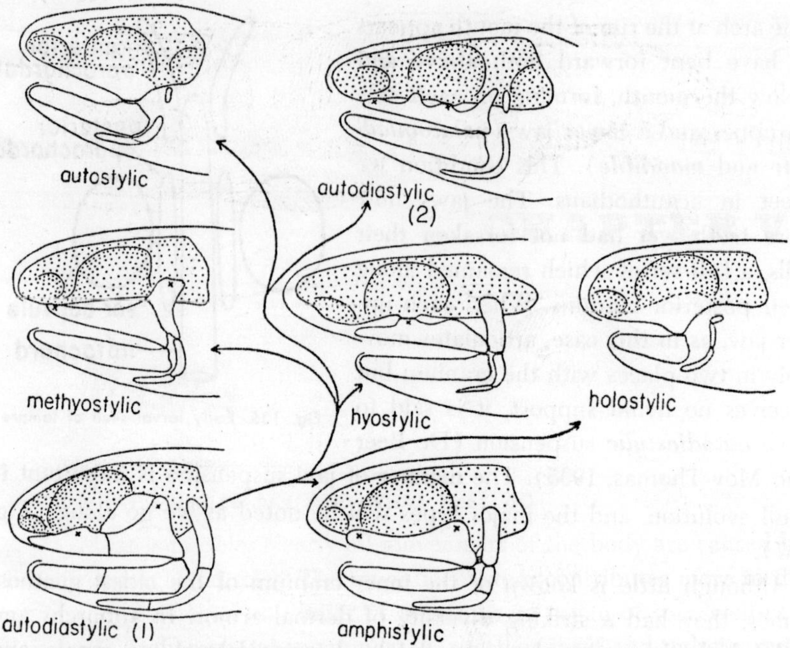

Fig. 137. Methods of jaw suspension in vertebrates. Most primitive, *autodiastylic (1)*, is that of acanthodians. *Amphistylic* occurs in some primitive sharks. *Holostylic* is found in Holocephali. *Hyostylic* is the type in modern sharks and rays, and *methyostylic* is that of Actinopteri. *Autodiastylic* (2) is characteristic of labyrinthodont Amphibia, while *autostylic* occurs in lungfishes, modern Amphibia, many reptiles, and mammals. See text for more details.

like). In most sharks and rays the upper jaw remains loosely articulated with the cranium at the basal process, just under the orbit. But the wide-open gill cleft which was present in acanthodians has closed from the bottom up, like a zipper pocket, leaving only a vestigial cleft, the *spiracle*, at the top, as if the zipper had jammed before it quite got there. Thus the angle of the jaw fits against, and is supported by, the upper piece of the hyoid arch (***hyomandibular***). This is the ***hyostylic*** type of jaw suspension (Fig. 137). Primitive sharks also have an articulation of the palatoquadrate with the otic region (*amphistylic*). In Holocephali (ghost fishes) the spiracular cleft closes entirely without involving the hyoid arch in support of the jaw, for the upper jaw unites solidly with the cranium; we call this jaw ***holostylic.***

Actinopteran Fishes

The Actinopteri or ray-finned fishes are an immense class, containing all higher fishes except Dipnoi (lungfishes) and Crossopterygii

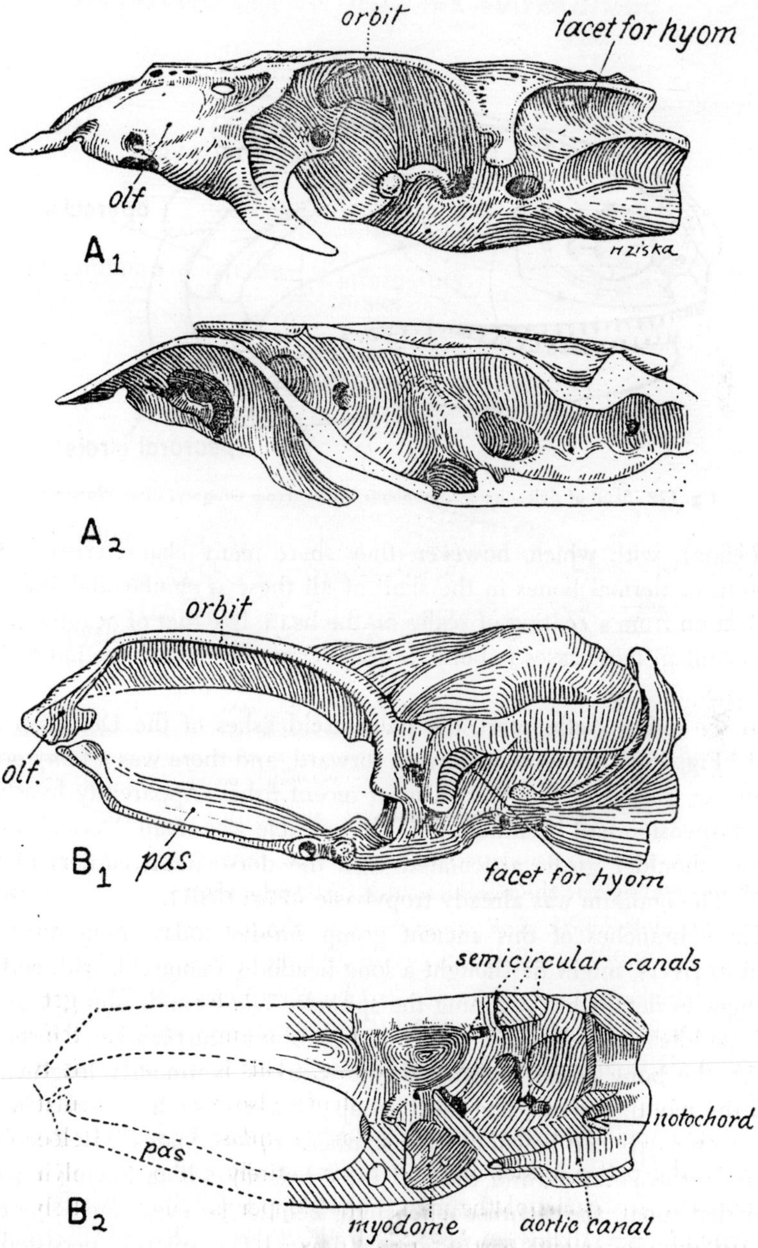

Fig. 138. Shark neurocranium (A_1) compared with paleoniscid neurocranium (B_1) in lateral view. A_2 and B_2 are sagittal sections of the same. (From Gregory, 1933, Transactions of the American Philosophical Society, vol. 23, after Allis and Watson.)

***hyom,* hyomandibular; *olf,* olfactory capsule; *pas,* parasphenoid.**

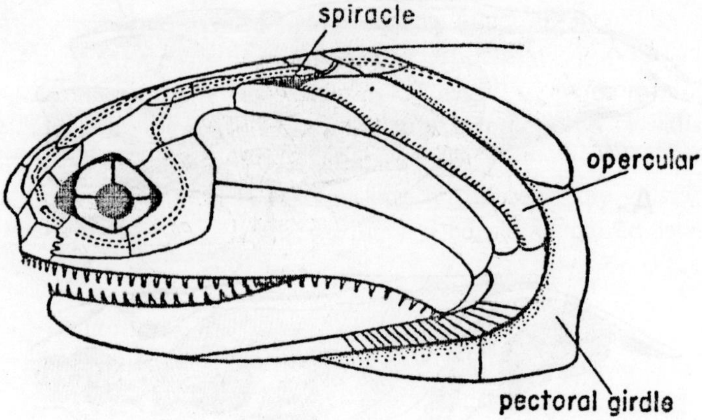

Fig. 139. Skull of paleoniscid, *Cheirolepis trailli.* (From Gregory, after Watson.)

(lobefins), with which, however, they share many characteristics. The pattern of dermal bones in the skull of all these is similar and suggests derivation from a coating of scales on the head, like that of acanthodians. The skull in most cases is bony, but a few kinds show reduction or loss of bone.

In the oldest Actinopteri, the paleoniscid fishes of the Devonian period (Fig. 139), the eyes were well forward, and there was a bony operculum or gill cover like that in most recent fishes. Apparently hyostylic jaw suspension had been adopted. The spiracle was small. Dermal bones of the shoulder girdle articulated with the dorsolateral corners of the skull. The cranium was already tropybasic (Fig. 138B).

Three branches of this ancient group survive today. *Polypterus,* of African rivers, might be thought a long-headed paleoniscid, with certain changes in details but retaining the spiracle. The Chondrostei, sturgeons and paddlefishes, declined to a mud-grubbing, cartilaginous form, in which the bony armor is reduced, the rostrum enormously lengthened, and the mouth ventral. The third group, Holostei, we have today in the freshwater gars and bowfins of the southern United States. The first ones came in the Permian, and differed from paleoniscids in having a more extended snout, the mouth not reaching behind the eyes, and some correlated rearrangements of the dermal bones. The spiracle persisted in some; it is found in the bowfin, *Amia.*

The great subclass Teleostei, containing all higher fishes, was derived in the Jurassic from Holostei. They have no spiracle, but it is difficult

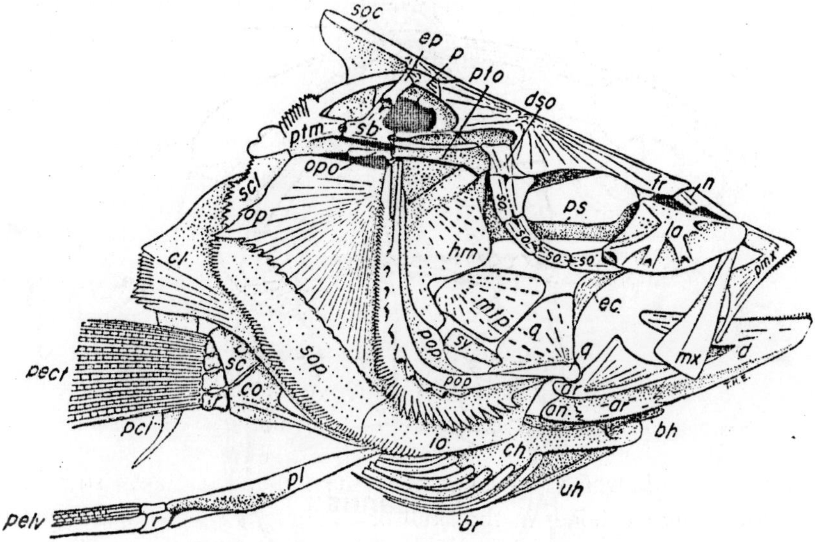

Fig. 140. Skull of yellow perch, *Perca flavescens*.
an, angular; *ar*, articular; *bh*, basihyal; *br*, branchiostegal ray; *ch*, ceratohyal; *cl*, cleithrum; *co*, coracoid; *d*, dentary; *dso*, dermosphenotic; *ec*, ectopterygoid; *ep*, dermal epiotic; *fr*, frontal; *hm*, hyomandibular; *io*, interopercular; *la*, lacrimal; *mtp*, metapterygoid; *mx*, maxillary; *n*, nasal; *op*, opercular; *opo*, opisthotic; *p*, parietal; *pcl*, postcleithrum; *pect*, pectoral fin; *pl*, pelvis; *pop*, preopercular; *ps*, parasphenoid; *ptm*, posttemporal; *pto*, pterotic; *q*, quadrate; *r*, radial; *sb*, scale bone; *sc*, scapula; *scl*, supracleithrum; *so*, suborbital; *soc*, supraoccipital; *sop*, subopercular; *sy*, symplectic; *uh*, urohyal.

otherwise to define the skull characters, for the diversity of form is almost unlimited. Ordinarily the cranium is overlain by numerous thin dermal bones, some of them carrying canals or grooves for the sensory canal system (Figs. 140, 141).

1. *Neurocranium.* The bulging posterior section contains the brain, and its cavity is confluent with those of the inner ears. Only a thin septum lies between the orbits, while the anterior section consists of a dorsal *ethmoid* and a ventral *vomer*, the latter usually with teeth.

2. *Dermal bones.* Roughly these are in four groups:

a. Roof: *Nasals, frontals, parietals, posttemporals*, etc.

b. Circumorbitals: A ring of small bones around each eye, bearing a sensory canal.

c. Dermal jaw bones: For the upper jaw, a small *premaxillary* on each side, followed by a larger *maxillary*. In advanced kinds the maxillaries recede from the margin of the mouth, and their

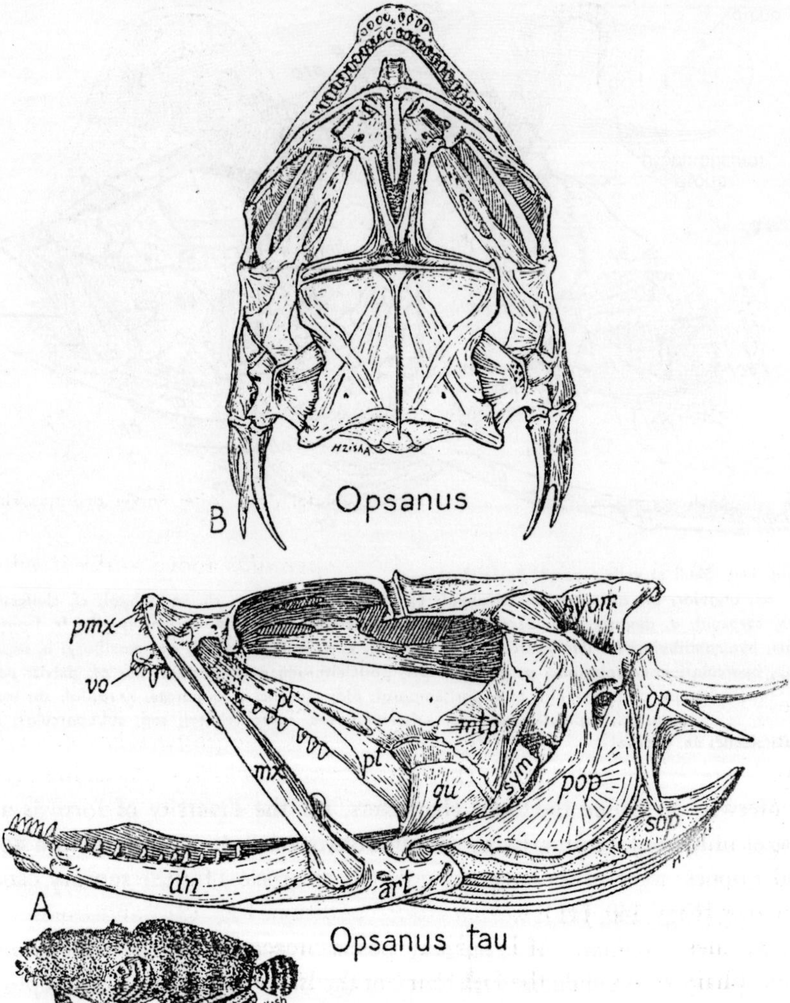

Fig. 141. Skull of toadfish, *Opsanus tau*. A, lateral; B, dorsal. (From Gregory, 1933, Transactions of the American Philosophical Society, vol. 23.)

art, articular; *dn*, dentary; *hyom*, hyomandibular; *mtp*, metapterygoid; *mx*, maxillary; *op*, opercular; *pl*, palatine; *pmx*, premaxillary; *pop*, preopercular; *pt*, pterygoid; *qu*, quadrate; *sop*, subopercular; *sym*, symplectic; *vo*, vomer.

places are taken by the enlarged, toothed premaxillaries, which become protrusible. In the lower jaw the *dentaries* form the tooth-bearing rim.

d. Opercular series: Just behind the jaws come the bones of the gill

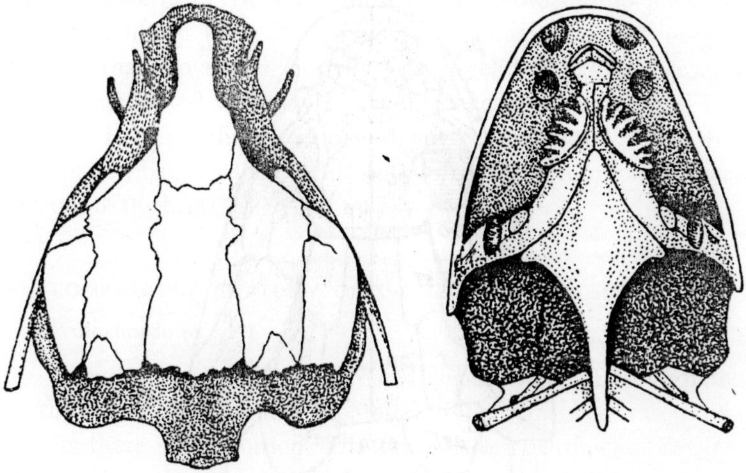

Fig. 142. Skull of lungfish, *Epiceratodus;* dorsal and palatal views. (After Parker and Haswell.)

cover, diminishing from the large ***opercular*** above to the slender ***branchiostegal rays*** beneath.

3. ***Primary jaws and suspensorium.*** The upper jaw, preformed in cartilage, usually ossifies in 4 or 5 pieces. The term ***suspensorium*** refers to these bones plus the upper pieces of the hyoid arch, all of which work as a mechanical unit in supporting the lower jaw. The primary lower jaw is represented in teleosts by (1) the ***articular*** bone, which replaces the base of ***Meckel's cartilage,*** and (2) the unossified distal part of Meckel's cartilage, hidden and enclosed by the dentary. The ***hyomandibular*** is large and important; it articulates dorsally on the side of the cranium, and supports the upper jaw (***methyostylic*** condition, Fig. 137) and the opercular bones. Five pairs of branchial arches succeed the hyoid in the wall of the pharynx, but only the first four bear gills.

Dipnoi (Lungfishes)

We remarked that the dermal skull bones of higher fishes are so much like scales as to suggest that they originated in an ancestor whose head was covered with small scales. In Westoll's (1943) review of the origin of tetrapods he showed that such an ancestor must have had something like 90 dermal bones (scales) in the roof of its skull alone, and that the lines of descent leading to Actinopteri, lungfishes, and lobefins consolidated these units according to different plans. Consequently it is almost impossible to give equivalent names to the roofing bones in these

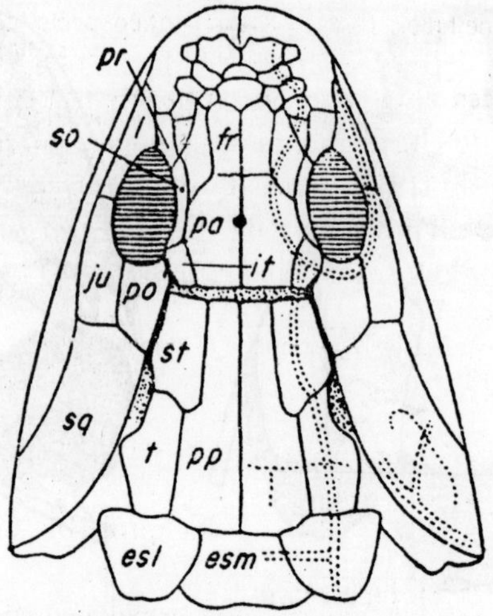

Fig. 143. Dermocranium of primitive lobefin, *Osteolepis*. (From Gregory, 1933, Transactions of the American Philosophical Society, vol. 23, after Watson.)

esl, lateral extrascapular; *esm*, median extrascapular; *fr*, frontal; *it*, intertemporal; *ju*, jugal; *l*, lacrimal; *pa*, parietal; *po*, postorbital; *pp*, postparietal; *pr*, prefrontal; *so*, supraorbital; *sq*, squamosal; *st*, supratemporal; *t*, tabular.

three groups, but there is every reason to believe that the roofing bones of a lobefin can be compared directly with those of tetrapods.

In the Australian lungfish, *Epiceratodus* (Fig. 142), the skull and shoulder girdle are still connected laterally, above the gill chamber. The palatoquadrate (primary upper jaw) fuses with the cranium, a condition called ***autosystylic;*** this releases the hyoid from any share in jaw support. Autosystyly here seems to be correlated with development of fan-like crushing dental plates instead of teeth.

Crossopterygian Fishes, and the Change to Tetrapods

Of the lobefinned fishes the only ones we shall consider are the Rhipidistia, which lived from the Devonian to the Permian period, for it is among them that the stem leading to Amphibia may be found (Fig. 143). During the change to land life the ***opercular series*** in the fish was lost (since gills were no longer useful to a terrestrial animal), the ***shoulder girdle*** was detached from the skull (allowing the head to bend freely), the ***nostrils*** rose above the rim of the mouth and became holes rather than notches, the whole ***nasal region*** lengthened out considerably, and the parts behind the eye became much shorter. In the notch at the

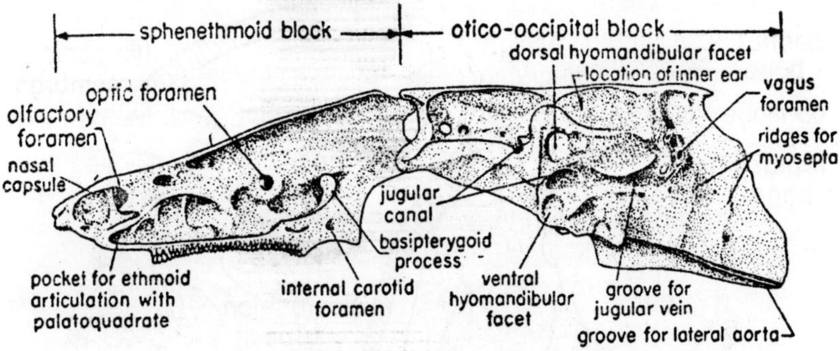

Fig. 144. Brincase of a crossopterygian, *Megalichthys (Ectosteorhachis)*. (After Romer.)

posterior corners of the skull there was probably an ***eardrum***, where formerly had been the spiracle.

But the old pattern of sensory canal grooves still appears in the early amphibians, suggesting that many of these were aquatic during a large part of their lives. The bones of the primary upper jaws, now forming most of the roof of the mouth, are surprisingly similar in the lobefin and early amphibian.

This brings us to the important question of the fate of the hyomandibular bone. In Rhipidistia the rather weak hyomandibular attached to the angle of the upper jaw, and no doubt gave it some support. As in no other known fishes, this hyomandibular had two points of articulation with the cranium, one above and one below the jugular vein on the side of the ear region (Figs. 144, 145). It also had a connection with the operculum, and with the lower part of the hyoid arch (five in all).

In Amphibia it seems as if the hyomandibular, discouraged by loss of the operculum (its chief responsibility), started to degenerate, but the long arm of coincidence reached out to save it, and gave it a wholly different function, to transmit sound waves from the air to the inner ear. Thus the hyomandibular of fishes evolved into the ***stapes***, a middle ear bone, in tetrapods (Fig. 145). But all five of the original articulations can still be recognized in a number of higher groups.

One more item should be noticed in Rhipidistia. The cranium was very long and consisted of two blocks, a sphenethmoid and an otico-occipital, between which was a movable joint, so that the head could be tilted slightly up and down. But it need not be thought that this peculiarity excludes them from possible ancestry to the Amphibia, as all other fea-

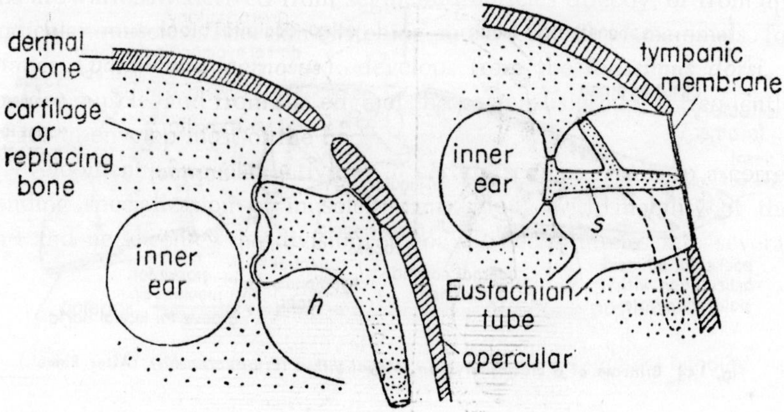

Fig. 145. Relationship between hyomandibular (h) in crossopterygian (left) and stapes (s) in tetrapod (right), diagrammatic.

tures point that way, and a fusion of the two blocks during development is all that is necessary.

Amphibians

The skull in primitive labyrinthodont Amphibia bore the characters we have just traced from lobe-finned fishes (Fig. 146). One tendency is conspicuous in labyrinthodonts from Lower Carboniferous to Triassic times; this is to enlarge and flatten the skull, to an almost unbelievable degree. Culminating in the huge Stereospondyli, these skulls were three or four feet long, about two feet wide, practically flat all over the upper surface, and completely decked with dermal bone except where nostrils, eyes, and pineal foramen opened upward. The braincase was ridiculously small, and cartilaginous. These animals died without further issue.

We show as an example the more conservative *Edops,* found in the Permo-Carboniferous of Texas (Romer and Witter, 1942), Fig. 147. The similar but slightly later *Eryops* is better known (Sawin, 1941). The skull of primitive labyrinthodonts articulated with the first vertebra by a ***single condyle*** just as in reptiles, which are descended from them; but among the later labyrinthodonts and modern Amphibia the condyle is double. A condyle implies free mobility of the joint, which had not been possible until the shoulder girdle detached from the skull; such condyles therefore occur in tetrapods but not in fishes.

Modern Amphibia represent three lines of descent from labyrinthodonts:

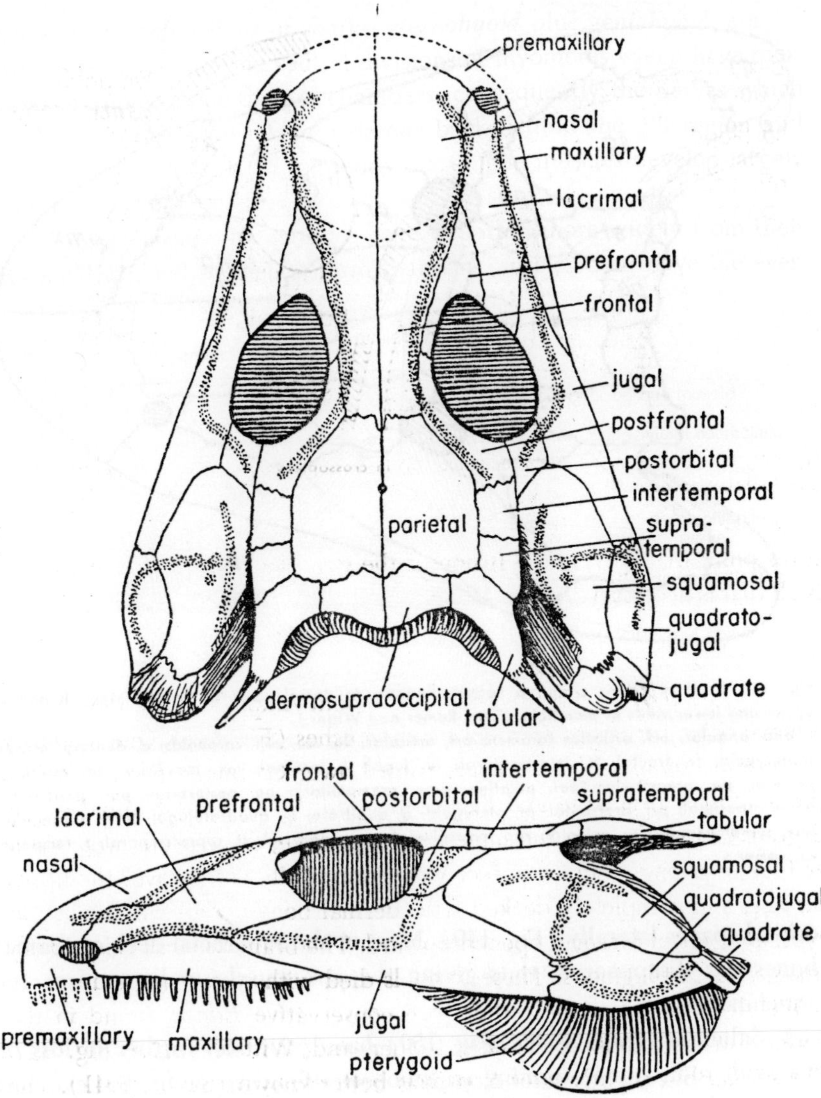

Fig. 146. Skull of *Paleogyrinus decorus*, a primitive labyrinthodont amphibian, in dorsal and lateral views. (After Watson.)

1. The limbless, burrowing Caecilia of the tropics show a partial fusion of cranial bones (Fig. 148), as well as a fairly complete deck. The eyes, being reduced, have very small orbits. The teeth suggest the labyrinthodont arrangement.

2. Caudata or salamanders have usually a narrow cranium with large

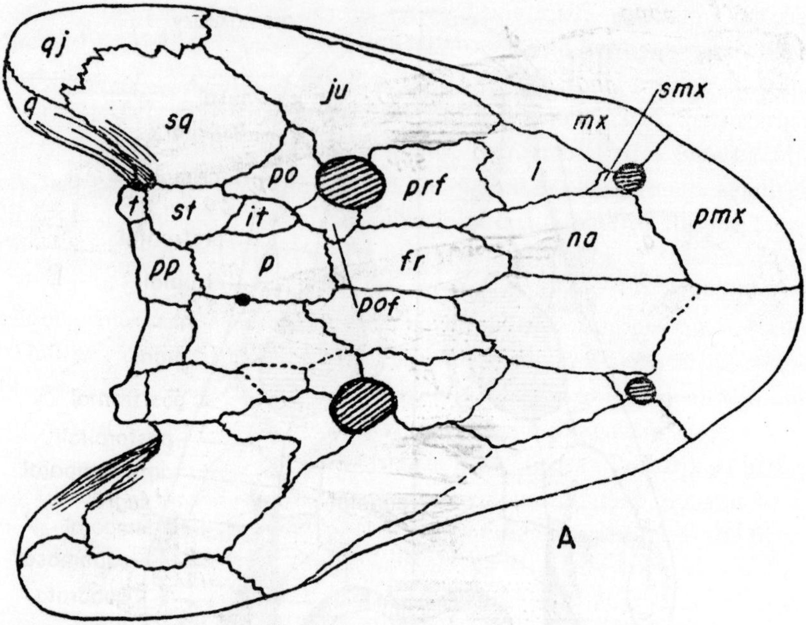

Fig. 147. Skull of *Edops craigi*, a Labyrinthodont. A, dorsal view; C, palatal view; B and D, upper and lower views of mandible. (From Romer and Witter.)

ang, angular; *arf*, articular foramen; *art*, articular; *co*, *coi*, *coll*, coronoids; *d*, dentary; *ect*, ectopterygoid; *fr*, frontal; *it*, intertemporal; *ju*, jugal; *l*, lacrimal; *mx*, maxillary; *na*, nasal; *p*, parietal; *pa*, prearticular; *pal*, palatine; *pmx*, premaxillary; *po*, postorbital; *pof*, postfrontal; *pp*, postparietal; *prf*, prefrontal; *pt*, pterygoid; *q*, quadrate; *qj*, quadratojugal; *sang*, surangular; *smx*, septomaxillary; *sp*, splenial; *spp*, postsplenial; *sq*, squamosal; *st*, supratemporal; *t*, temporal; *vo*, vomer.

orbital spaces laterally (Fig. 149). Most of the superficial deck of dermal bones has disappeared, thus giving the jaw muscles a larger area for attachment.

3. Salientia (frogs, toads, Fig. 150) have gone still further in loss of the skull roof, and in forming an immense orbital space for the eye and jaw muscles.

These trends characterize the recent groups of Amphibia: reduction of the number of skull bones; firm union of primary upper jaws with the cranium (***autostyly***); shortening of the posterior part of the cranium so that the last two pairs of cranial nerves (of the 12 pairs found in labyrinthodonts) now emerge from the spinal cord, behind the skull, leaving only 10 pairs of actual cranial nerves. The gill arches are usually reduced because of loss of the gills in most adult Amphibia (Fig. 151).

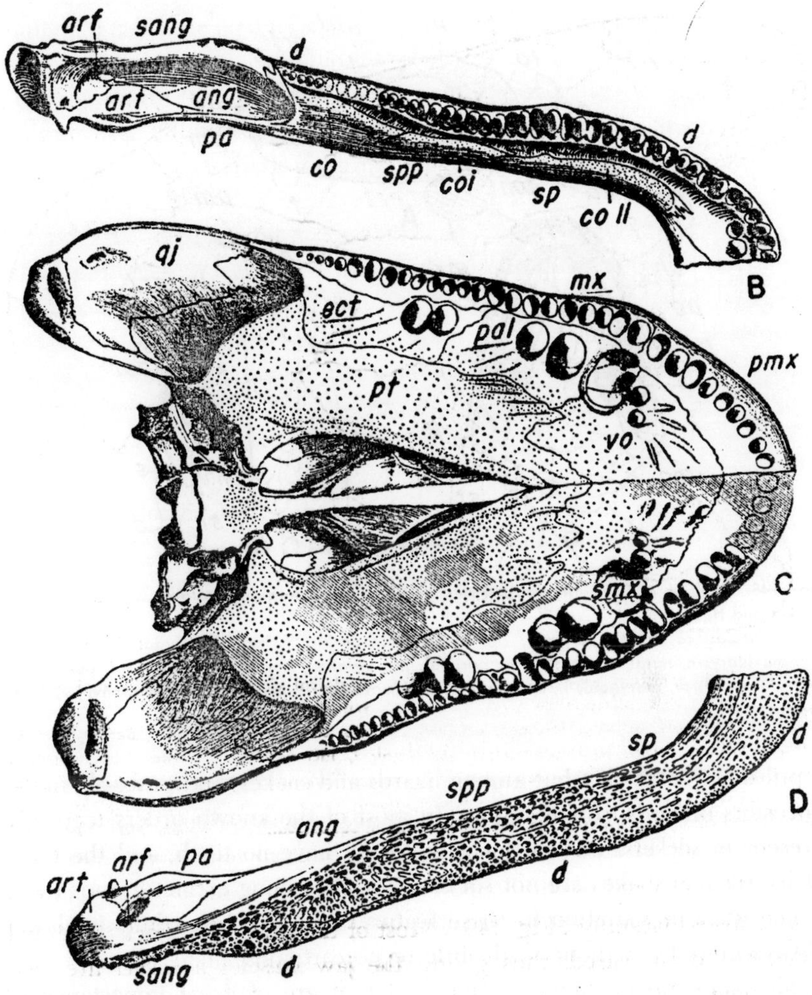

Reptiles

Living reptiles are but a scanty vestige of the far richer population of Mesozoic times. It is important in discussing the origin of birds and mammals, as well as in understanding the scope of this remarkable class, to mention skull characters of some of the orders.

Most reptiles have a single occipital condyle, received from early amphibians, but in the mammal-like reptiles (Therapsida) of the Triassic, it became double, and remained so in mammals. Most of the orders of

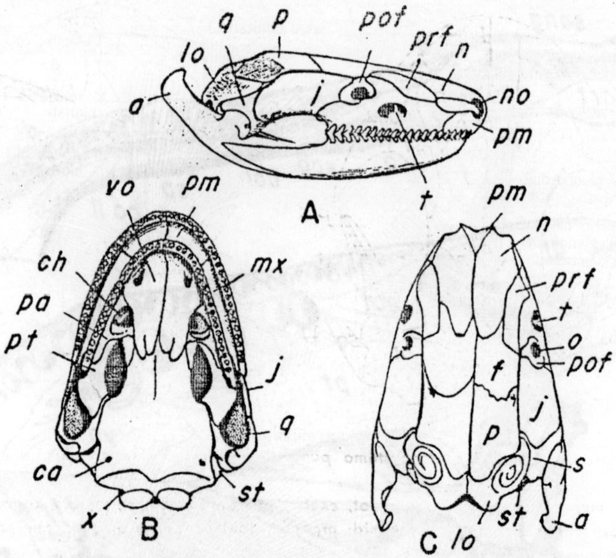

Fig. 148. Skull of a Caecilian, *Ichthyophis glutinosa*. A, lateral; B, palatal; C, dorsal. (From Parker and Haswell.)

a, articular; *ca*, carotid foramen; *ch*, inner nasal opening; *f*, frontal; *j*, jugal; *lo*, exoccipital; *mx*, maxillary; *n*, nasal; *no*, nostril; *p*, parietal; *pa*, palatine; *pm*, premaxillary; *pof*, postfrontal; *prf*, prefrontal; *pt*, pterygoid; *q*, quadrate; *s*, squamosal; *st*, stapes; *t*, tentacular groove; *x*, vagus foramen.

reptiles are autostylic, but among lizards and snakes, and probably in the ancestors of birds, this is not so. In most of the known orders teeth are present in sockets. But turtles (Chelonia) have no teeth, and the teeth of lizards and snakes are not socketed, while among extinct groups there is considerable variation in these features. The skull in adults is almost always bony throughout, with little or no cartilage.

The temporal openings in the skull roof are of great importance in visualizing relationships. The oldest order, Cotylosauria (Fig. 152), like the labyrinthodonts, had no such openings, and are called Anapsida in reference to this. The Chelonia (turtles and tortoises, Fig. 153) appear to have branched directly from Cotylosauria, and their skull roof has melted away, so to speak, from the posterior edges, leaving large temporal openings in most kinds; but since this was a receding of the edge rather than a perforation, this type of skull is usually called anapsid likewise. In the Therapsida (mammal-like reptiles) a single large temporal opening occurred on each side, bounded below by an arch formed mostly

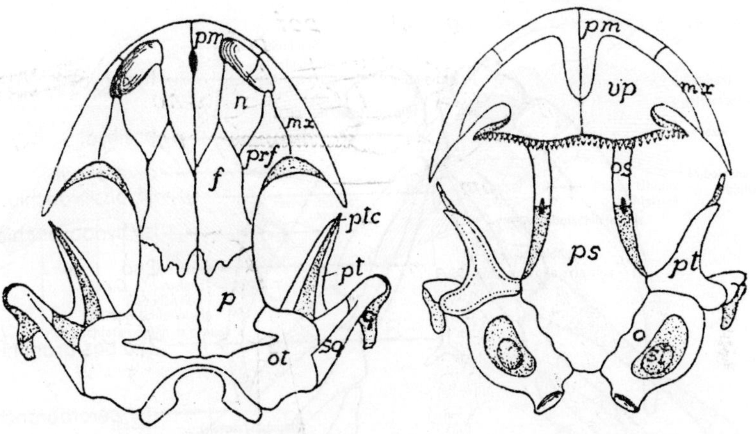

Fig. 149. Skull of salamander, *Ambystoma punctatum;* dorsal on left, ventral on right. (From Kingsley.)

f, frontal; *mx,* maxillary; *n,* nasal; *o, ot,* exoccipital; *os,* sphenethmoid; *pm,* premaxillary; *prf,* prefrontal; *ps,* parasphenoid; *pt,* pterygoid; *ptc,* pterygoid cartilage; *q,* quadrate; *sq,* squamosal; *st,* stapes; *vp,* vomer.

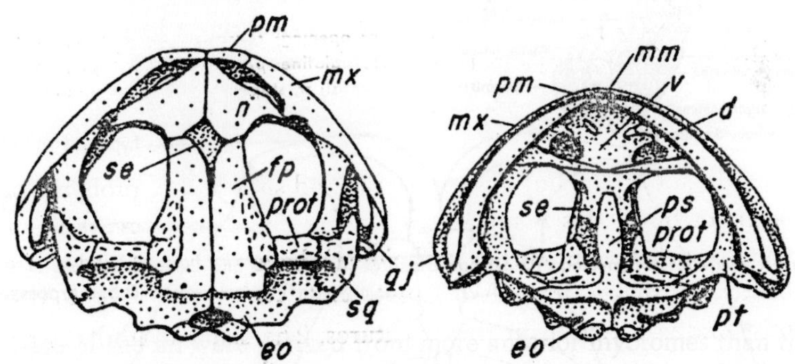

Fig. 150. Skull of Toad, *Bufo americanus:* dorsal on left, ventral on right. (Redrawn from Kingsley, *Outlines of Comparative Anatomy of Vertebrates,* 3rd ed., Blakiston.)

d, dentary; *eo,* exoccipital; *fp,* frontal; *mm,* mentomeckelian; *mx,* maxillary; *n,* nasal; *pm,* premaxillary; *pl,* palatine; *prot,* prootic; *ps,* parasphenoid; *pt,* pterygoid; *qj,* quadratojugal; *se,* sphenethmoid; *sq,* squamosal; *v,* vomer.

of the jugal (zygomatic) bone (Fig. 154A). This arrangement, clearly the same as in mammals, is the synapsid type. Several less familiar orders, such as the marine, whale-like Ichthyosauria, show the parapsid arrangement, with a single opening on each side farther dorsad than the preceding. Finally, the great majority of advanced reptiles are of the

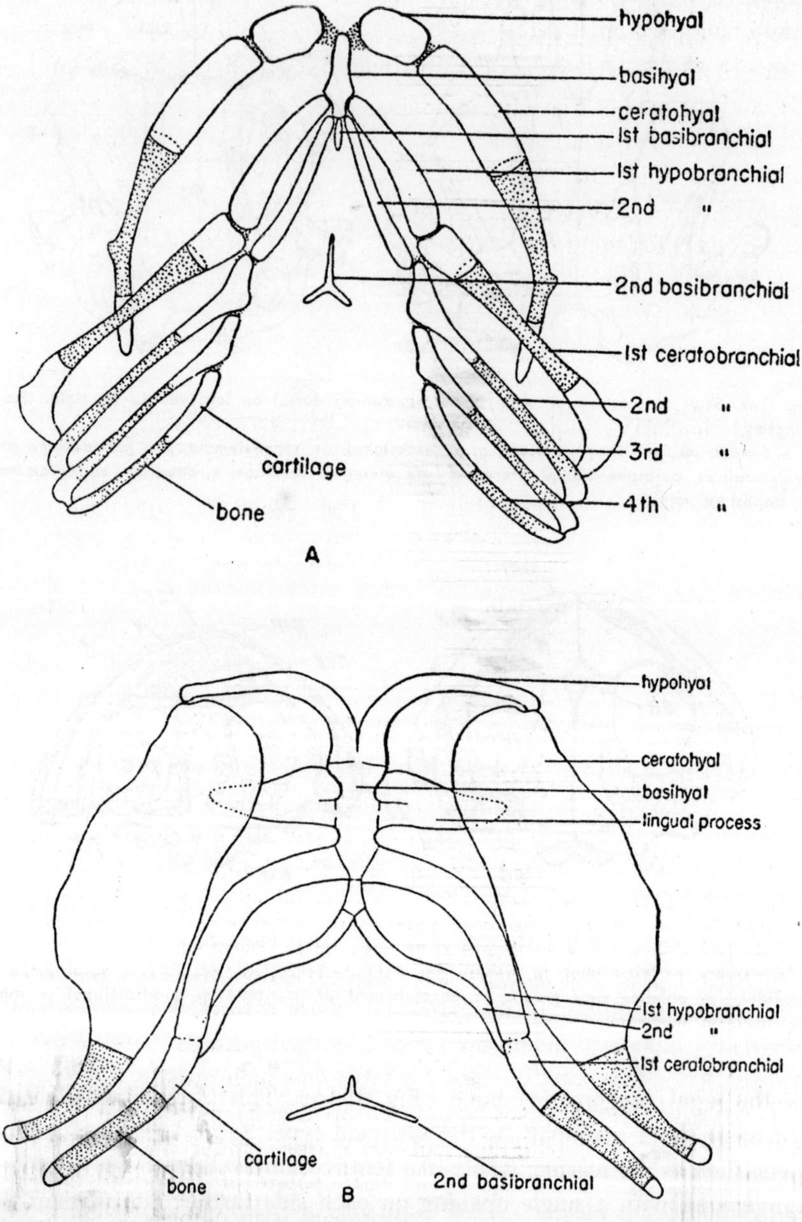

Fig. 151. Hyoid and branchial arches of a salamander (A) larva, and (B) adult, *Dicamptodon ensatus*; ventral view; bone stippled, cartilage clear.

diapsid type, with two pairs of temporal openings (unless secondarily changed). The Rhynchocephalia, represented today by the "living fossil," *Sphenodon* (Fig. 155), and the extinct Eosuchia and Thecodontia (somewhat alligator-like), are probably the oldest Diapsida. Crocodiles, dinosaurs, birds, flying reptiles (Pterosauria), all are thought to have originated from the thecodonts, while Squamata (lizards and snakes) probably came from early Eosuchia. In the Squamata the diapsid condition is modified by a breakdown of the lower temporal arch (lizards) and then of the upper as well (snakes), so that only by courtesy to their ancestors may we call them Diapsida.

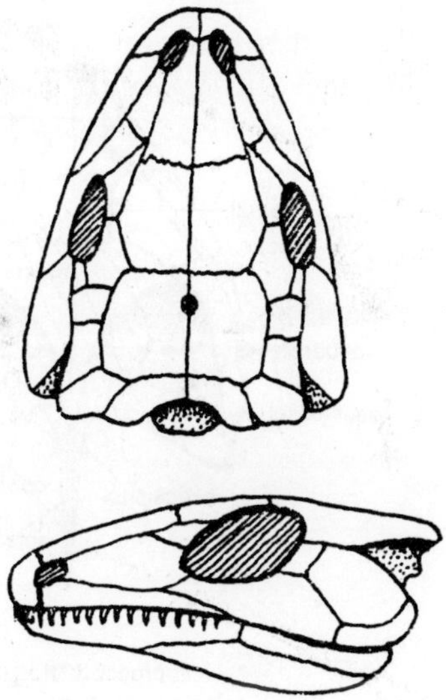

Fig. 152. Skull of *Seymouria*, dorsal and lateral, to illustrate Cotylosaur type. (Redrawn from Watson in *Encyclopedia Britannica*.)

Here we should note the ***kinetic*** condition of certain reptile skulls. This is a freedom of movement of certain bones. The anterior part of the cranium in some dinosaurs, lizards, snakes, and presumably in ancestors of birds, is capable of slight up-and-down movement against the posterior part. In most lizards (Fig. 156) and in snakes (Fig. 154D) the quadrate bone hinges loosely against the squamosal above and the mandible below, making it possible to thrust the lower jaw forward; some bones of the upper jaw also attach to the quadrate and can slide a short distance forward. This double-jointed quadrate mechanism, called ***streptostyly,*** also occurs in birds; we have noted previously that these groups are not autostylic like most reptiles.

One more point remains, as to the palate. Usually this differs but little from that of Amphibia, but in two groups there is formed a second roof to the mouth, the hard palate, by which the nasal passages become separate from the mouth cavity. Crocodilia are one such group (Fig. 154B, C). The device here makes it possible to breathe by putting the nostrils just out of water, while the mouth is wide open under water, perhaps holding and drowning the prey. The hard palate of Therapsida originated quite independently of this, and indeed is incomplete in most of the known kinds; it anticipates the mammalian condition, presumably being a device to aid breathing while chewing, for the teeth suggest that

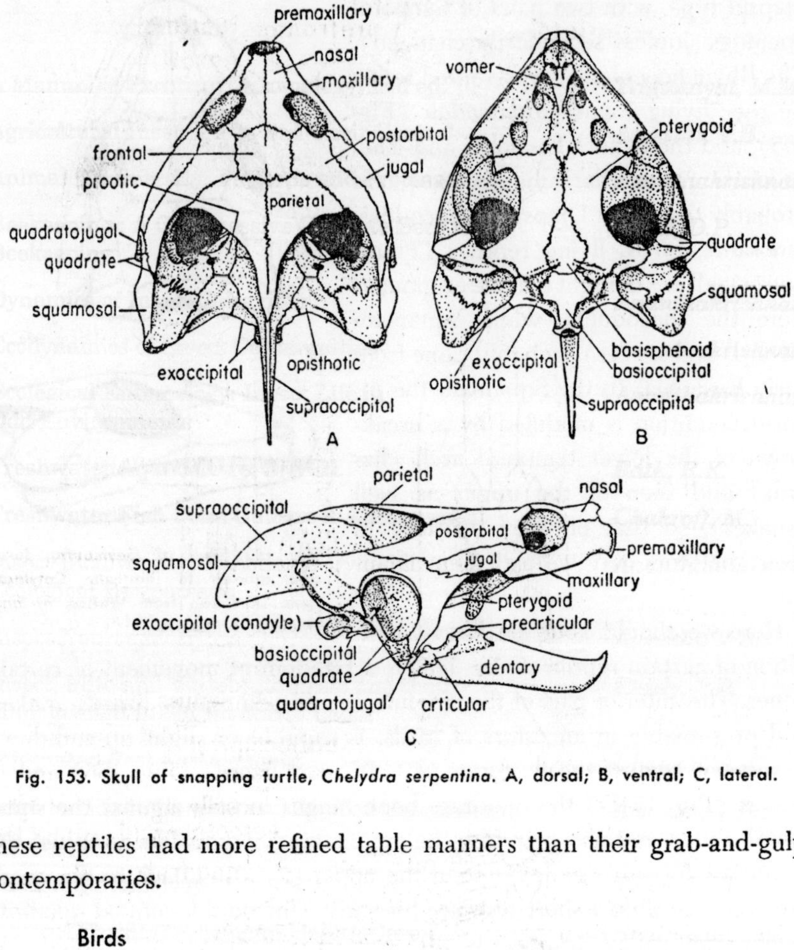

Fig. 153. Skull of snapping turtle, *Chelydra serpentina*. A, dorsal; B, ventral; C, lateral.

these reptiles had more refined table manners than their grab-and-gulp contemporaries.

Birds

In sharp contrast with the variety of important characters in reptile skulls, those of birds show little fundamental variation (Fig. 157). The more uniform traits are: expanded and rounded cranium (brain large), with a thinning of bones (flying calls for reduced weight); extension of premaxillaries, and often maxillaries and nasals, into a beak, toothless in all modern birds; movable quadrate, hinged dorsally on the squamosal but ventrally on (1) mandible, (2) pterygoid, and (3) a lateral rod consisting of quadratojugal and jugal, which forms the lower margin of the temporal fossa; frequently but not always some up-and-down hinge movement of the anterior parts of the skull upon the posterior, aided by the quadrate and attending muscles; single occipital con-

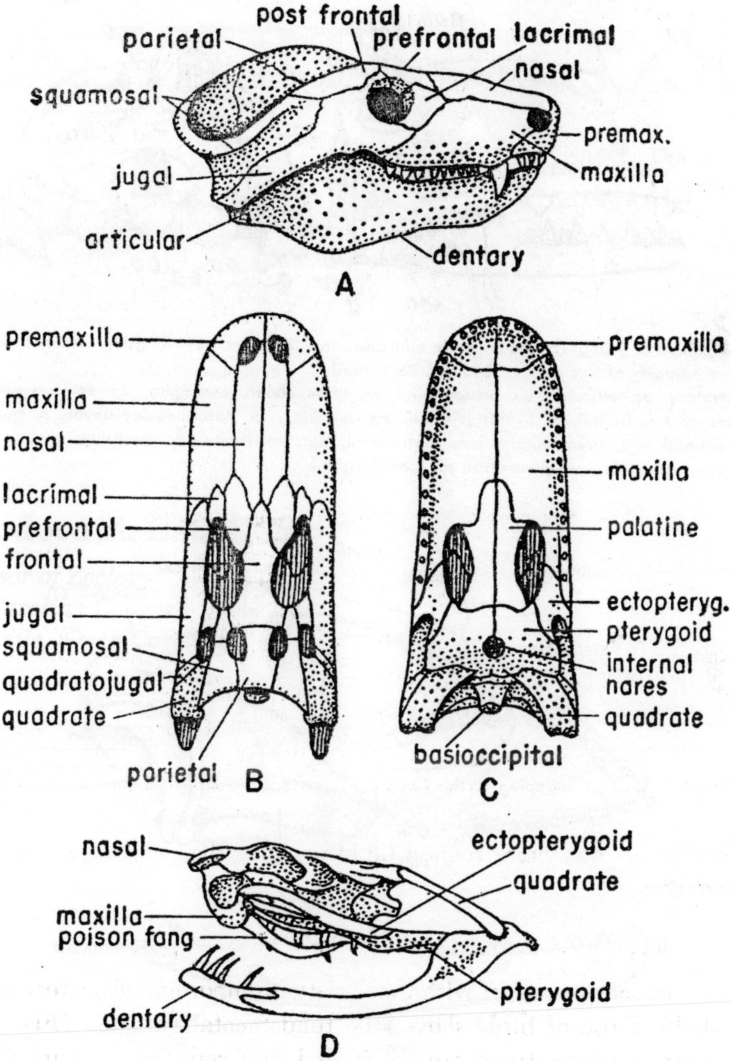

Fig. 154. Skulls of (A) mammal-like reptile, *Cynognathus;* (B, C) alligator, dorsal and ventral (after Andrews in *Encyclopaedia Britannica*); (D) rattlesnake (after Watson in *Encyclopaedia Britannica*).

dyle; extensive fusion of skull bones with one another except at movable articulations.

Comparing the palatal view of a bird skull (Fig. 158) with that of a reptile (e.g., lizard), the differences at first appear great, but they are mostly differences of proportions, for the bird skull is much longer in

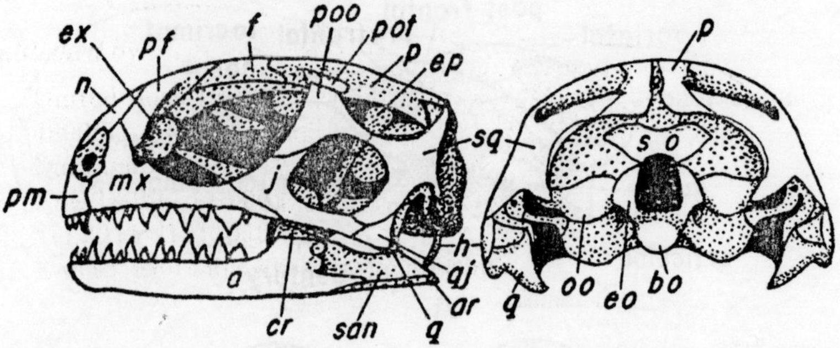

Fig. 155. Skull of young *Sphenodon*, side and posterior. (Redrawn from Kingsley, *Outlines of Comparative Anatomy of Vertebrates*, 3rd ed., Blakiston.)

a, dentary; *ar*, articular; *bo*, basioccipital; *cr*, coronoid; *eo*, exoccipital; *ep*, epipterygoid; *ex*, "extranasal"; *f*, frontal; *h*, hyoid; *j*, jugal; *mx*, maxillary; *n*, nasal; *oo*, opisthotic; *p*, parietal; *pf*, prefrontal; *pm*, premaxillary; *poo*, postorbital; *pot*, postfrontal; *q*, quadrate; *qj*, quadratojugal; *san*, surangular; *sq*, squamosal; *so*, supraoccipital.

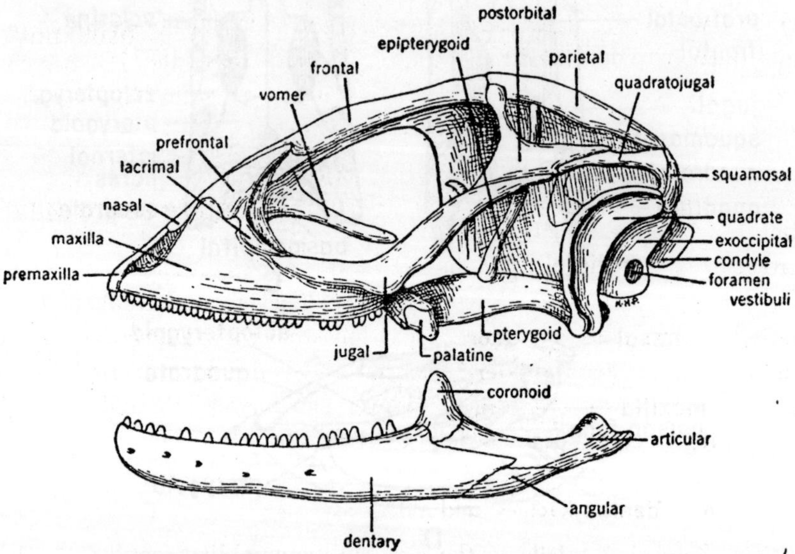

Fig. 156. Skull of lizard, *Sceloporus*. (Reproduced by permission from *Comparative Anatomy* by L. A. Adams & S. Eddy, published by John Wiley & Sons, Inc., 1949.)

front of the internal nares, on account of the beak. The originally separate vomers become one, this in many cases forming a thin internarial septum, or disappearing. The illusion of an elongated hard palate is furthered by the greatly lengthened premaxillaries in most birds, so that the nasal region is left far behind, yet it represents approximately the front of the skull in a reptile.

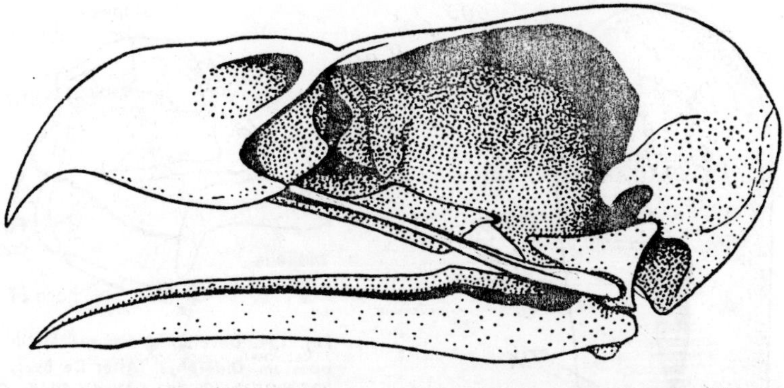

Fig. 157. Skull of eagle. (Redrawn from Beebe, *The Bird*, Henry Holt.)

The basisphenoid (floor of the braincase) and parasphenoid, just anterior to it, have fused, and the latter makes a so-called rostrum against which the bones of the palate fit, usually with a loose contact which permits them to slide. Since the work of T. H. Huxley (1867) the several ways in which the details of the palate differ in various groups have counted heavily in classifying birds.

The hyoid apparatus of birds is much modified in the service of the tongue. Besides some slender median rods, it has a pair of delicate "horns" which are actually the first branchial bars, and which extend posterolaterally, serving for attachment of the muscles that thrust out the tongue. In long-tongued birds, like woodpeckers and hummingbirds, these horns become so extraordinarily long that they must curve up over the back of the head (under the skin) and come down in front of the eyes, and they may even (in the flicker) go forward through the right nostril into the cavity of the beak.

Mammal-like Reptiles, and Mammals

Returning briefly to the Triassic mammal-like reptiles, we shall follow some of the changes in skull structure which led to mammals. This is one of the two or three greatest transitions in vertebrate history. Already showing in therapsid reptiles are the following:

1. Double occipital condyle, which remains in mammals.
2. Heterodont teeth (distinguishable as incisors, canines, molars).
3. Single pair of temporal openings, as in mammals.
4. Hard or secondary palate in various stages of completion, by the extension inward of palatal processes of the premaxillaries and maxil-

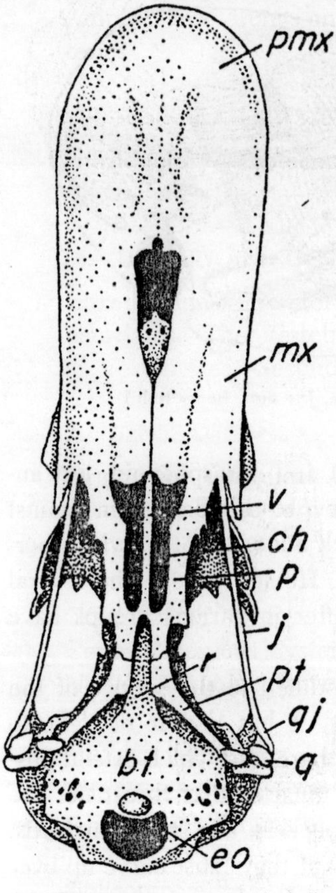

Fig. 158. Palatal view of duck skull. (Redrawn after Kingsley, *Outlines of Comparative Anatomy of Vertebrates*, 3rd ed., Blakiston.)

bt, basisphenoid; *ch*, choana; *eo*, exoccipital; *j*, jugal; *mx*, maxillary; *p*, palatine; *pmx*, premaxillary; *pt*, pterygoid; *q*, quadrate; *qj*, quadratojugal; *r*, parasphenoid; *v*, vomer.

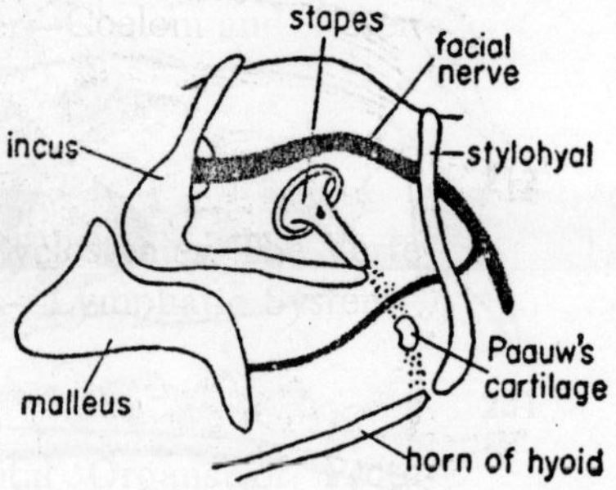

Fig. 159. Developing ear ossicles in young opossum, *Didelphys*. (After De Beer, *Development of the Vertebrate Skull*, Oxford University Press.)

laries, so that a sheet of bone forms the roof of the mouth and at the same time a floor for the nasal passages. The small paired vomers, commonly called prevomers, fuse together in some mammal-like reptiles and support the palate, between the two nasal cavities, just as the single vomer does in mammals; the "prevomers" in all vertebrates from the lobe-finned fishes up are evidently vomers. (The theory prevalent for many years was that the mammalian vomer was the reptilian parasphenoid, and that the reptilian vomers disappeared.)

5. In advanced Therapsida the glenoid fossa, where the lower jaw articulates, is composed of two bones, internally the quadrate (which was the only one in that position before), and externally the squamosal. Immediately behind this fossa is the cavity of the middle ear. In the course of the change to mammals the quadrate leaves the glenoid fossa, becomes connected with the ***stapes*** in the ear, and thus makes a sound-transmitting bone, known as the ***incus*** (Fig. 159).

6. At the same time the articular facet on the lower jaw, in some Therapsida, is composed of both dentary and articular bones. When the quadrate becomes an ear bone the articular does also, in this way keep-

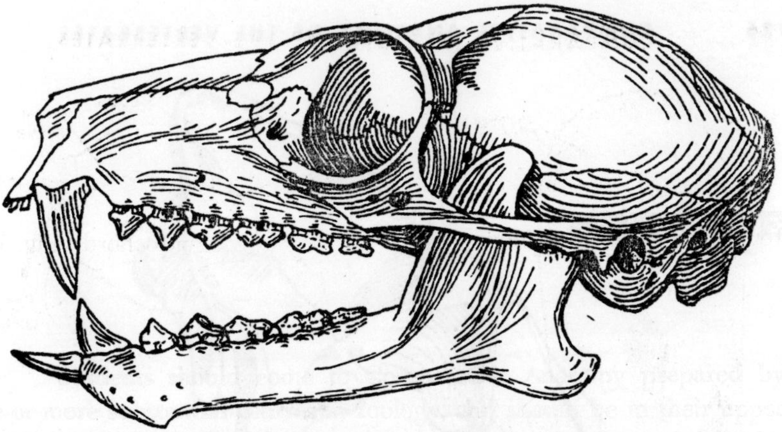

Fig. 160. Skull of lemur. (Redrawn from Parker and Haswell, *Textbook of Zoology*, 6th ed., Macmillan.)

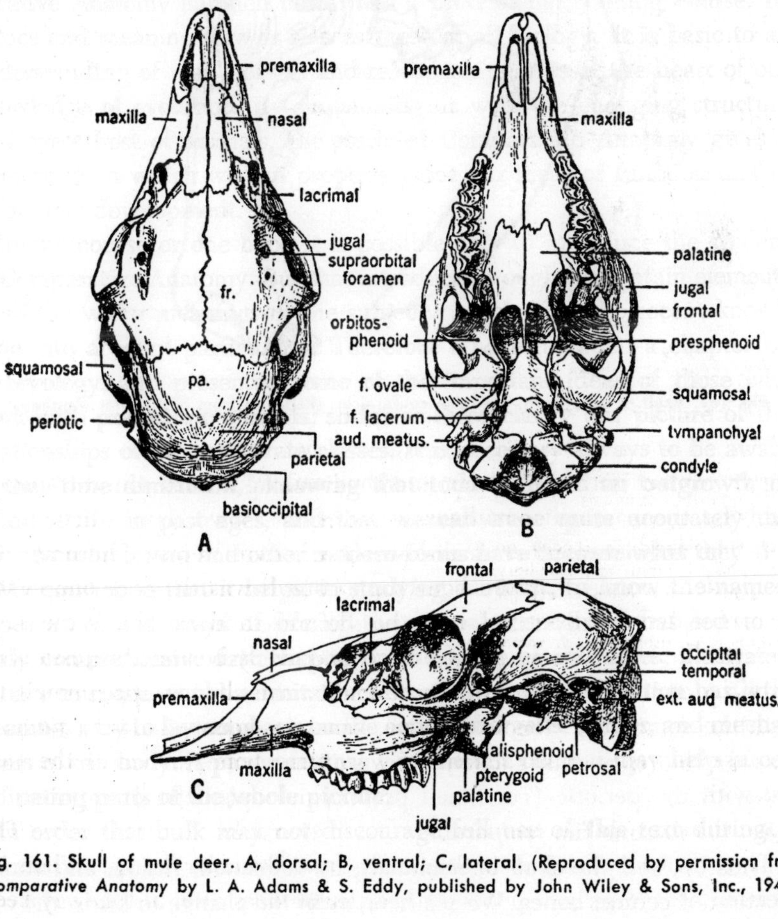

Fig. 161. Skull of mule deer. A, dorsal; B, ventral; C, lateral. (Reproduced by permission from *Comparative Anatomy* by L. A. Adams & S. Eddy, published by John Wiley & Sons, Inc., 1949.)

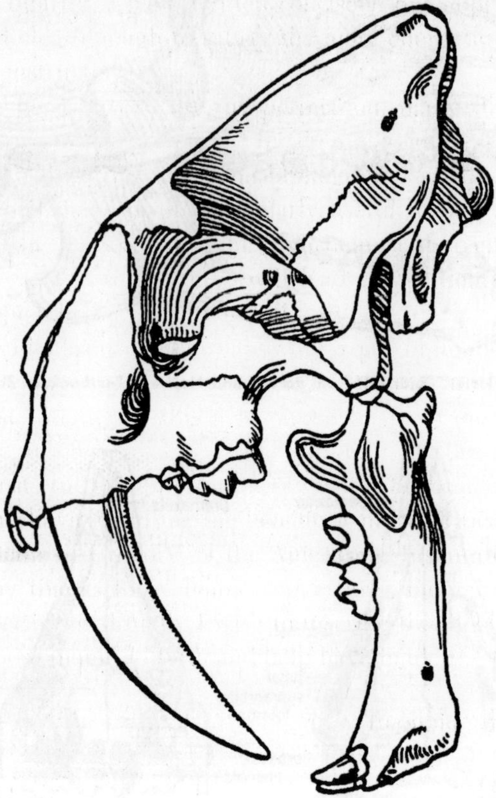

Fig. 162. Skull of sabertooth, *Smilodon californicus*. (Redrawn from Scott, after Matthew.)

ing its contact with the former, and thus making a third sound-transmitting bone, the mammalian ***malleus.***

7. The angular, another small lower-jaw bone of reptiles, becomes the ***tympanic,*** a crescent-shaped bone supporting the eardrum or tympanum in mammals (usually fused with the squamosal, however). With these changes, and the disappearance of other lower-jaw units, we have left only the dentary as the lower jaw of mammals, and it articulates only with the squamosal. The latter term, however, is not used in mammals, because the reptilian squamosal fuses with the tympanic, and frequently too with the periotic (combined prootic and opisthotic of reptiles) to make the mammalian ***temporal.***

Thus evolved the skull of mammals, by reduction, fusion, and translocation of certain bones. We see here, as in the change to birds, the operation of Williston's law, that the number of separate elements in the skull tends to decrease during evolution.

Usually the braincase is conspicuously larger in mammals than in reptiles, because of the larger brain (Fig. 160). The dentary bone develops a high ***coronoid process*** in most mammals, for insertion of the temporal and masseter muscles. Ordinarily the jaws and snout extend far forward of the eyes. Several thin scroll-like ***turbinal*** bones are present in the nasal passages on the sphenoid and ethmoid bones (Fig. 289), and the olfactory epithelium is thus greatly increased in area. The ***hyoid*** connects with the ear capsule by a ligament to the ***styloid process,*** while its middle portion, plus some vestiges of the branchial arches, takes part in forming the ***larynx.***

Beyond these general features we find an endless variety of skull structure in mammals (Figs. 160, 161, 162). Some of the more striking changes seen in particular cases are: 1. Loss of some or all teeth. 2. Increase in bulk of skull, the bones becoming extremely massive; this is correlated with bearing great weight or providing attachment for powerful muscles. In such skulls the brain cavity and other passages do not increase in proportion with the whole skull. 3. Reduction to much less than average bulk, as in the paper-thin skull of a shrew. Here the orbits, ear capsules, and brain cavity are of larger than average proportions, and the bones do not develop conspicuous crests for muscle attachment.

CHAPTER V

Skeleton (Continued)

NOTOCHORD

The ***notochord*** is the one distinctive feature of the skeleton of all chordates, and therefore of all vertebrates. Among most of the latter it is succeeded in development by a series of ***vertebrae,*** which partly or wholly replace it. They are not formed by the notochord, but develop out of small cartilages (***arcualia***) adjacent to it, frequently supplemented by bone or cartilage deposited in or around the sheaths of the notochord. Among several of the most primitive groups of vertebrates no actual vertebrae exist, but only arcualia in various degrees of elaboration, lying against the notochord. Vertebrae (as structures which take the place of the notochord) originated several times in the course of evolution, and by means of different patterns of development.

The notochord typically has two sheaths, an outer ***elastica externa,*** and an inner ***fibrous sheath.*** The ***somites,*** or body segments of the embryo which are to form longitudinal muscles, lie in series along each side. Between each somite and the next one there forms a connective tissue ***myoseptum,*** which reaches inward to the surface of the elastica externa. Along the inner face of each somite in early stages, against the notochordal sheaths, there appears a cluster of cells called a ***sclerotome*** (Fig. 163), one of these on each side of the notochord, per segment. Sclerotomes are therefore ***segmental*** in position, and septa ***intersegmental.***

Each sclerotome, in many of the lower vertebrates at least, gives rise to four small cartilage pieces, two above and two below. These are arcualia. Their names and positions, in relation to the original sclerotome, are:

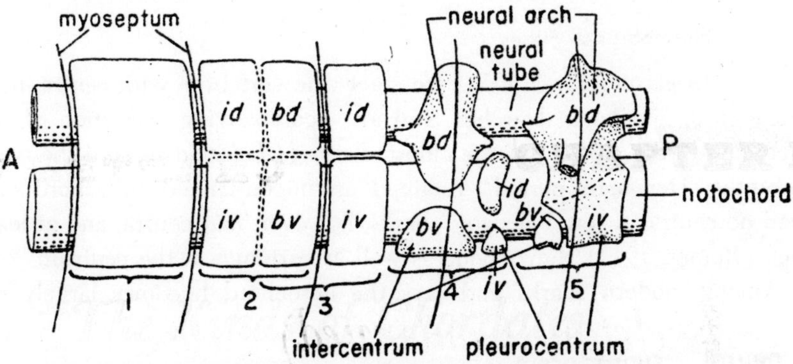

Fig. 163. Stages in development of vertebrae from sclerotomes, diagrammatic.
1, sclerotome; 2, four procartilage arcualia formed by one sclerotome; 3, basidorsal (*bd*) and basiventral (*bv*) of one segment unite with interdorsal (*id*) and interventral (*iv*) of next posterior segment to make a vertebra; 4, a rhachitomous vertebra (*Chelidosaurus*) in which the intercentrum is larger than the pleurocentrum; 5, a vertebra of the *Seymouria* type in which the pleurocentrum is larger, as in reptiles (after Romer).

Basidorsal—posterior, dorsal
Basiventral—posterior, ventral
Interdorsal—anterior, dorsal
Interventral—anterior, ventral

As these develop in most animals, the interdorsals and interventrals unite with the basidorsals and basiventrals of the next preceding segment, not with those of their own. Thus a completed vertebra is intersegmental, with the corresponding myoseptum meeting its lateral surface, while the muscle segments themselves alternate with the vertebrae, as they must do to cause movement of the column. Now we must look at some of the modifications of this general scheme:

Cyclostomes

In cyclostomes (living) the notochord is present throughout life. There are no ventral arcualia, but in each segment (in lampreys) two pairs of dorsals, of which the more anterior is presumably interdorsal and the posterior basidorsal; hagfishes have none. From the notochordal canal in the skull of ostracoderms it is known that they also had a fully developed notochord, but arcualia are not indicated by the fossils, being no doubt of soft cartilage.

In early gnathostomes the notochord must have persisted throughout life, but little else can be said about it.

Elasmobranch Fishes

In elasmobranch fishes we meet true vertebrae with centra, neural arches, and haemal arches, and an accompanying reduction of the notochord (Fig. 164). But the oldest sharks (*Cladoselache* and the pleuracanths, Devonian period) retained an unconstricted notochord and had no centra. A few Paleozoic sharks, however, had centra, and at least one (Romer, 1942) shows bone as well as cartilage in the centrum.

Among modern sharks and rays the notochord becomes largely re-

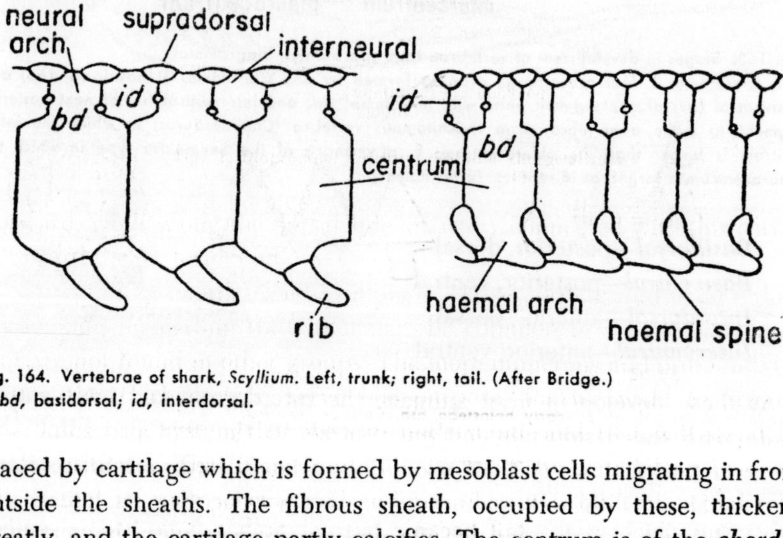

Fig. 164. Vertebrae of shark, *Scyllium*. Left, trunk; right, tail. (After Bridge.) *bd*, basidorsal; *id*, interdorsal.

placed by cartilage which is formed by mesoblast cells migrating in from outside the sheaths. The fibrous sheath, occupied by these, thickens greatly, and the cartilage partly calcifies. The centrum is of the ***chordal*** type; it is concave or funnel shaped at each end (***amphicoelous,*** Fig. 166). If, as in *Squalus,* a cross section of the centrum shows a single ring of calcification surrounding the pinched notochord, the vertebra is ***cyclospondylous,*** but there are other patterns of calcification as well.

Turning now to the arcualia, the two basidorsals, resting on the centrum outside the notochordal sheaths, develop into a ***neural arch*** by meeting and fusing above the neural canal. Just behind, fitting over the joint between two centra, the interdorsals do likewise, making an ***intercalary*** arch. On the lower side of the centrum each basiventral forms a lateral flange or ***parapophysis,*** to which, in the trunk, ribs attach. But in the tail the two basiventrals of each vertebra unite beneath the dorsal aorta and caudal vein to make a ***haemal arch,*** practically an inverted image of the neural arch above. The tail vertebrae are double, that is, for each segment there are two centra. This is *diplospondyly.* It occurs in some other fishes, as we shall see, evidently increasing the flexibility of the tail.

Actinopteri

Among Actinopteri, the ray-finned fishes, the earliest retained a complete notochord with no centra, but above the notochord lay a series of neural arches, usually prolonged into high neural spines, and in corresponding positions below were subnotochordal pieces (no doubt basi-

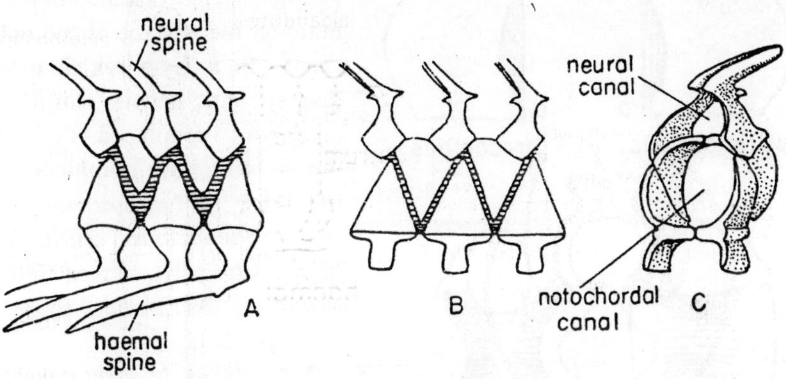

Fig. 165. Vertebrae of an early holostean fish, *Caturus furcatus.* A, caudal; B, C, trunk. (After Zittel.)
bd, basidorsal; *id,* interdorsal; *bv,* basiventral.

ventrals) which in the tail became haemal arches with haemal spines (Fig. 165).

Polypterus, which we have mentioned before as a modern side-line descendant of the paleoniscids, has complete bony amphicoelous centra, with neural arches leading into sloping neural spines dorsally, and prominent parapophyses for the ribs, in the trunk, while in the tail haemal arches and spines replace these by a gradual transition. The superficial resemblance of these vertebrae to those of higher fishes is no doubt secondary, since evidence from the skull and elsewhere indicates that the latter arose along an independent line.

Chondrostei (sturgeons, etc.) retain a complete notochord, and may be regarded as either primitive or, possibly, neotenic (keeping immature traits in maturity), in having all four pairs of arcualia distinct in the adult. They are cartilaginous.

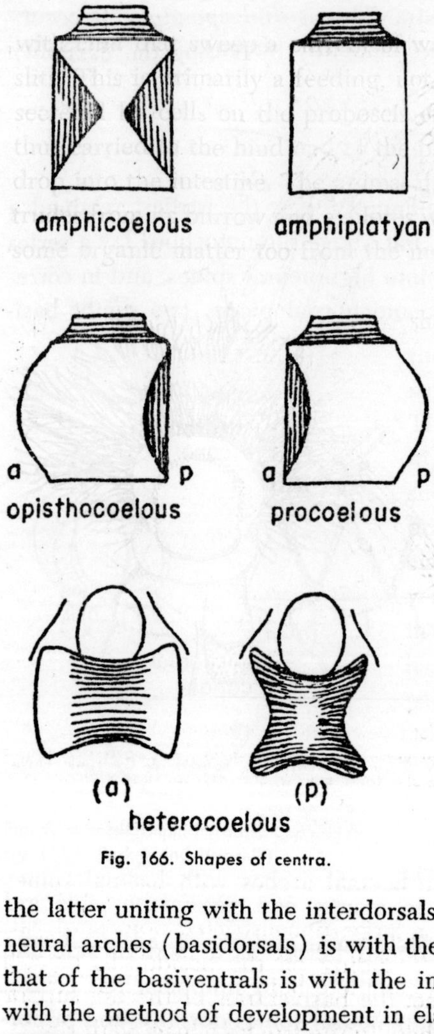

Fig. 166. Shapes of centra.

Holostei show through their history a gradual increase of completeness of vertebrae and replacement of the notochord. The centrum is ***perichordal*** (Fig. 165), and remains so in teleosts; this means there is no penetration of the notochordal sheaths by cartilage-forming or bone-forming cells, but the skeletogenous layer outside deposits bone on the surface of the outer sheath in a layer which grows deeper and heavier until the notochord is replaced. Typically the vertebrae are ***amphicoelous,*** but those of *Lepisosteus* (gar) are ***opisthocoelous*** (centra concave posteriorly, convex anteriorly, Fig. 166), unlike those of any other fish.

Amia shows, in early development, the four pairs of arcualia, but the interventrals soon disappear by fusion with basiventrals, and the centrum is formed by the latter uniting with the interdorsals. Curiously, the relationship of the neural arches (basidorsals) is with the interdorsals *in front* of them, and that of the basiventrals is with the interventrals in front; this contrasts with the method of development in elasmobranchs and tetrapods, and is another indication that vertebrae have appeared independently in these groups. *Amia* is diplospondylous in the tail.

Vertebrae of teleosts develop in fundamentally the same manner as in *Amia* (Figs. 167, 168). The growing centrum, cylindrical in form, tends to engulf the arcualia. Anterior and posterior processes develop from the upper corner of each centrum, interlocking with those of the preceding and following centra to strengthen the whole column. These are ***zygapophyses,*** comparable to but not homologous with those in tetrapods.

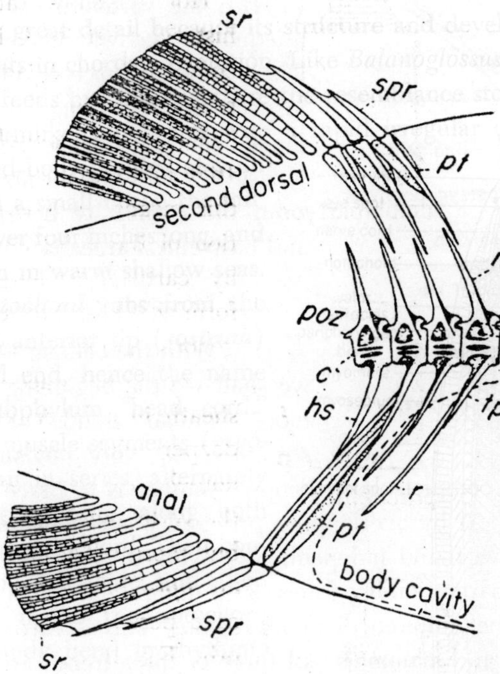

Fig. 167. Portion of vertebrae and fin skeleton of yellow perch, *Perca flavescens.*
c, centrum; *dspn,* dorsal foramina for spinal nerves; *hs,* haemal spine; *poz,* postzygapophysis; *prz,* prezygapophysis; *pt,* pterygiophore; *r,* rib; *spr,* spiny ray; *sr,* soft ray; *tp,* parapophysis; *vspn,* ventral foramina for spinal nerves.

Dipnoi

Dipnoi or lungfishes have a large, unconstricted notochord with incomplete vertebrae (Fig. 169). Standing on the basidorsals of each segment is a neural arch and spine; the basiventrals of the tail support haemal arches and spines, while those in the trunk connect with ribs.

Crossopterygii

The crossopterygian notochord was also large, but anteriorly ringlike centra constricted it somewhat (Fig. 170). Farther back in the trunk the centra separated into right and left crescents and became thin; finally in the tail none are represented, unless they were of cartilage. This is the situation in *Eusthenopteron* according to Gregory and Raven

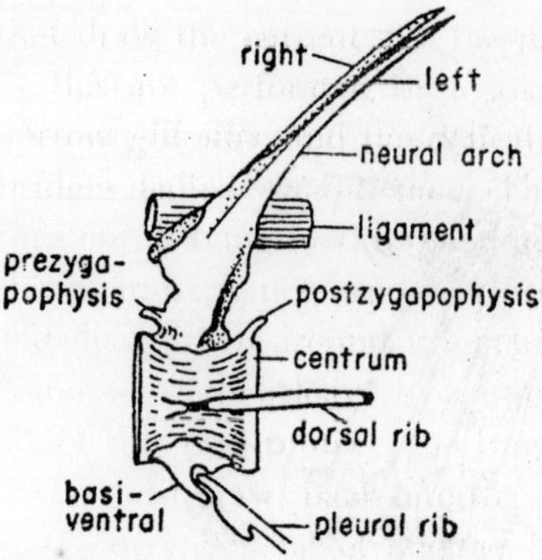

Fig. 168. Trunk vertebrae of pike, *Esox lucius*. (After Goodrich.)

(1941); the vertebral column is not adequately known in other kinds. There is no evidence of ribs. We may suspect that the ring- or crescent-shaped centra had their origin in mesoblast cells invading the fibrous sheath from the regions of both basidorsal and basiventral arcualia.

Tetrapods

With the coming of tetrapods we shall see the beginning of certain more familiar trends of vertebral structure, not very obvious at first. One is to develop complete bony centra. Another is to brace each vertebra against the next by overlapping *pre-* and *postzygapophyses;* both of these features we found independently evolved in the teleost fishes. A third, not so apparent among fishes, is to differentiate the column according to functional differences in its regions. This is carried out impressively in birds and mammals. All three of these trends are strongest in terrestrial animals because of the more complex mechanism needed to carry and move the body out of water. In secondarily aquatic groups of long standing, as whales and ichthyosaurs, there is a return to a less differentiated column.

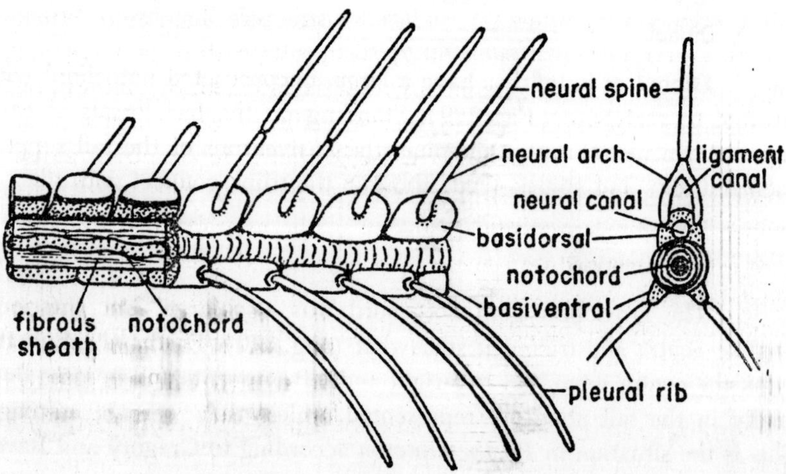

Fig. 169. Vertebrae of lungfish, *Epiceratodus*. (After Goodrich.)

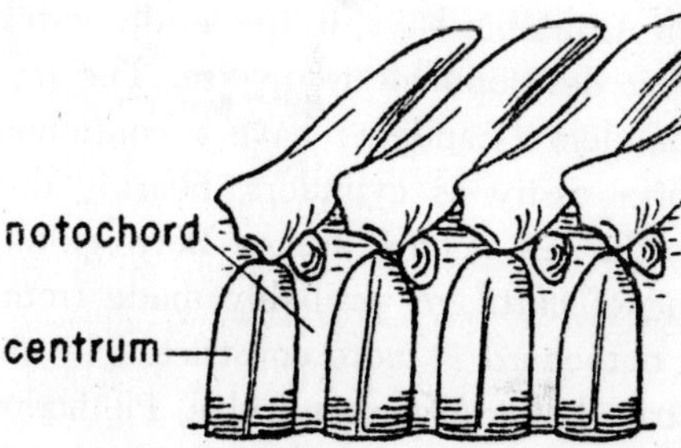

Fig. 170. Trunk vertebrae of crossopterygian, *Eusthenopteron.* (After Gregory and Raven.)

Some of the oldest known amphibian vertebrae, belonging to *Eogyrinus* (an embolomerous labyrinthodont), had a double centrum (Fig. 171), against both parts of which the neural arch fitted. The arch already bore zygapophyses, and laterally it had a transverse process to meet the tuberculum (upper head) of the rib on each side. There was a small facet on each side of the anterior section of the centrum (***hypocentrum***) to articulate with the lower head of the same rib. Presumably, although the situation is still not clear, the neural arch and spine came from the basidorsal arcualia, the hypocentrum from the basiventral, and the second section (***pleurocentrum***) from interdorsal, plus, possibly, interventral. Both hypo- and pleurocentrum were almost complete, leaving only a minute passage for the notochord.

We meet a puzzling situation in the second suborder, Rhachitomi (Fig. 163, nos. 4, 5), for here the notochord was much less interrupted and the vertebrae looked more primitive. The whole structure suggests retention of larval form, but the parts were bony and the animals more terrestrial than aquatic. Recent finds suggest that the rhachitomous type occurs as early as the embolomerous.

The Stereospondyli, last and largest of the labyrinthodonts, had an amphicoelous centrum bearing a strong arch formed evidently from basiventrals and basidorsals, respectively, but with no indication of interdorsals or interventrals. The centrum sometimes had a perforation for the notochord, but usually not.

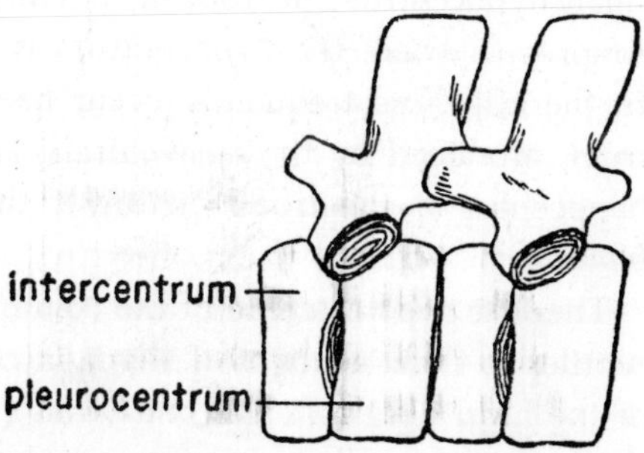

Fig. 171. Embolomerous vertebrae of *Cricotus.* (After Noble.)

Practically all modern Amphibia have, in the adult, vertebrae of a single piece each, but their development is diverse. The more primitive, aquatic kinds of salamanders (Caudata) have a continuous notochord around which the centra ossify as cylinders, bearing dorsally neural arches and ventrally, in the tail, haemal arches. Between any two centra is an intervertebral ring of cartilage, probably made from interdorsals and interventrals. The notochord is more constricted by the intervertebral cartilage in terrestrial kinds (Salamandridae, Plethodontidae), and the joints are opisthocoelous (ball and socket, the socket being on the posterior face of each centrum). The ***sacrum*** in Caudata, Salientia, and labyrinthodonts (Caecilia have none) consists of one or two scarcely modified trunk vertebrae from which short sacral ribs extend laterad to meet the ilium on each side.

Frogs and toads (Salientia) show a great variety and high specialization of vertebrae. The number is nine or less, in two cases being only six; this is the least known in any vertebrate animal. Reduction is partly by loss and partly by fusion of formerly separate vertebrae, as in the rodlike ***urostyle*** at the posterior end. The joint between vertebrae is formed by a ball that ossifies from, probably, interdorsal cartilages and fastens to either the anterior end of a centrum, making it opisthocoelous, or the posterior end, making it procoelous (concave in front); the latter condition exists in our native frogs and toads. A tadpole has no vertebrae in its tail. Those of the trunk ossify at the time of transformation.

Reptiles

Vertebrae of reptiles are as various as their other characters. Generally the centrum is a pleurocentrum; that is, it develops from interventral arcualia, while basiventrals may in some primitive kinds form a small hypocentrum in front of it. This is contrary to the arrangement in Amphibia, where the hypocentrum is primary, or may be the only part. In the tail there frequently occur haemal arches distinct from the centrum, produced by the basiventrals, and therefore homologous with the hypocentra of the trunk (if any); these are familiarly called ***chevron bones.***

The first two vertebrae of the column, the atlas and axis, begin in early reptiles to show changes of shape in connection with increasing mobility of the head (Fig. 172). From Evans (1939) we may summarize some of these changes, which are not completely realized until we reach mammals: "the atlas spinous process is lost and the opposite halves of the neural arch fuse with each other and with the atlas intercentrum to form a ring; with the rise of the reptiles the anterior facets on the atlas become markedly concave for the reception of the ball-like occipital condyle; the atlas centrum becomes separated from its neural arches . . . and finally fuses with the axis centrum to form the odontoid; and prominent lateral processes grow out from the sides of the atlas." The odontoid process of the axis (second vertebra) acts in a sense as a pivot around which the ringlike atlas can rotate; up-and-down movements of the skull take place at the joint where the condyles fit against the atlas. The development in mammals of the transverse processes of the atlas, to which Evans refers, is related directly to the appearance of strong muscles which reach from them to the skull. Other points of general interest in the vertebrae of reptiles are these:

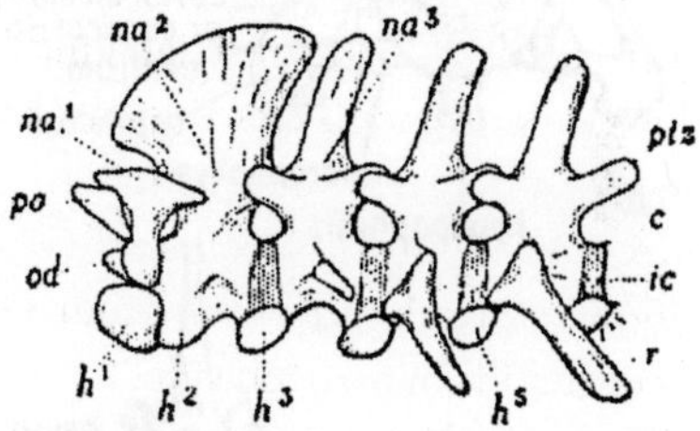

Fig. 172. Anterior vertebrae of *Sphenodon*, left side, showing proatlas (*pa*), atlas (*na*[1], *h*[1]), axis (*na*[2], *h*[2]). (From Goodrich, *Studies on the Structure and Development of the Vertebrates*, Macmillan.)

c, pleurocentrum; *h*, hypo- or intercentrum; *ic*, intervertebral disc; *na*, neural arch; *od*, odontoid process (first pleurocentrum); *pa*, proatlas; *ptz*, postzygapophysis; *r*, rib.

1. Most reptiles have ribs all the way from the third vertebra of the neck to the sacrum (Fig. 173). They attach by a double articulation: the lower head (capitulum) of the rib meets either a simple facet or a process, the ***parapophysis;*** the upper articulation of the rib (tuberculum) is to a more prominent process, the ***diapophysis;*** sometimes para- and diapophysis are on one common transverse process. In several orders the cervical (neck) ribs have been reduced to short spines, either articulated or fused with their vertebrae; the small canal which was present between the upper and lower articulations still remains as the ***vertebrarterial canal,*** and distinguishes these reduced ribs from actual transverse processes belonging to the vertebrae alone.

2. In the ***sacrum*** two vertebrae usually take part. Each bears a pair of sacral ribs, short peglike bones extending from the centrum to the adjacent ilium. But reptiles which have no hind limbs and therefore no pelvis also lack a sacral region. Contrasted with this is the great enlargement of the sacrum in certain dinosaurs. For example it included eight or nine vertebrae among the trachodonts or duckbills, and seven to ten in the quite different Ceratopsia. The neural canal was so much enlarged in the sacrum of the sauropod dinosaurs that it must have housed a ganglion

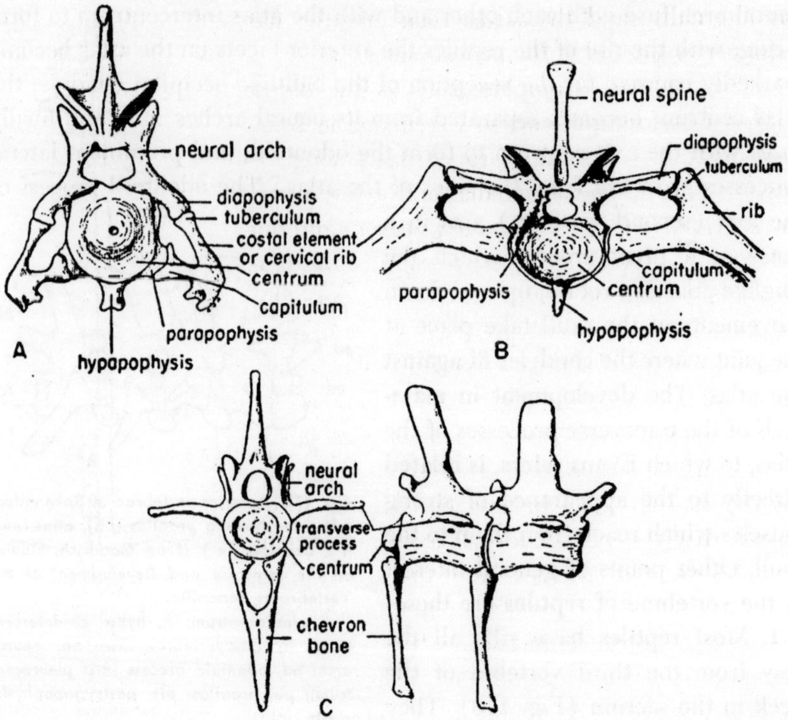

Fig. 173. Vertebrae of crocodile. A, cervical; B, thoracic; C, caudal. (Reproduced by permission from *Comparative Anatomy* by L. A. Adams & S. Eddy, published by John Wiley & Sons, Inc., 1949.)

several times the size of the brain. This has inspired a much-quoted poem about the dinosaur's displaced intellect: "If something slipped the forward mind, 'twas rescued by the one behind," and so on. Among flying reptiles, Pterosauria, the sacral vertebrae numbered four to seven, and in such advanced types as *Pteranodon* of the Cretaceous period there was also a fusion of thoracic vertebrae with a bony shield overlying them, forming a ***notarium,*** to the sides of which the scapulae articulated. This extraordinary arrangement, reminiscent of that in rays, doubtless strengthened the delicately built skeleton, for supporting the bases of the wings.

3. The carapace of turtles holds all the trunk vertebrae in a solid grip, so that they make little more than a narrow ridge along its inner face. The neural spines fuse with (but do not themselves form) the flat neural plates which lie in series along the center of the carapace. Short transverse processes from the centra and neural arches meet the bases of the

ribs internally, and the latter then join the flat costal plates, laterad of the neural plates.

4. The centrum is amphicoelous in the oldest reptiles, Cotylosauria, and so it remains in some Therapsida, in geckos and Uropeltidae among lizards, in some Rhynchocephalia, and a few more. The centra become ***amphiplatyan*** (flat on anterior and posterior faces) in a few aquatic reptiles, as the plesiosaurs, one family of Rhynchocephalia, and some extinct Crocodilia. They are amphiplatyan also in a few dinosaurs. The procoelous condition comes in turtles, most lizards, mosasaurs, snakes, modern Crocodilia, and pterosaurs. There are opisthocoelous vertebrae in some dinosaurs and turtles. This may be less bewildering if we suggest that the amphicoelous vertebrae are the primitive type in reptiles, and that from them arose, according to functional adaptations in various orders, the other kinds of centra. Hence the presence of any one type does not necessarily imply direct relationship among those animals that have it.

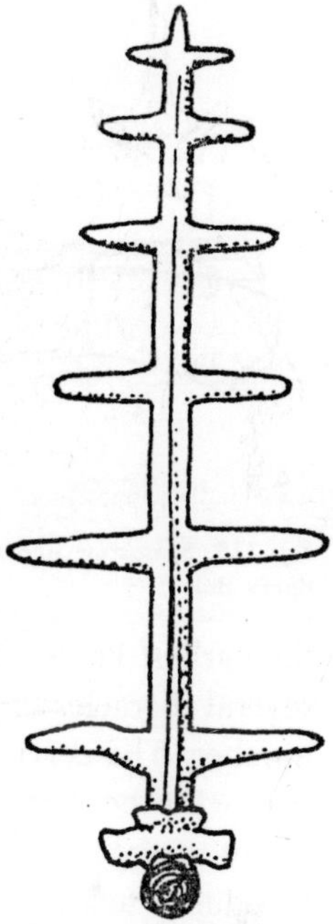

Fig. 174. Dorsal vertebra of a pelycosaur, *Naosaurus*, showing high neural spine. (After Goodrich.)

5. In Pelycosauria, which were early relatives of the mammal-like reptiles, occurred a strange modification. The neural spines of *Dimetrodon* rose like the masts of a ship far above the dorsal surface of the body; there is evidence that they supported an upright sheet of skin. In *Edaphosaurus* (appropriately called "Daffy" by some paleontologists) and in *Naosaurus* (Fig. 174) these spines also bore numerous horizontal bars, suggesting yardarms! To the best of our knowledge they were not used to hang mutineers.

Birds

In correlation with the great expansion of the pelvis in birds, from twelve to sixteen or more vertebrae and their ribs commonly fuse to form a ***synsacrum*** (Fig. 175); this means that not only are the two primitive reptilian sacral vertebrae included, but also some that would be (if they lacked ribs) lumbar vertebrae, one rib-bearing thoracic vertebra, and some of the caudals. The thoracic region is short, with a maximum of eleven or twelve vertebrae bearing fully developed ribs in *Archeopteryx*,

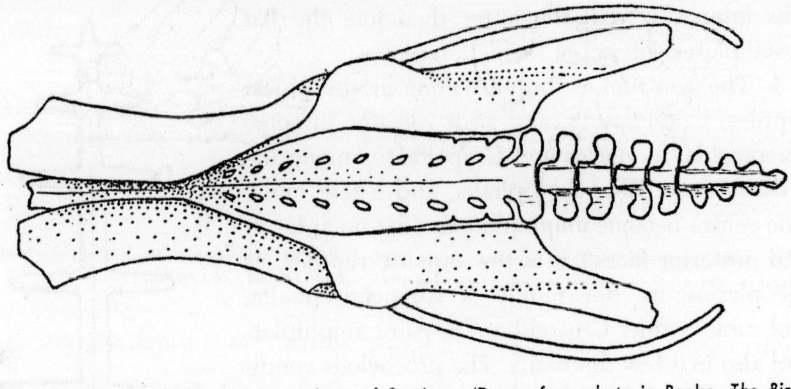

Fig. 175. Synsacrum and caudal vertebrae of flamingo. (Drawn from photo in Beebe, *The Bird*, Henry Holt.)

the earliest known bird, to five or six in the more advanced kinds. Often several thoracics fuse in the shoulder region. The neck contains a variable number. The cervical ribs articulated by two heads in *Archeopteryx* but fuse with the vertebrae in modern birds. The tail, finally, was long, with freely movable vertebrae, in *Archeopteryx*, but in Neornithes (modern) is reduced to a few free units followed by a block, the ***pygostyle***, containing several fused caudals.

The ends of the neck vertebrae are saddle shaped (***heterocoelous***, Fig. 166). This combines a strong joint with great flexibility. If you sat in the saddle on the front of a centrum you would face left or right; if you sat in the posterior saddle you would face dorsally or ventrally. Only the atlas vertebra is procoelous, to fit the condyle of the skull. But in *Archeopteryx* and in *Ichthyornis*, a toothed bird of the Cretaceous period, the vertebrae were amphicoelous, a heritage from the reptiles.

Mammals

Most mammals (Fig. 176) have five regions in the vertebral column: cervical, thoracic, lumbar, sacral, and caudal. The neural arch and spine develop from basidorsal arcualia, and the centrum from interdorsals and perhaps also interventrals. It is a pleurocentrum, as with birds and most reptiles, while the hypocentrum is here seldom present except as the ***chevron bones*** (***haemal arches***) of the tail in several of the older orders. The articular surfaces of the centra are slightly concave, or may be flat (***amphiplatyan***) as in whales. They carry throughout their growth a disc, the ***epiphysis***, between which and the body of the centrum

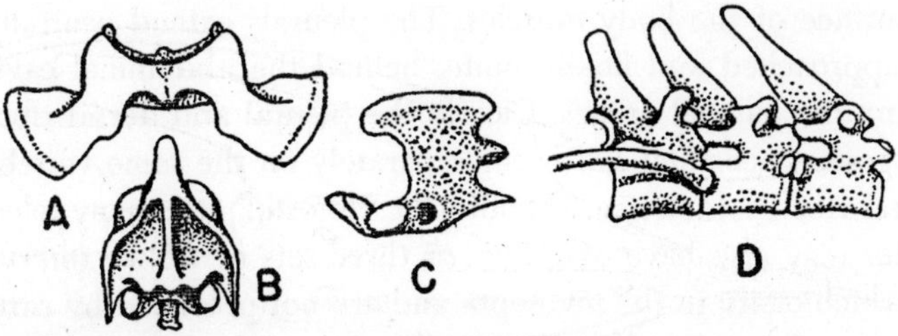

Fig. 176. Vertebrae of hare, *Lepus*. A, atlas, ventral; B, axis, ventral; C, axis, lateral; D, thoracic, lateral. (After Parker and Haswell, *Textbook of Zoology*, 6th ed., Macmillan.)

growth mainly takes place, until at maturity the epiphysial joints close and the vertebra is complete.

The cervicals bear ribs which fuse with the centra and are recognizable only by the persistent ***vertebrarterial canal*** which they enclose. The number is nearly always seven (exceptions: six in manatees and two-toed sloth; nine in three-toed sloth).

The thoracics, most often thirteen in number, carry transverse processes for the tubercula of the ribs, while the capitulum usually meets a facet on two adjoining centra. Neural spines develop in proportion to the bulk of dorsal muscles; therefore in those animals, as the larger ruminants or the elephants, which bear a heavy head, they are particularly long and powerful. In the lumbar vertebrae of most mammals the neural spines and transverse processes are long and project obliquely forward. The sacrum is usually small, containing three or four fused vertebrae, but in whales and Sirenia, which have no hind limbs, a sacrum cannot be distinguished.

RIBS AND STERNUM

Ribs are by no means universal among vertebrates, nor are the kinds of ribs in various groups all derived from a single ancestry. Those of sharks and rays are short rods of cartilage attached to the centra of corresponding vertebrae in the trunk and extending laterally for a short distance in the horizontal septum between epaxial and hypaxial muscles. They develop in continuity with the basiventral arcualia, at least in the procartilage stage. This type of rib may be called ***dorsal.***

In actinopterous fishes there are sometimes ***dorsal*** ribs and ***ventral,*** or ***pleural,*** ribs (Fig. 177); both occur in *Polypterus*. The dorsals project out from the centra in the horizontal septum; the pleurals lie along the inner surface of the body muscles. The pleurals extend ventrally as the tail is approached and finally unite, behind the abdominal cavity, thus becoming the haemal arches. Clearly the pleural and dorsal ribs are not homologous, since they can occur separately on the same vertebra.

Pleural ribs characterize Chondrostei, Holostei, and many teleosts, but the latter may also have one, two, or three sets of riblike ***intermuscular bones*** which ossify in the myosepta and are not preceded by cartilage. A sardine is well equipped with these, but among higher teleosts they appear less often, and in several groups even the pleural ribs are lost.

Lungfishes have bony pleural ribs, articulated with the basiventral

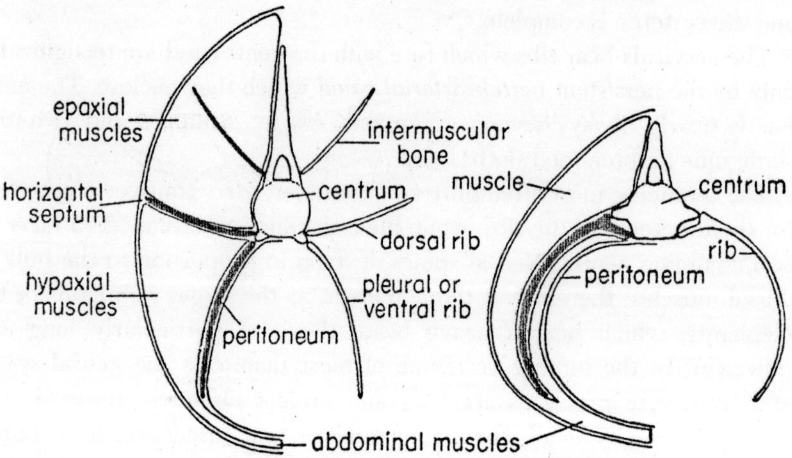

Fig. 177. Types of ribs in fish (left) and tetrapod (right).

cartilages. We can probably say that this type goes back to the common ancestry of lungfishes and Actinopteri. But in crossopterygians no ribs have yet been found.

Coming to tetrapods (Fig. 177) we find a dorsal rather than pleural type of rib; that is, they lie between layers of muscles in the body wall. We may infer a third origin of ribs, perhaps this time in connection with support of the body wall in animals which had become terrestrial. This is especially shown by the two-headed form of the earliest tetrapod ribs, for the dorsal head (*tuberculum*) anchors the rib in its extended position, while the ventral one (*capitulum*) acts as a brace against the centrum. Typically a short *parapophysis* receives the capitulum and a longer *diapophysis* the tuberculum (Fig. 173). One or both may be borne on a more extensive *transverse process.* The ribs may become single headed, as in a number of reptiles, while frogs and toads have (except for certain primitive species) lost them, and much lengthened transverse processes take their place.

Among modern Amphibia ribs do not reach the sternum but in other ribbed tetrapods some, at least, do. The attachment usually is by a flexible band of cartilage or a separately ossified part of the rib, called the ***sternal rib.*** The remaining part, associated with the vertebra, is the ***vertebral rib.*** Any which do not meet the sternum are ***floating ribs.*** Rhynchocephalia, Crocodilia, and many kinds of birds carry an ***uncinate process*** on the posterior margin of each rib, which overlaps the rib behind. We have noted already in connection with vertebrae the differences between cervical, thoracic, and sacral ribs.

Abdominal ribs or ***gastralia*** are slender dermal bones not related to

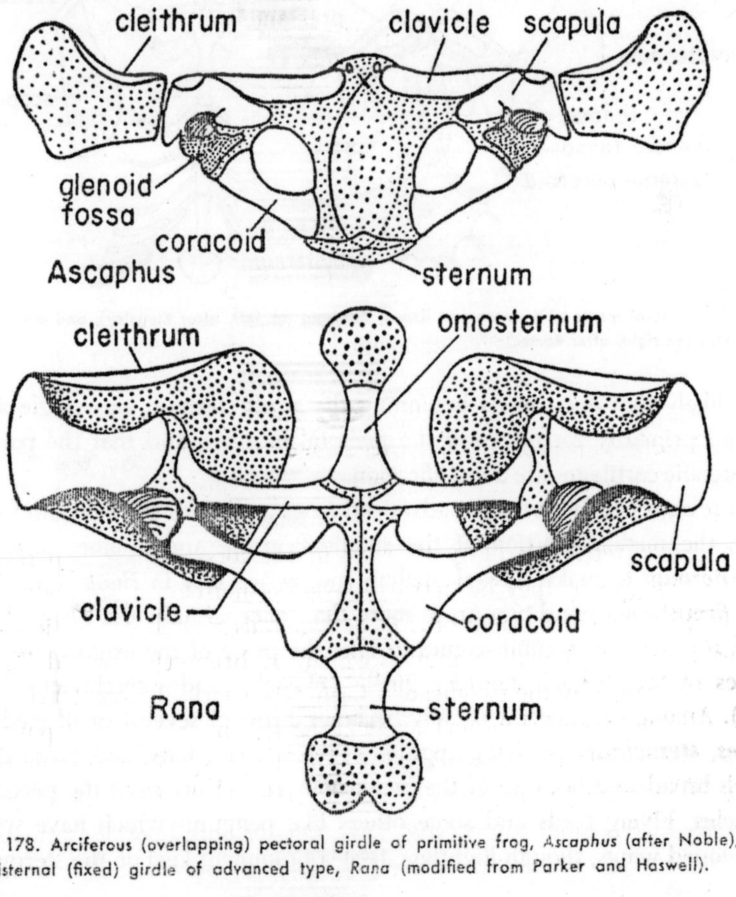

Fig. 178. Arciferous (overlapping) pectoral girdle of primitive frog, *Ascaphus* (after Noble), and firmisternal (fixed) girdle of advanced type, *Rana* (modified from Parker and Haswell).

true ribs, set in two or three series in the ventral wall of the abdomen. They seem to have originated in the early labyrinthodont Amphibia as oblique rows of scales or scutes. They did not persist in higher Amphibia but were inherited by a number of orders of reptiles. *Archeopteryx*, the earliest known bird, had twelve or thirteen pairs of abdominal ribs, but they do not appear in any later birds, or in mammals.

The ***sternum*** is a plate or rod of cartilage or bone between the ventral ends of the thoracic ribs. It occurs in tetrapods but not fishes, is much reduced in modern Amphibia and absent in Caecilia, amphisbaenids, snakes, plesiosaurs, and ichthyosaurs. It develops in reptiles, birds, and mammals from a pair of separate longitudinal cartilage bars which attach to the lower ends of the ribs and later fuse medially with each other. But

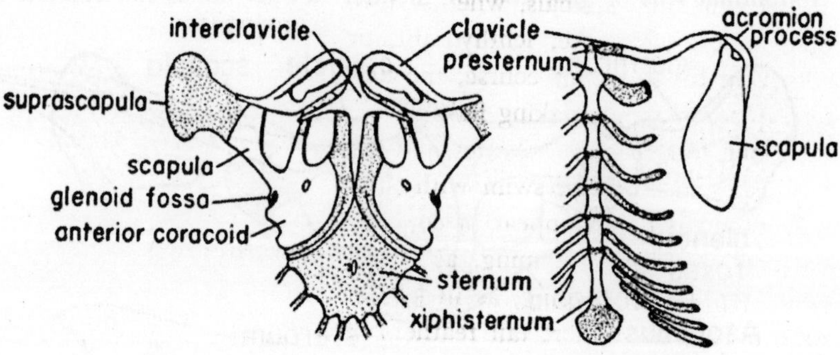

Fig. 179. Pectoral girdle and sternum in lizard, *Xantusia* (on left, after Kingsley), and marsupial, *Didelphys* (on right, after Romer).

it is likely that even in its beginning the adult sternum was single and served primarily for origin of the pectoral muscles, and that the paired embryonic cartilages are a modification.

In several families of more advanced frogs and toads the sternum fuses with the median portion of the shoulder girdle and a more anterior ***omosternum*** to make a fixed median rod, as we find in *Rana*. Probably this ***firmisternal*** condition arose more than once in Salientia (Fig. 178).

In reptiles the sternum comes to take the place of the original breast-plates of the dermal shoulder girdle (clavicles and interclavicle, Fig. 179). Among mammals it usually becomes a row of several small median bones, ***sternebrae.*** In flying reptiles (Pterosauria), bats, and birds it is much broadened because of the increase in area of origin of the pectoral muscles. Flying birds and some others like penguins which have well-developed wings, though flightless, bear a prominent keel on the sternum, which adds still more to the pectoral muscle area. The Ratitae (ostriches, etc.) lack such a keel, perhaps through loss.

TAIL AND MEDIAN FINS

The *tail* is that part of an animal, posterior to the body cavity, borne primarily by the vertebral column; this means in most cases that it starts behind the anus and the pelvic limbs, but there are exceptions. In fishes, again with exceptions, the tail is the primary locomotor organ, because in the flexing of the body the most effective action is at the posterior end. As a rule, then, a flattened surface, the *caudal fin,* develops there in any group of tail-swimming animals, whether fish, salamander, sea snake, ichthyosaur, or whale; in the latter, of course, the flattening is horizontal, making flukes.

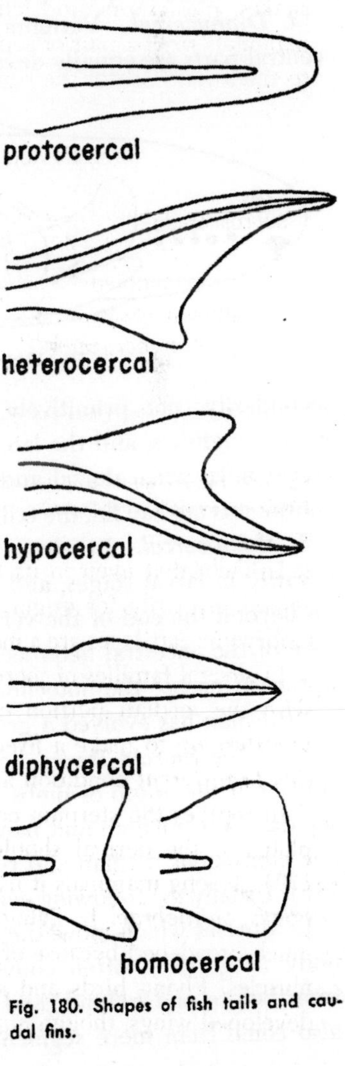

Fig. 180. Shapes of fish tails and caudal fins.

Among terrestrial vertebrates and the few fishes that do not swim with their tails, other functions appear: a counterbalance used while running, as in bipedal reptiles, or leaping, as in kangaroos; a base to support tail feathers of birds; a prehensile organ, as in sea horses, chameleons, and several kinds of New World monkeys; a defensive weapon, as in stingrays and porcupines; a reservoir of food or water, as in Gila monsters; or, finally, to swish away insects, as in most grazing animals. If none of these functions exist, the tail may disappear; so it has done in frogs, sloths, bears, apes, and man.

As to the *caudal fin,* there are several kinds, whose relationships we may show (Fig. 180):

1. ***Protocercal.*** Possibly ancestral type of fin, with notochord (or vertebral column) ending straight and with fin equally extended above and below; pres-

ent in many young larval fishes but in no adults, so far as known; those of lampreys and hagfishes undoubtedly diphycercal (see below).

2. ***Heterocercal.*** Many fishes of the older groups have this, in which the vertebral column ends in the elongated upper lobe. Some ostracoderms and nearly all the older gnathostome groups were heterocercal.

3. ***Hypocercal,*** or "reversed heterocercal." Vertebral column ends in lower lobe of caudal fin; occurred in Anaspida and some other ostracoderms, and in ichthyosaurs, where a tail fin evolved with a down-swept vertebral column upon their return to marine life.

4. ***Diphycercal.*** Vertebral column goes to end of fin, and dorsal and ventral parts are equally developed; but the term applies to fins which are

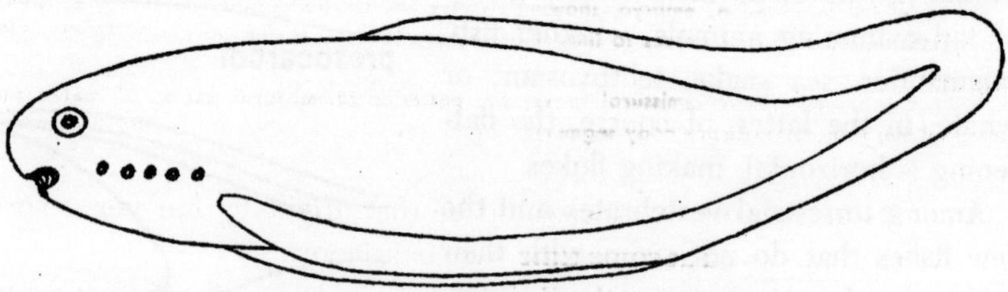

Fig. 181. Imaginary primitive fin folds.

secondarily, not primitively, thus. Modern cyclostomes, Holocephali, living lungfishes, and the later crossopterygians all have (or had) diphycercal tails; when dorsal and anal fins unite with the caudal to make a continuous median fin, the tail is ***gephyrocercal.***

5. ***Homocercal.*** In teleost fishes the heterocercal condition shows temporarily in larval stages, and is succeeded by a fully symmetrical caudal fin beyond the end of the vertebral column; this appears to be the lower lobe of the ancestral heterocercal fin, for it grows up and out behind the upturned end of the notochord as the latter degenerates. Upon this successful plan has evolved a great variety of homocercal fins.

The Fin-Fold Theory. Since this theory concerns the fundamental subject of the origin of limbs, we must examine it carefully. It is that the separate fins, median and paired, originated in some early ancestor from one continuous median and two ventrolateral folds along the body (Fig. 181). Usually the following evidence is given:

1. Muscles which move the fins are derived from many segments of the body (Fig. 182), often converging towards the developing limb bud from before and behind; and the spinal nerves going to these muscles also come from more segments of the body than the adult fin or limb

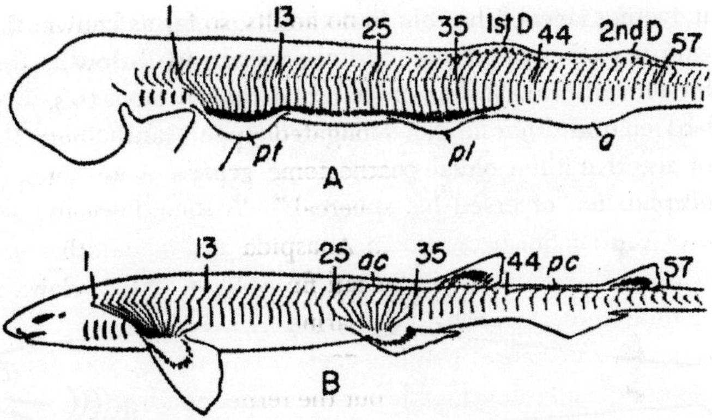

Fig. 182. *Scyllium canicula.* A, embryo, showing somites, muscle buds, and limb buds; B, adult, showing relationship of spinal nerves to fins. (From Eaton, 1945, *J. Morphology*, vol. 76 (3): 209, after Goodrich.)

a, anal fin; *ac*, anterior commissural nerve; *pc*, posterior commissural nerve; *pl*, pelvic fin; *pt*, pectoral fin; numerals indicate body segments.

covers at its base; the inference would be that originally fins were more extensive than now, and at some time were continuous.

2. In the very ancient acanthodian fishes (Fig. 183A) there were up to seven pairs of ventrolateral fins, which suggests breakup of two folds.

3. In the Devonian shark *Cladoselache* (Fig. 184) the paired fins were broader at the base than distally; their skeletal supports ran out parallel and quite uniformly from the base; this again might imply departure from a once continuous fold on each side.

4. Even *Branchiostoma* (Amphioxus) is often brought into the picture because its metapleural folds seem to match the hypothetical plan (Fig. 183, C, D).

Here, however, are some contrary arguments: In the study of living and fossil fishes one does not find any truly primitive group in which the fins are continuous (Fig. 185). They fuse and make continuous median folds in a number of cases, including lampreys and the very ancient pleuracanth

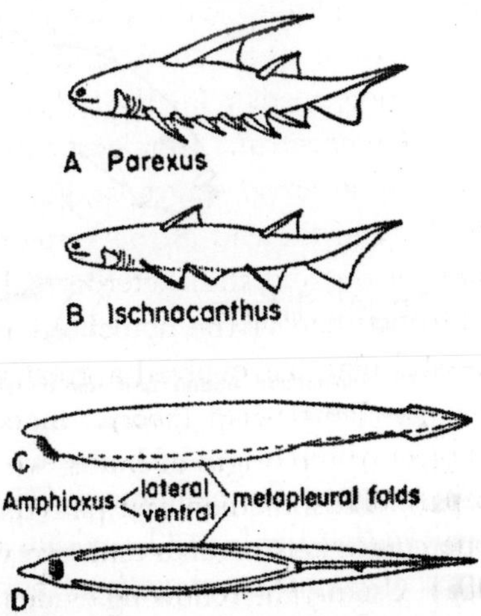

Fig. 183. Two acanthodians. A, *Parexus* with five pairs; B, *Ischnacanthus* with two pairs of ventral folds; C, D, lateral and ventral views of Amphioxus, showing metapleural folds.

sharks; but fossil evidence reveals that these are specialized and came from still earlier kinds in which the fins were separate. Indeed in the oldest of all known vertebrates, the ostracoderms, the median and paired fins were short-based and often no more than spines. In acanthodians the author cannot see that the several paired "fins" represent anything but a special multiplication of defensive spines, which sometimes but not al-

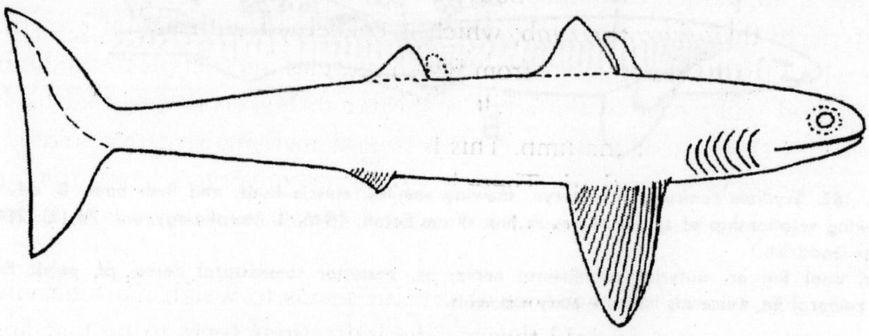

Fig. 184. Primitive shark, *Cladoselache*. (Based on Romer.)

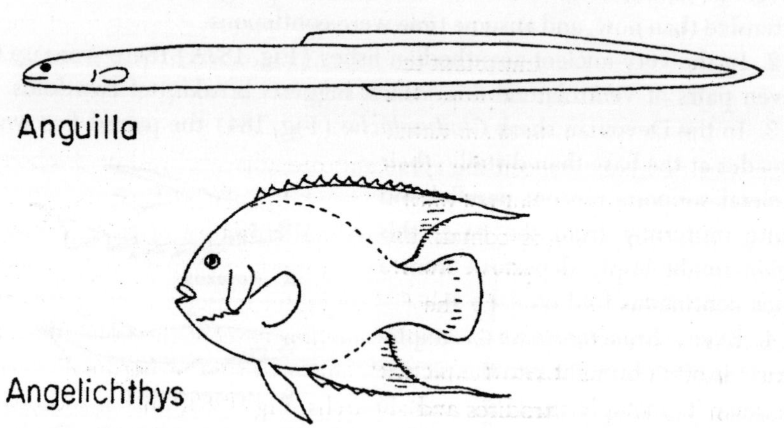

Fig. 185. Specialized bodies and fins in teleosts. Above, an eel, *Anguilla;* below, a coral-reef fish, *Angelichthys*.

ways support fin membranes. Gregory and Raven (1941) give reasons to believe that the broad-based fins of *Cladoselache* were not primitive but specialized, and analogous to those of rays. Amphioxus would seem to be irrelevant, since its metapleural folds develop in correlation with, and on the outside of, the atrium; several writers now incline to the view that Amphioxus is not at all an ancestral chordate.

As to the argument that segmental muscles and nerves converge towards the growing limb from points before and behind its base, it is not difficult to show that this "convergence" is an illusion. When a limb bud first appears, it is, like some other embryonic organs, relatively large (Fig. 182). The body segments from which it receives muscles are those that lie adjacent to it. After the muscle buds and their accompanying nerves enter the limb bud they adjust their rate and direction of growth to the limb, rather than the body. But the body grows faster antero-posteriorly than does the limb, which is concerned with lateral growth. Therefore the body segments from which muscles and nerves of the limb originated will cover, in the adult, a greater distance forward and back than does the base of the limb. This is true of median, as well as paired, fins, in, for example, a shark. There has been no convergence towards the limb base, but a divergence of nerves and muscle segments within the body.

Thus the balance of evidence at present seems to weigh more heavily against than for the fin-fold theory. The indications seem to be that fins arose in connection with median and paired spines, already established in ostracoderms (Figs. 10, 11). Perhaps the defensive function of such spines was at first more important than the locomotor. The pectoral spines were in a position to act as props, keeping the body upright when resting on the bottom. It would seem natural that a membrane or fold of skin should extend out between the spine and the adjacent wall of the body, as in many cases it did, but not until we reach the placoderms, apparently, did such a fold begin to contain an internal structure of rays. Without such rays, and deeper pieces for their support, no effective locomotor action could be produced by the fins, since fin muscles need places for insertion. The evolution of fins for locomotion (except the tail) seems to have been slow, haphazard, and with curious experiments on the way (see, for instance, arthrodires and antiarchs, Fig. 186), until a successful "basic patent" was finally attained in fishes above the placoderms.

Dermal Fin Rays. Whether median or paired, fins are nearly always supported by parallel rods just beneath the epidermis. Such dermal rays may be jointed or not. To their deep ends attach the muscles that move the fins, or sometimes to bones or cartilages that in turn hold the dermal rays. In elasmobranch fins the dermal rays are nonjointed flexible rods called ***ceratotrichia.*** In Actinopteri, the ray-finned fishes, jointed rays derived from rows of scales replace ceratotrichia; they are called ***lepidotrichia*** (Figs. 118, 187). Other groups of fishes show one or both of these types in differing combinations.

Fig. 186. An antiarch, *Pterichthyodes milleri.* (After Moy-Thomas.)

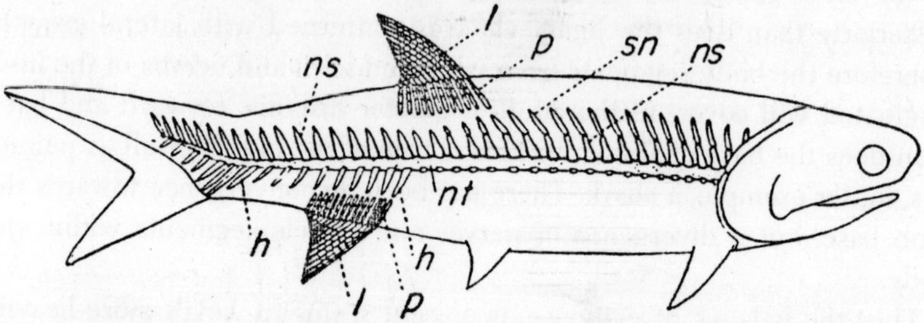

Fig. 187. Diagram of axial skeleton and fin supports in a primitive actinopteran. (From Eaton, 1945, *J. Morphology,* vol. 76 (3): 202.)
h, **haemal arches; *l,* lepidotrichia; *ns,* neural spines; *nt,* notochord; *p,* pterygiophores; *sn,* supraneurals.**

Pterygiophores. These are the deeper bones or cartilages on which dermal rays rest (Fig. 187); the word means "fin bearers." Each consists of a single piece, or two or three; the single ones are in more advanced fishes and are the result of fusion. The separate elements, if present, are called ***radials;*** if, as in sharks, some are much larger than others, they are called ***basals.***

PAIRED LIMBS AND GIRDLES

Paired Limbs

There is no clear evidence that the paired fins of fishes were primitively continuous, and there is excellent evidence that the two girdles had very different beginnings. Nevertheless, the pectoral and pelvic fins, or limbs, usually resemble each other in surprising detail.

The *pectorals* lie close behind the head. Their girdle consists of (1) a paired *scapulocoracoid* portion of cartilage or of replacing bone, with which the fins or limbs articulate, and (2) a superficial, usually more elaborate, series of *dermal bones* primitively connected with the skull.

The *pelvic* appendages in most cases lie next to the anus and the posterior end of the body cavity. Their girdle consists of paired cartilages or replacing bones, often joined medially, and in most tetrapods attaining a dorsal connection with the vertebral column by way of the sacral ribs.

Although no trace of paired fins remains in lampreys and hagfishes, most ostracoderms had pectorals, either in the form of flaps preceded by spines (Fig. 10), or spines alone. Some Anaspida (Fig. 11) had paired rows of spines along the abdomen; these, with the frequent addition of fin membranes behind them, we see again in acanthodians (Figs. 12, 183A, B). Most placoderms had spines rather than fins, either movably or (at first) immovably fixed on the sides of the thoracic armor. These facts seem to indicate (1) that a dermal spine acted as the first skeletal support of a fin, (2) that such spines may have occurred before any fins, (3) that

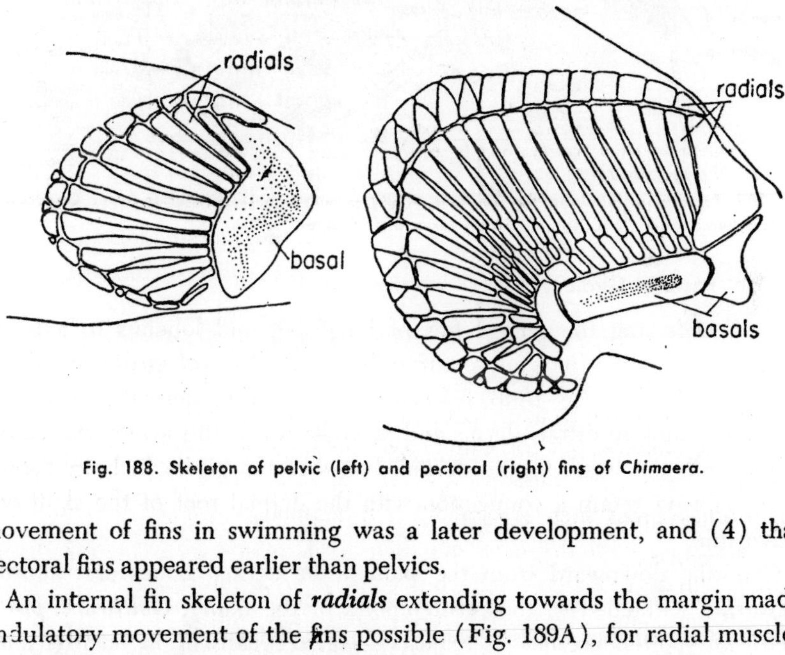

Fig. 188. Skeleton of pelvic (left) and pectoral (right) fins of *Chimaera.*

movement of fins in swimming was a later development, and (4) that pectoral fins appeared earlier than pelvics.

An internal fin skeleton of ***radials*** extending towards the margin made undulatory movement of the fins possible (Fig. 189A), for radial muscles could raise and depress them. This mechanism came in acanthodians, if not earlier. Still more effective action resulted from enlargement or fusion of some inner radials, now called ***basals.*** Gregory and Raven (1941) very plausibly suggest that the scapulocoracoids, and also the pelvis, are simply further enlarged and fused basals.

Apart from the girdle, the basal elements of the fin proper in elasmobranchs are usually three, ***propterygium, mesopterygium,*** and ***metapterygium*** (Figs. 188, 189A). A similar, although doubtless independent, structure is found in *Polypterus* (Fig. 189B). Once more may we remark that *Polypterus* is not a crossopterygian? Beyond these large pieces extend radials, usually many jointed and sometimes, as in rays, forked at the tips. Among Actinopteri it is usual for the internal skeleton of the paired fins to be much reduced, for dermal rays bear the brunt of fin support and movement. Usually not more than two rows (salmon) or one row (most other teleosts) are left, articulating with the scapulocoracoid or the pelvis.

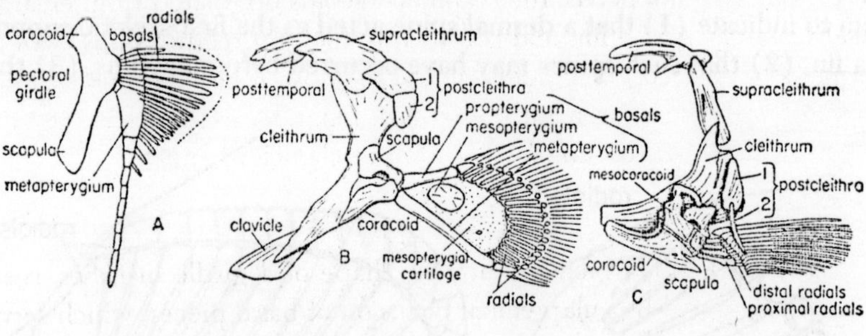

Fig. 189. Pectoral girdles and fin skeletons of (A) primitive shark, *Cladodus* (after Goodrich); (B) *Polypterus* (after Goodrich); (C) *Salmo fario* (after Bridge).

Pectoral Girdle

Deferring the paired fins of lungfishes and lobefins to a later paragraph, we may look more carefully at the *pectoral girdle* of fishes. Its dermal bones (Fig. 189B, C) can be traced to the dermal plates covering the trunk in ostracoderms and placoderms. As this armor was associated with the head, so the dermal bones of the girdle in bony fishes nearly always retain a connection with the dermal roof of the skull on either side.

Counting downward from the skull, these dermal bones are: ***supracleithrum*** (attached to the posttemporal of the skull), ***cleithrum, postcleithrum*** (often present in teleosts), ***clavicle*** (absent in teleosts and other advanced Actinopteri, but present in lungfishes and lobefins). As elasmobranchs have no bones they lack a connection between the skull and girdle; the rays, however, have a peculiar dorsal plate developed from the neural spines of several anterior dorsal vertebrae, to articulate laterally with the upper end of each scapulocoracoid. The connection between the posttemporal and supracleithrum in crossopts was apparently lost even in the early Amphibia; both of these bones promptly disappeared so that the head became more readily movable on the neck.

To trace the origin of the endodermal (cartilage or replacing bone) part of the girdle, namely the scapulocoracoid, is more difficult, since we cannot easily choose between two alternatives. In the ostracoderms Stensiö (1927) found a large internal wall at the back of the skull, which he called postcranial because it lay behind the gill region and must have provided part of the base for attachment of the pectoral fins. This wall might, by detaching from the skull and reducing, become the scapulocoracoids. The other and more likely alternative is (Gregory and Raven, 1941) that, like the pelvis, the scapulocoracoids originated from enlarged basal elements of the fin, and secondarily took up relationships with the dermal bones of the shoulder girdle. In most fishes, but not elasmobranchs, these pieces remain paired and separate, except that they may be linked indirectly by union of cleithra or clavicles beneath.

Pelvis

The *pelvis*, which has not the shape of a girdle in fishes, seems to have originated by enlargement or fusion of basal pieces, which served

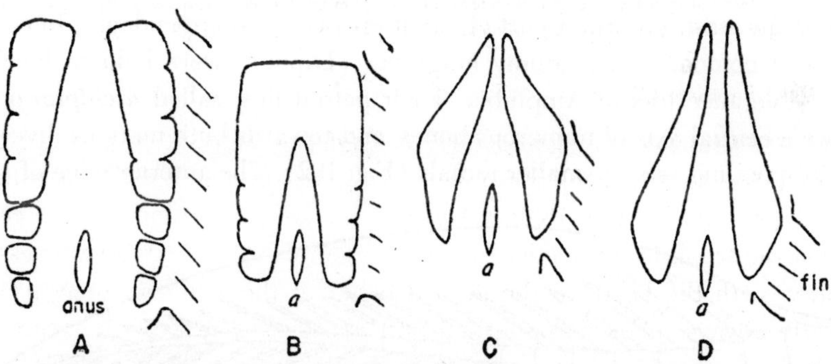

Fig. 190. Pelvis of fishes, showing presumed origin of deep basal cartilages, diagrammatic. A, primitive shark; B, modern shark; C, primitive actinopteran, lobefin, etc.; D, teleost.

for origin of radial muscles of the pelvic fins. The pelvis is therefore paired in primitive fishes (early sharks, Holocephali, crossopterygians, *Amia*, etc., Fig. 190).

In the modern sharks and rays the two pieces fuse to make a single crossbar in front of the anus. Likewise the halves fuse, although differently, in modern lungfishes, and again still differently, in teleosts. Thus the pelvis, at first paired, has become a single structure not less than four times among vertebrates, if we add the tetrapods to those already mentioned.

Among many higher teleost fishes the pelvic girdle and fins forsake their traditional place near the anus (Fig. 191) and become ***thoracic;*** in doing so, the anterior end of the pelvis attaches to the ventral tip of the united cleithra. A few groups carry the process even further; codfishes and blennies have ***jugular*** pelvics, that is, beneath the throat. Not uncommonly they may be lost altogether (eels), or modified in peculiar ways, as the circular sucking disc of the lumpsuckers (Fig. 207).

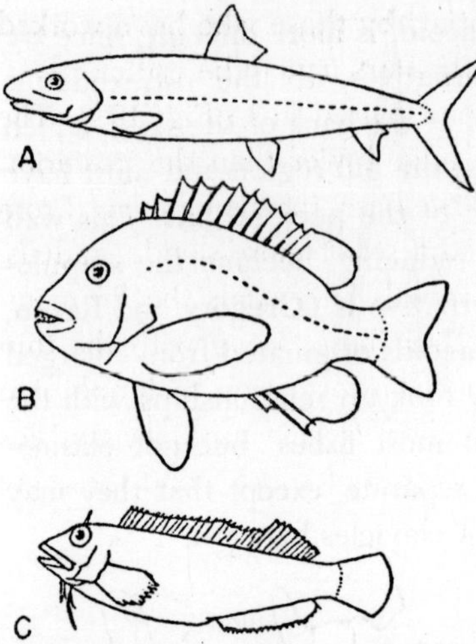

Fig. 191. Positions of pelvic fins in teleosts. A, abdominal, in *Albula;* B, thoracic, in *Archosargus;* C, jugular, in *Clinus.*

Paired Fins

The nature of dipnoan and crossopterygian paired fins, and derivation of tetrapod limbs from the latter, is the most important part of our study of paired appendages. In early days of comparative anatomy lungfishes (Dipnoi) were held to be the probable ancestors of Amphibia. Their paired fins, called ***archipterygia,*** have a central axis of numerous bones, and towards both margins diverge corresponding sets of smaller radials (Fig. 192). The anterior edge of the

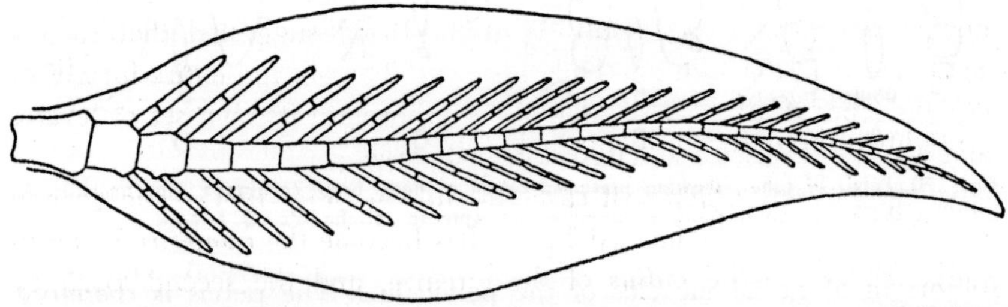

Fig. 192. Skeleton of archipterygial fin of lungfish, *Epiceratodus.* (After Parker and Haswell.)

pectoral is held higher than the posterior, but both pairs trail in the water. The attempt to imagine this structure changed into that of a salamander's leg and foot met insurmountable difficulties, for nothing intermediate could be found.

With better understanding of crossopterygians and of the early Amphibia (labyrinthodonts) the claim of lungfishes to tetrapod ancestry collapsed. Lungfish skulls and teeth, as well as fins, are far too highly specialized, but those of crossopts meet the requirements much better.

The view given here is accepted in essentials by those who have worked on the problem recently, but in some particulars it may be criticized.

Early crossopts had broad paddle-shaped paired fins (Fig. 193). The pectorals were carried with the anterior edge lower than the posterior. A row of diminishing pieces ran out from the base, forming an axis. From the anterior (preaxial) side there diverged a set of more slender radials, but not from the posterior. Beyond the radials and axials the fin membrane was supported by dermal fin rays, but these, along with the thin margins of the fins, were lost during the transformation to amphibian limbs. We may take it that in the most generalized crossopterygian fins each *axial* piece was attended by a *primary radial* and beyond these lay

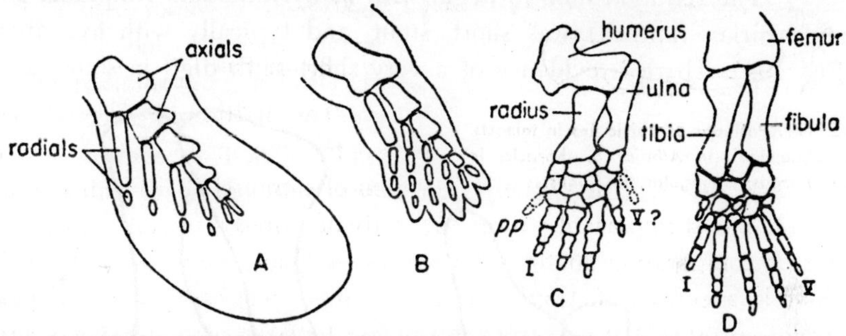

Fig. 193. Diagram of (A) crossopterygian fin skeleton; (B) possible transition to tetrapod limb; (C) the forelimb of *Eryops*; and (D) hind limb of *Trematops*.
***pp*, prepollex; I, V, numbered digits.**

one or two ***secondary radials.*** Westoll (1943) suggested that such a series corresponds with one body segment. Thus in the entire fin five or six segments are indicated, each presumably with its own set of levator and depressor muscles, and its own spinal nerve.

In the limb of *Eryops,* an early amphibian (Fig. 193C), the general pattern is similar. The first axial piece has become the ***humerus,*** while its radial has made the ***radius*** of the forearm, and the second axial, by lengthening, is recognizable as the ***ulna.*** The small ***carpal*** bones of the foot (wrist) make a succession of diagonal rows slanting towards the preaxial edge, just as their counterparts did in the fin, but the bones of the ***digits*** are new; they were not represented at all in the fin. According to this scheme, the primary and secondary radials of the fish fin, plus a few more elements in series with them (tertiary and quaternary radials, if you like), can account for essentially the whole foot, and also for the obliqueness of the rows of carpal bones. This is consistent with Westoll's segmental idea, and fits, with relatively little adjustment, the plan of foot muscles in Amphibia.

We must notice the changes in position and function of the limb. In order to bear the weight of the body in walking on land, the flat ventral surface had to be placed on the ground with its tip forward. This required (1) a more flexible shoulder joint, and (2) an elbow, so that the lower arm and hand could be flexed. In the hind limb (Fig. 193D) the change was similar except that a knee joint, directed forward, developed between the rounded end of the *femur* and the heads of the *tibia* and *fibula*. The wrist and ankle seem to have developed their flexibility later and more gradually.

Tetrapod Limb

The tetrapod limb to begin with (labyrinthodont Amphibia and cotylosaurian reptiles) was short, stout, and typically with five digits (Fig. 194). There is evidence of a very short sixth digit in some laby-

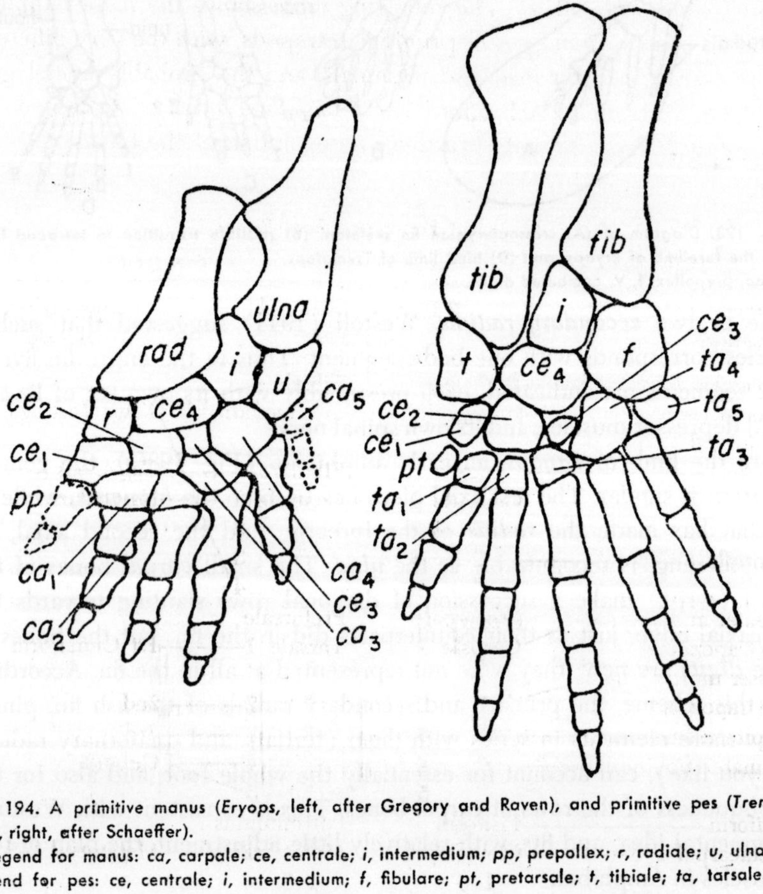

Fig. 194. A primitive manus (*Eryops*, left, after Gregory and Raven), and primitive pes (*Trematops*, right, after Schaeffer).

Legend for manus: *ca*, carpale; *ce*, centrale; *i*, intermedium; *pp*, prepollex; *r*, radiale; *u*, ulnare. Legend for pes: *ce*, centrale; *i*, intermedium; *f*, fibulare; *pt*, pretarsale; *t*, tibiale; *ta*, tarsale.

rinthodonts, the *prepollex* (mediad of the thumb) or *prehallux* (mediad of the first toe in the hind foot). No modern amphibians have more than four digits in the front foot (*manus*), but nearly all have five in the hind foot (*pes*). The famous two-toed "fossil footprint" named *Thinopus antiquus*, supposed to be that of an Upper Devonian amphibian, is figured in many books and discussed widely, but unless its validity is confirmed, it can have no weight in the problem of tetrapod origin; it is perhaps a coprolite (impression of feces).

The oblique pattern of the small foot bones, which we have traced from the crossopterygian fin, was already somewhat obscured in labyrinthodonts by the close and many-sided articulation of these pieces with one another, for the foot now had to be firm enough to support the animal's weight on land. Eventually the oblique pattern disappeared as new methods of jointing developed, or as individual bones atrophied or consolidated with others. The following table shows the names applied to bones of manus and pes of primitive tetrapods, with the very different set of names used for man and mammals, and the probable homologies between the two sets. Schaeffer (1941) has done much to straighten out the confusion that existed regarding the evolution of the tarsus.

Names of Hand and Foot Bones

Hand (manus): Mammalian names	Hand (manus): Primitive Tetrapod names	Foot (pes): Primitive Tetrapod names	Foot (pes): Mammalian names
Navicular (scaphoid)	Radiale	Tibiale	
Lunate (semilunar)	Intermedium	Intermedium ?	Talus (astragalus)
Triquetral (cuneiform)	Ulnare	Fibulare	Calcaneus
	Centrale 1	Centrale 1	Navicular
	" 2	" 2	
Centrale (in mole)	" 3	" 3	
	" 4	" 4	
	Precarpale	Pretarsale	
Greater multangular (trapezium)	Carpale 1	Tarsale 1	1st Cuneiform
Lesser multangular (trapezoid)	" 2	" 2	2nd "
Capitate (magnum)	" 3	" 3	3rd "
Hamate (unciform)	" 4	" 4	Cuboid
	" 5	" 5	
Pisiform	Pisiform	Postminimus	
Metacarpal 1–5	Metacarpal 1–5	Metatarsal 1–5	Metatarsal 1–5

Modifications of the primitive plan are innumerable. Tetrapod limbs display a wonderful range of structural differences. The fact most obvious in surveying them is that each adaptive type appeared more than once, in unrelated groups (e.g., ***cursorial*** or running, ***graviportal*** or weight bearing, ***leaping, climbing, gliding, flying, burrowing, swimming***). This often results in a surprising similarity of body or limbs in animals of diverse ancestry, that is, convergent evolution (Figs. 195, 196).

In Amphibia the limbs are lost in certain eellike ancient amphibians,

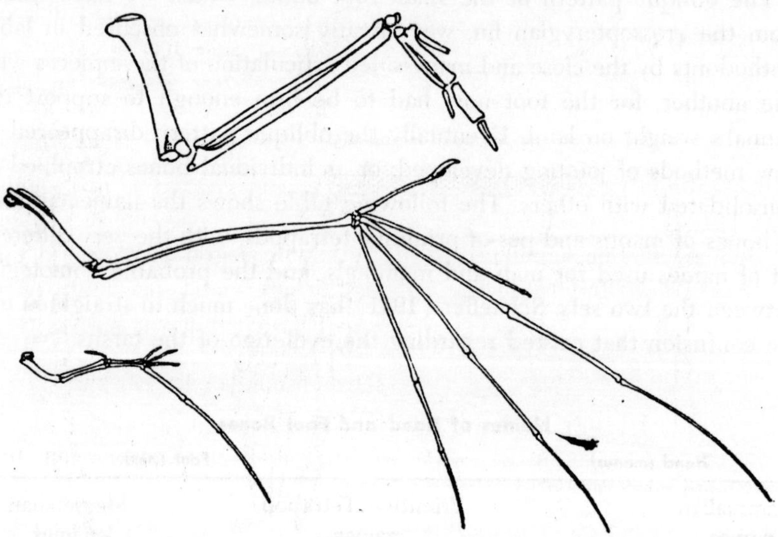

Fig. 195. Comparison of three unrelated types of wing skeleton. A bird, the condor (top, redrawn from Beebe); a fruit bat, *Pteropus* (redrawn from Parker and Haswell); a flying reptile, *Pterodactylus* (bottom, after Parker and Haswell).

in the burrowing wormlike Caecilia, and nearly lost in *Siren* and *Amphiuma,* which are eel-shaped Caudata. Otherwise, salamanders have four toes in front, and five or rarely four on the hind feet, with carpal and tarsal bones varying from six to twelve. The Salientia, adapted usually for leaping, have the radius and ulna fused, with four toes in front, plus sometimes a prepollex. In the hind leg the tibia and fibula are fused, the tibiale and fibulare elongated and often partly fused (increasing the effective length of the foot), while the digits (five) are webbed.

In the tarsus of most reptiles the astragalus and calcaneus (fibulare) form a proximal row, followed by a transverse joint, and then the tarsalia plus perhaps one or two centralia; also the limbs are generally held laterally, as with Amphibia, so that the body is on the ground when at rest and may even drag along during locomotion. The Therapsida (mammal-like) held their limbs more nearly beneath, and the astragalus and calcaneus were mammalian in form, with the main joint of the hind foot proximal to them. In the flipperlike limbs of ichthyosaurs and other marine reptiles the humerus and femur were short and heavy, the more distal bones being small and often greatly increased in number, arranged in parallel rows. Dinosaurs, as we noted before, were primitively bipedal but in more advanced kinds often quadrupedal. The carpals and tarsals

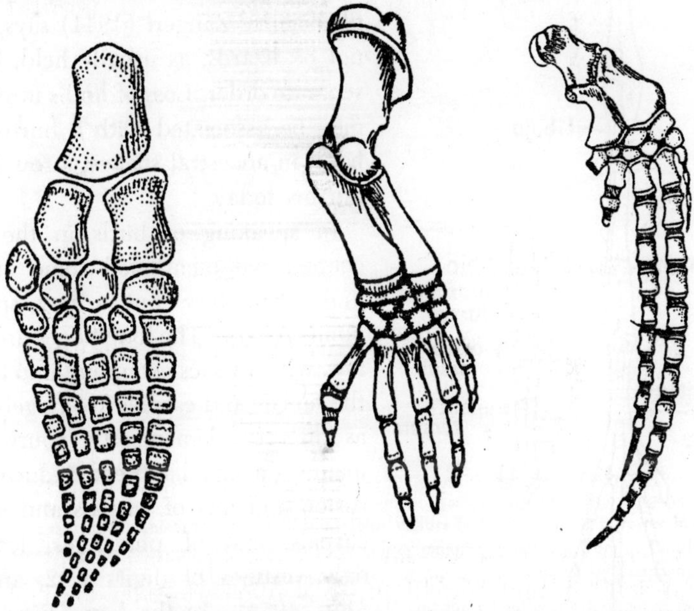

Fig. 196. Three unrelated types of flipper (forelimb) in aquatic animals: a porpoise, *Globicephalus* (right, after Parker and Haswell); a marine reptile, *Ichthyosaurus* (left); a dugong (middle, after Parker and Haswell). All are external aspects of right flippers.

were much reduced, while the metacarpals and metatarsals were heavily built and supported the weight of the body; they and the digits in ornithischian dinosaurs were sometimes reduced to three in the hind feet, making enormous birdlike footprints in the Mesozoic mud.

The pterosaurs we have described briefly, but may add that the long bones, like those of birds, were hollow, apparently with air sacs in them, while the radius and ulna were fused, metacarpals fused and lengthened, and of course the phalanges of the fourth finger extraordinarily lengthened; in the hind leg the tibia and fibula were fused, the tarsals much reduced. All these features go with making a light, strong skeleton.

Crocodilia have reduced carpals and tarsals, five digits in the manus and four in the pes. In Chelonia the feet are quite primitive, but webbed (especially the pes) in many freshwater turtles, clublike in the heavy land tortoises (*Testudo*), and powerful flippers in the sea turtles. The astragalus and calcaneus in *Sphenodon* are fused and the tarsals reduced, but otherwise its limbs are generalized; those of lizards are similar but often with elongated digits. The burrowing, generally limbless group Amphisbaenia, Zangerl (1944) says, may not be lizards, as usually held, but a separate order. Loss of limbs in snakes may be associated with a burrowing habit in ancestral snakes; a few kinds burrow today.

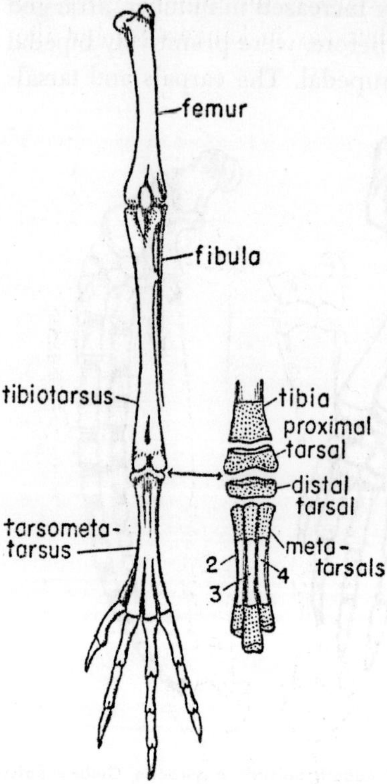

Fig. 197. Leg and foot bones of pigeon, *Columba*. Adult, left; embryo, right. (After Parker and Haswell, *Textbook of Zoology*, 6th ed., Macmillan.)

In speaking of birds in the first chapter we mentioned some of the limb characters, but a few special points remain. The long bones are hollow, with air sacs extending into them; the tarsals and carpals are largely lost as discrete elements. Of course the manus (wing) has been reduced by fusion and loss of carpals and metacarpals, and of phalanges, leaving only vestiges of digits 1, 2, and 3. (Fig. 195A). In the legs (Fig. 197) the fibula is vestigial, and, most important, the tibia fuses with the proximal tarsal bones making a ***tibiotarsus*** (the lower leg bone), while the distal tarsals fuse with the metatarsals making a ***tarsometatarsus*** (the ankle bone). The digits are usually four, sometimes three (first and fifth lost); there are only two (third and fourth) in the ostrich.

The appendages in Mesozoic mammals are not known. Those of monotremes are somewhat specialized for digging or swimming. For the most primitive mammalian type we must therefore look to the generalized marsupials and insectivores. The fore feet of marsupials have five simple digits, but on the basis of hind foot structure there are two groups: Didactyla (including opossums) in which the hind foot also is simple and five-toed, and Syndactyla (the majority of marsupials) in which the 1st digit of the hind foot is thumblike or lacking, second and third small and not separated from each other, while the fourth and fifth are normal or enlarged. The latter situation may have resulted from an early arboreal adaptation in which the foot was used for grasping, as it still is in many marsupials, followed by a return of some types to the ground, when the fourth and fifth digits enlarged still further (as in kangaroos) but the useless thumb became vestigial.

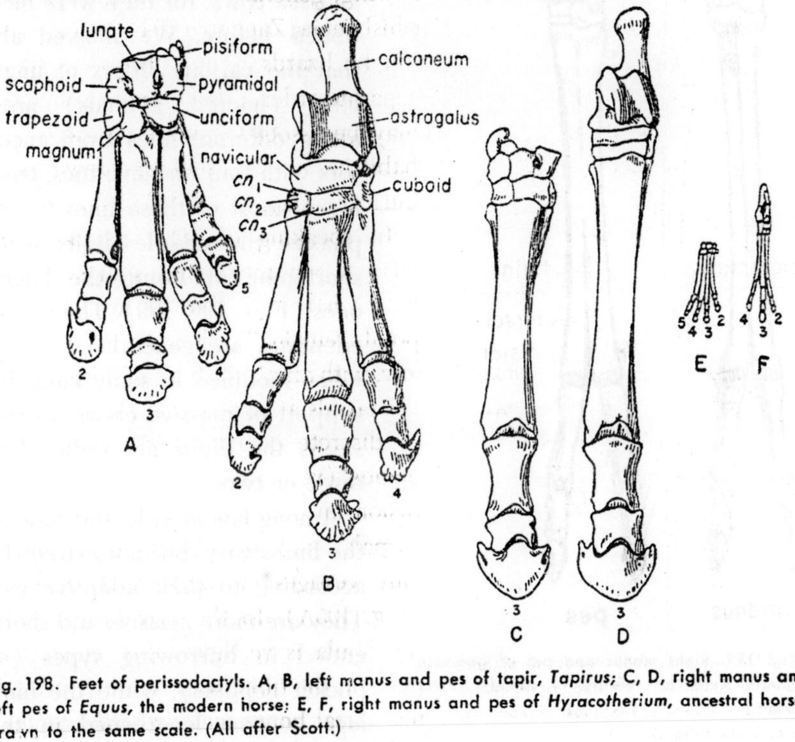

Fig. 198. Feet of perissodactyls. A, B, left manus and pes of tapir, *Tapirus;* C, D, right manus and left pes of *Equus,* the modern horse; E, F, right manus and pes of *Hyracotherium,* ancestral horse, drawn to the same scale. (All after Scott.)

Shrews, especially tree shrews, have the most primitive, plantigrade (walking on the soles), five-toed feet that we can find among placental mammals. In the burrowing insectivores (moles) the bones of the fore limb and hand are heavy and broad, and movement is lateral. The arboreal, gliding Indo-Malayan Dermoptera (the so-called flying lemurs, which are not lemurs and do not fly), are related to both insectivores and primates; their feet are of a primitive sort, but a broad fold of skin extends between fore and hind legs, as in the flying squirrels. We have dealt in the first chapter with the hands and feet of Primates, noting that a grasping thumb and big toe are characteristic; in a few cases the fingers of the hand act together as a hook in seizing branches of trees while the animal travels rapidly through the forest; thus the gibbons have reduced, and spider monkeys completely lost, the thumb. It is an example of sacrificing a primitive character in favor of a high specialization.

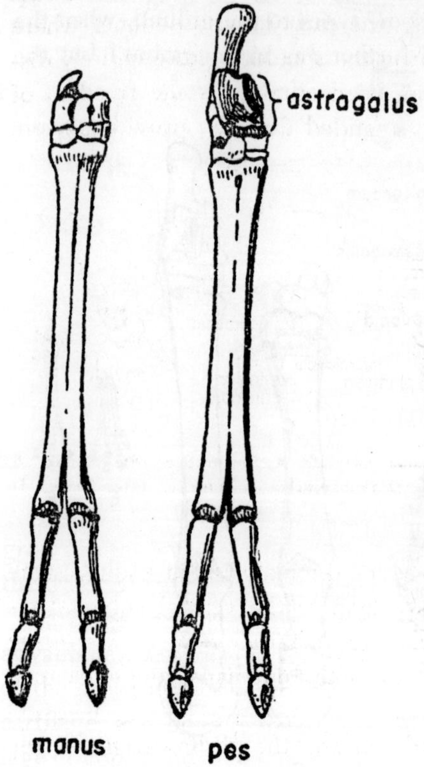

Fig. 199. Right manus and pes of guanaco (South American relative of camel). (After Scott.)

We need not review the limb characters of carnivores, whales, and sea cows, for they were mentioned before. We showed also that the various orders of ungulates (hoofed animals) arose along different lines from ancestors with four and five toes, tending in most of those lines to emphasize the middle digits while shortening or losing the lateral ones (Figs. 198, 199). The feet of edentates are generalized except when modified in some kinds for support of massive claws; in tree sloths the digits are reduced to three or two.

Among lagomorphs and rodents the limbs vary, but not extremely, according to their adaptive use. They are more massive and short-ended in burrowing types (as pocket gophers), while the hind legs become lengthened in the runners and leapers; in more than one family of rodents a bipedal habit has arisen (jerboas, kangaroo rats), with the result that the forelegs are very small and the animal is essentially a miniature kangaroo in shape and behavior. At the other extreme among terrestrial mammals come the graviportal Proboscidea, with massive pillarlike limbs to support their weight, digits shortened, and foot bones in parallel series.

The Pectoral Girdle in Tetrapods

We saw that in higher fishes the girdle consists of a deeper ***scapulocoracoid,*** preformed in cartilage, and a more superficial ***dermal se-***

ries, linked dorsolaterally with the roof of the skull, and meeting at the midventral line. In the fish the girdle is a stable base for articulation of the pectoral fins.

In tetrapods the girdle becomes more and more freed from the rest of the skeleton, and progressively simplified by loss of parts. The first step in both of these processes occurred in the early labyrinthodonts, where the *supracleithrum* disappeared, leaving the shoulder girdle entirely disconnected from the skull, so that the head obtained some freedom of movement. The *scapula* meanwhile expanded dorsally, providing more

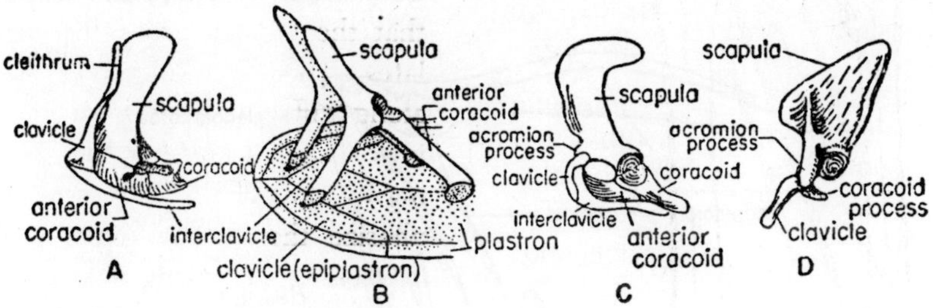

Fig. 200. Pectoral girdles of reptiles and mammals, left side. A, *Dimetrodon* (after Romer); B, turtle, pectoral girdle in relation to plastron; C, *Ornithorhynchus* (monotreme, after Romer); D, gorilla.

room for the enlarging limb muscles which originated on it, and the same factor probably accounts for the expansion of the ventral coracoid region. The dermal parts of this girdle were a rather reduced *cleithrum* and a *clavicle* (on each side, of course), with a median diamond-shaped breastplate, the *interclavicle.*

Modern Amphibia have no dermal bones in the girdle except for the narrow cleithrum and clavicle of Salientia. In Caudata the girdle consists only of a scapulocoracoid ossification surmounted by suprascapular cartilage, and extending ventrad as a coracoid plate of cartilage; this usually overlaps its opposite in the median line, and meets, behind, a small cartilaginous sternum. In Salientia there is a separate coracoid bone, properly called *anterior coracoid.* Toads, the primitive frog *Ascaphus* (Fig. 178), and tree frogs (*Hyla*) have the right and left halves of the girdle overlapping at the median line; this loose sort of girdle is called *arciferal.* Advanced frogs (*Rana* and others) enjoy a solid median fusion of the girdle, called the *firmisternal* type.

Reptiles, except those that have lost the pectoral girdle, keep the anterior coracoid (Fig. 200A), but behind this, in mammal-like reptiles, appears the *coracoid* proper; we shall follow it presently into mammals.

The clavicle and interclavicle remain in most reptiles, but the cleithrum persists only in cotylosaurs and some mammal-like reptiles. Turtles show an extraordinary condition: the pectoral girdle, starting to develop anteriorly to the ribs, becomes internal to them (Fig. 200B). That part to which the humerus is attached is a triple prong, consisting of a slender V-shaped scapula and the bladelike anterior coracoid. The clavicles and interclavicle persist in turtles as paired and median plates of the plastron (ventral shield). Clavicles are absent, however, in Crocodilia.

Birds (Fig. 201) have a strong anterior coracoid reaching from the sternum to the head of the humerus (glenoid cavity), and the scapula

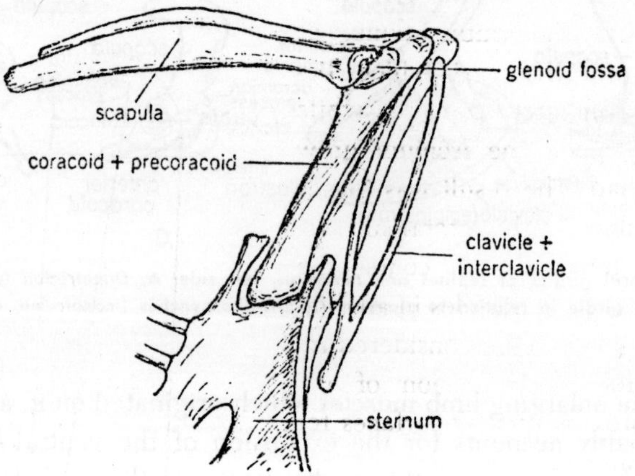

Fig. 201. Pectoral girdle of turkey, right side. (After Adams and Eddy.)

extends back like a knife blade. The clavicles fuse at their ventral tips with the interclavicle and form the "wishbone," usually attached to the sternum.

The primitive reptilian girdle is passed on, little changed, to the monotreme mammals (Fig. 200C), where the interclavicle makes a T-shaped breastplate. Higher mammals, of course, have no interclavicle or anterior coracoid; the coracoid itself is present only as a hooked process (Fig. 200D) on the scapula, while the clavicle, if full sized, goes from the sternum to the acromion process of the scapula. It may, however, be vestigial. The last stage in reduction and release of the pectoral girdle comes by total loss of the clavicle in ungulates, elephants, many carnivores, and the marine groups, where nothing but the scapula is left, and nothing but muscles serve to connect it with the axial skeleton. Thus the scapula enters freely into movements of the limb, and is no longer a fixed base.

Pelvic Girdle in Crossopterygian Fishes

The *pelvic girdle* in crossopterygian fishes was no more than a pair of small simple bones lying on either side of the cloaca and serving for articulation of their respective fin skeletons. In *Eusthenopteron,* according to the restoration by Gregory and Raven (1941), these two bones were in contact at their anterior ends, and each bore an iliac process projecting dorsally but not nearly reaching the vertebral column (Fig. 202). The acetabulum (socket) was at the posterior end. According to these authors the pelvis was then equivalent to the pubis alone, of tetrapods, and the part which became the ilium grew upward, while that which made the ischium grew posteromediad. The result was to put the acetabulum on the lateral surface, as it is in tetrapods. Of course the enlargement of leg muscles in adaptation to land life may be considered responsible for the extension of the bones, to provide greater surfaces for their attachment.

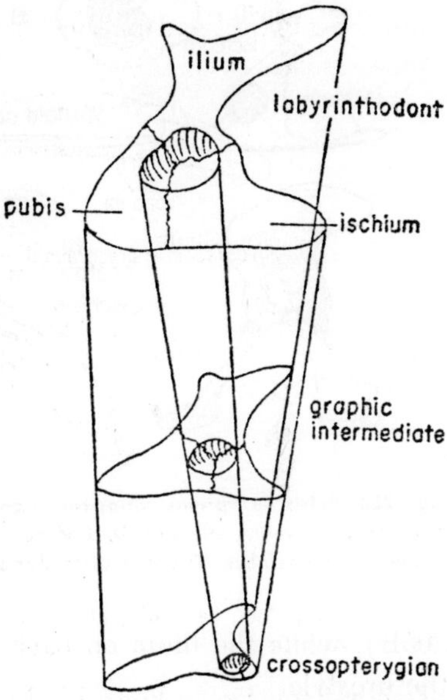

Fig. 202. Probable relationship between pelvis of crossopterygian fish (bottom) and that of labyrinthodont (top), as shown by graphic intermediate form. Left lateral views. (After Gregory and Raven.)

The next step, found in early Amphibia, was to make a symphysis (union by suture) between the two pubic bones; that of the ischia came somewhat later, but presently a broad flattened ventral plate was established, from which the ilia rose vertically on either side to meet the ends of one pair of ribs and thus produce a true girdle and a sacrum for the first time. The earliest labyrinthodonts did not yet have the sacroiliac connection. A single pair of ribs is involved in Amphibia, but the sacral region contains two pairs in most reptiles. The acetabulum, in all but a few tetrapods, is the meeting place of all three pelvic bones.

The pelvis of Caudata is degenerate, consisting usually of an ischiopubic cartilage with centers of ossification ventrally only for the ischia; the ilia are partly ossified. Sutures may be lacking. In lung-breathing salamanders (Fig. 203A) a Y-shaped (ypsiloid) cartilage extends forward from the pubes in the median line and is movable; by pulling it in (dorsad) the salamander thrusts air forward in the lungs, causing the head to rise as it floats in the water. The ypsiloid cartilage is therefore part of a special hydrostatic mechanism. In Salientia, as part of the leaping adaptation, the pubis and ischium may be much reduced (Fig.

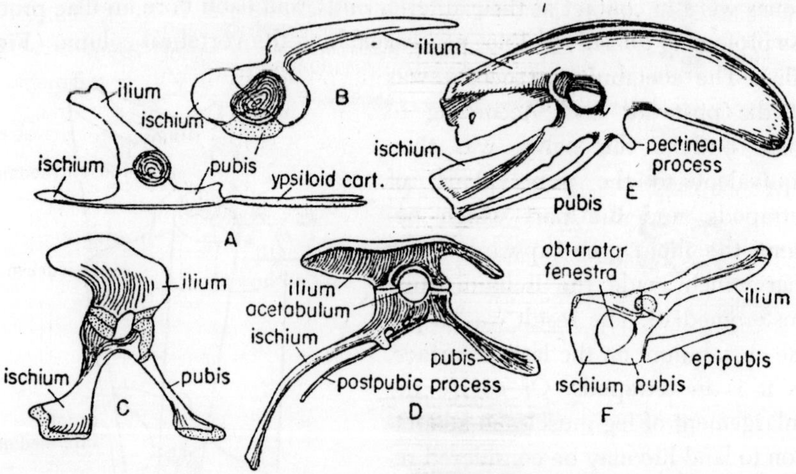

Fig. 203. Pelvis in various tetrapods, right lateral views. A, salamander, *Cryptobranchus;* B, frog, *Rana;* C, young alligator; D, dinosaur, *Iguanodon,* showing bird type; E, ratite bird, *Apteryx;* F, opossum, *Didelphys.* (D and E after Goodrich; F after Romer.)

203B), while the ilium on each side extends far forward, parallel with the urostyle.

Early reptiles had a complete ventral plate like that of labyrinthodonts, with the ilia attached to two pairs of sacral ribs. A small foramen in each pubis gave passage for the obturator nerve, and continues to do so in most reptiles, but the obturator foramen as seen in mammals and the Therapsida is the result of union of this pubic foramen with a larger opening which presently developed between the pubis and ischium (properly called puboischial fenestra). The fenestra is wide in most reptiles, so that the pubic symphysis does not connect with the ischial symphysis, and as seen from the side the pelvis is triradiate (Fig. 203C).

Ornithischian dinosaurs and birds share certain peculiarities of the pelvis that possibly indicate relationship (Fig. 203, D, E). The ischium and pubis are greatly prolonged caudad, parallel with each other, and do not form symphyses (there is a pubic symphysis in *Archeopteryx*, but in no higher birds). Ornithischia also have a large "prepubic process" forward and down from each pubis; this corresponds to the "pectineal process" of birds, shorter but in the same position. The ilium, of course, is greatly expanded and thinned, as mentioned already in connection with the synsacrum of birds. The acetabulum is partly open, probably reducing weight by eliminating superfluous bone.

There are not many remarkable features of the pelvis in mammals. Monotremes and marsupials have a pair of ***epipubic*** or ***marsupial*** bones in the abdominal wall just anterior to the pubes, from which they develop embryonically (Fig. 203F). Usually the ventral symphysis is short, made only by the pubes in some insectivores, carnivores, and Primates. It is absent in other insectivores and in Chiroptera. The pelvis as a whole degenerates to a vestige or disappears in Cetacea and Sirenia, in connection with loss of their hind legs.

CHAPTER VI

Muscles

WHAT MUSCLES ARE AND WHAT THEY DO

Meat is muscle. Nearly all movements of the body are caused by the contraction of muscles. The muscular system contributes more to the bulk of the body than any other. A muscle is an organ of soft, well-vascularized tissue which, when stimulated, shortens itself and thereby pulls toward one another the parts to which it is attached. When the stimulus ceases, the muscle relaxes and regains its original shape passively. Muscles pull; they never push.

In spite of this highly plastic form, the positions and anatomical relations of muscles are as constant in any species of animal as are those of other organs. Therefore it is no more difficult to understand the anatomy of the muscular system than to grasp that of the skeleton or the blood vessels. Yet muscles are rather commonly neglected in books of comparative anatomy.

The reason for this neglect may be that anatomists differ on the homologies and names of many muscles among the classes of vertebrates. Few authors have attempted to follow muscles carefully through a series from fish to mammal. This is not an easy task for two reasons: the fossils give little direct evidence of muscular anatomy, so that we must depend largely on modern types; and only a handful of workers have used evidence from embryology, although this is quite readily obtainable.

Of the three kinds of muscle tissue found in vertebrates (Fig. 204), two can scarcely be treated in a comparative anatomy course because they do not constitute organs in their own right, or because their anat-

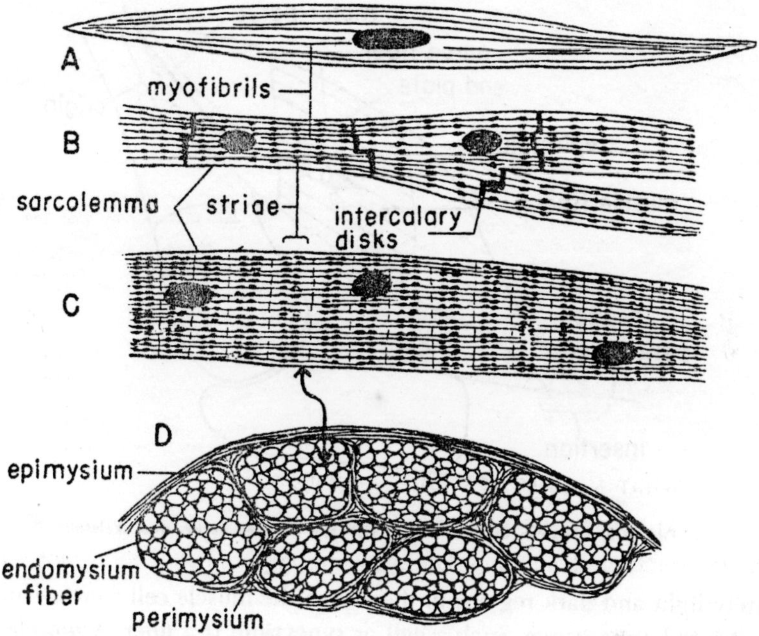

Fig. 204. A, smooth muscle cell; B, cardiac muscle fiber; C, skeletal muscle fiber; D, bundles of muscle fibers in a muscle, cross section.

omy is microscopic and shows little significant change. These two are ***smooth*** and *cardiac* muscle.

Smooth muscle occurs in the walls of the digestive tract as circular and lengthwise layers, in the walls of arteries and arterioles, circularly, in the passages of the urogenital system, in the deep layer of the skin (dermis) and elsewhere. The cells are spindle-like, with a single nucleus, relatively slow to contract, but powerful and activated by the autonomic nervous system; this means that with one or two exceptions they are involuntary.

Cardiac muscle forms the bulk of the heart wall, as well as a part of the pulmonary artery and aorta. The cells are generally branched, interlacing, with many nuclei, and partly striated (cross-banded). Of course they are involuntary and under autonomic regulation, but contraction is initiated *within the muscle itself*, not by a nervous impulse.

The third kind of muscle, ***striated*** (skeletal or "voluntary"), with which we are concerned from here on, is made of large slender many-nucleate cells (more properly called ***syncytia***), in which occur a great number of lengthwise ***myofibrils***, the actual contractile threads; alter-

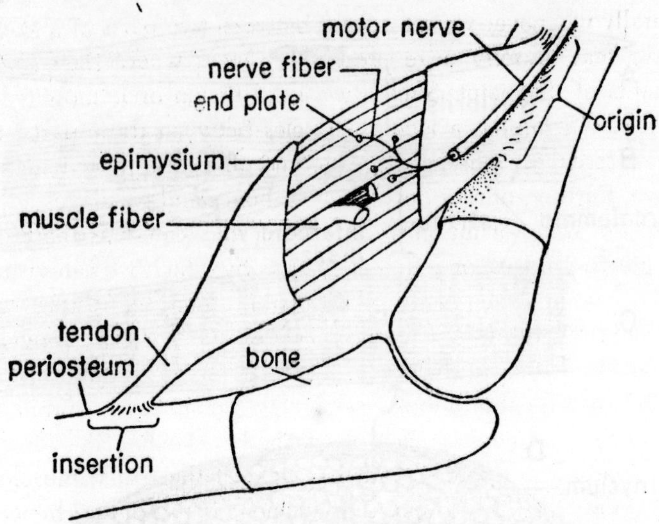

Fig. 205. Diagram of relationship between muscle, nerve, and skeleton.

nately light and dark regions in these give the muscle cell a cross-banded or striated appearance. Such a cell or syncytium is a ***fiber.*** A ***muscle*** is a great bundle of these fibers parallel to one another (Fig. 205). Small groups of fibers within the muscle are more or less set apart by a thin connective tissue wrapping, the ***perimysium,*** while the common sheath of connective tissue around all is the ***epimysium.*** By means of the connective tissue in these sheaths the muscle is fastened at either end to, usually, some part of the skeleton. A ***tendon*** is simply a thick cord or band continuous with the epimysium and serving to attach the muscle to a part which it may not reach by itself. A ***fascia*** is a sheet of such tissue.

Striated muscles are activated by the "voluntary" nervous system. Its ***motor nerves*** send branches to all muscle fibers (Fig. 205). Normally any contraction of a muscle is a response to a barrage of impulses from the central nervous system. The direction in which the fibers lie is, of course, the direction in which they will pull. The sheath or tendon of the muscle is continuous with the connective tissue sheath of the bone (***periosteum***) or cartilage (***perichondrium***).

The more stationary attachment of a muscle is its ***origin;*** the more movable is its ***insertion.*** Our chewing muscles, then, ***originate*** on the cranium and ***insert*** on the jaw. Usually muscles work in antagonistic pairs or groups, one pulling a bone one way, the other in the opposite direction.

Naturally whenever we see a joint between two parts of a skeleton we can infer that muscles were present in places where they could make movements of that joint. Similarly a solid fusion or immobility of parts of the skeleton implies a lack of muscles between those parts; it would be useless to have a muscle begin and end on a solid piece which it could not move. Further, during development, bones and cartilages adjust their shape to the stress of muscles upon them, and skeletons therefore carry ridges, crests, grooves, or roughened areas by which we can visualize the extent of muscle attachments. It is possible, then, to "restore" a part of the muscular anatomy of animals known only from fossil bones.

Several terms used in the study of muscles should be introduced here, so that we may be able to talk more freely. To ***adduct*** is to pull a certain part towards an adjoining part of the body, as when the lower jaw is pulled towards the upper, or the arm against the trunk; muscles acting thus are ***adductors.*** The opposite movement is to ***abduct;*** hence ***abductors.*** To bend a joint (increase the sharpness of its angle) is to ***flex*** it; muscles bending the knee, fingers, etc., are ***flexors.*** Antagonistic to flexors are others which straighten or ***extend,*** the ***extensors.*** Muscles which raise something are ***levators;*** those that lower are ***depressors.*** Any so placed as to squeeze a cavity, giving it smaller circumference, are ***constrictors. Sphincters*** are those which constrict an opening, as of the rectum or bladder. There are also muscles which rotate a part, ***rotators;*** which turn an appendage upside down or supine, ***supinators;*** and which turn it back again so that the normal lower surface is down or prone, the ***pronators.*** These last terms are used mostly in connection with rotating the arm and hand.

A particular muscle is almost invariably supplied by a particular motor nerve. During the course of evolution, just as during individual development, the same nerve continues to supply that muscle even if the latter should shift its position, change its action, split into separate parts, or fuse with some other muscle. We therefore need to know the innervation (nerve supply) of muscles in order to be sure that we recognize corresponding muscles when we compare different animals. For example, the muscles supplied by the seventh cranial nerve in man have no resemblance to those so supplied in a fish, yet a study of developmental stages or of the anatomy of intermediate types shows very clearly that they are the same, albeit with numerous changes of position and function, splitting into parts, and fusion with others.

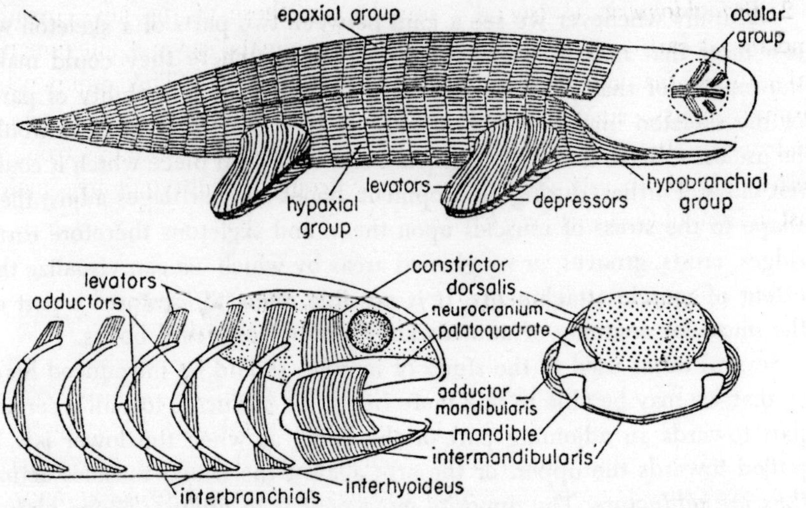

Fig. 206. Diagram of segmental muscles and their derivatives in a generalized vertebrate, at top. Lower left, diagram of primitive branchiomeric muscles (simplified). Lower right, transverse section through jaws of preceding.

ARRANGEMENT AND DIVISIONS OF THE SKELETAL MUSCLES

1. *Segmental* (somatic or myotomic) muscles are those which come from muscle segments or myotomes of the embryo, and are at first segmentally arranged (Fig. 206); in higher vertebrates they quickly lose this pattern. They are the muscles of the trunk, including neck and tail, and the following groups derived from these: hypobranchial (longitudinal throat) muscles; integumental muscles of the body in mammals, birds, and some reptiles; the muscle of the diaphragm in mammals. Also derived from myotomes are the muscles which move the eyeball (extrinsic ocular muscles). The muscles of the limbs come directly from myotomes in fishes and are therefore primitively segmental, although they develop within the limbs themselves in all higher groups.

In the neck, trunk, and tail, muscles of the right and left sides are separated dorsally by a ***dorsal septum*** of connective tissue, and ventrally by a corresponding ***linea alba.*** Along the body and tail in those animals whose muscle segments show clearly (except cyclostomes) there is also a ***horizontal septum*** separating the dorsal (***epaxial***) half from the ventral (***hypaxial***); this horizontal septum is usually at the level of the lateral line in fishes, and inwardly it meets the centra of the vertebrae. In birds and mammals the segmentation, as well as the separation of epaxial from hypaxial muscles, is largely lost by fusion.

2. ***Branchiomeric*** ("visceral") muscles are those of the jaws, hyoid, and throat cartilages (Fig. 206), plus (in mammals) several integumental muscles that move lips, nostrils, ears, eyelids, etc. This group was primitively associated with jaws and gill arches, and is supplied by cranial nerves. It has nothing to do with myotomes, but develops from a series of paired bands which in turn grow from the embryonic hypomere just in front of the heart (see Chapter II).

COMPARATIVE ANATOMY OF THE SEGMENTAL MUSCLES

Protochordates

Longitudinal muscles occur in the body wall of *Enteropneusta*, and in the tail of larval tunicates parallel to the notochord. But in neither of these is there segmentation. The tail muscles of tunicate larvae may be a reduced vestige of segmental muscles of an Amphioxus-like ancestor. *Branchiostoma* (Amphioxus) itself has 60 or more pairs of >-shaped myomeres (muscle segments) along the sides of the body; the point of each > is forward. Myomeres of right and left sides are placed alternately in Amphioxus and hagfishes, instead of opposite, as in all other Chordata. In the wall of the atrium lies a transverse sheet of muscle.

Vertebrates

Longitudinal Muscles of Neck, Trunk and Tail. In nearly all fishes, amphibians, and to a lesser extent reptiles, one can see a segmental arrangement of the longitudinal muscles. Myomeres (muscle segments) are separated from each other by myosepta (connective tissue walls), so that the muscle fibers, running longitudinally, reach only from one myoseptum to the next. The septa attach to the sides of their corresponding vertebrae; hence the myomeres must alternate with the vertebrae. (For development, see Chapter II.)

The mass of longitudinal muscles extends from skull to tail (Fig. 207), and, as seen in cross section, from middorsal to midventral line. In the tail the upper and lower portions are equal, but in the trunk there must be space for the body cavity and visceral organs below the level of the vertebrae, so that only a thin layer of muscle can surround the abdomen, leaving the bulk of body muscle on the dorsal side (Fig. 208).

Much distortion of this fundamental pattern has taken place. In birds and mammals, which no longer travel by wriggling undulations of the body, the embryonic myotomes blend, myosepta largely disappear, and muscle fibers extend for several segments. Segmentation still shows, of course, in the vertebrae, ribs, spinal nerves, and blood vessels.

Although it may be that the series of myomeres in some remote vertebrate ancestor extended along most of the head, uninterrupted, this is not true in any known vertebrate, because the skull occupies too much of the available space in the head. Therefore the first pair of trunk myomeres attaches to the back of the skull (occiput). In most fishes this

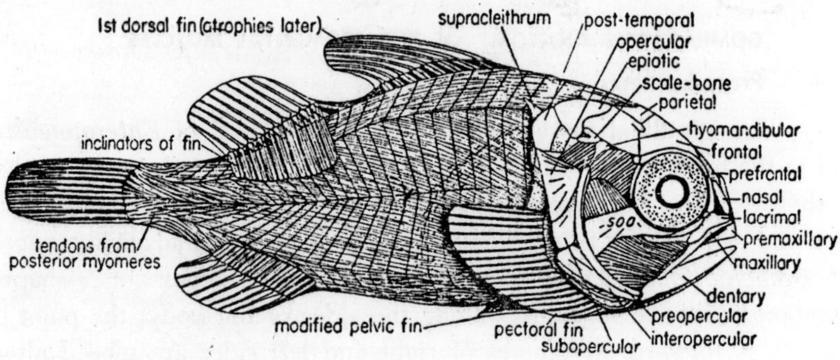

Fig. 207. General pattern of muscles in a fish, young *Cyclopterus lumpus* (lumpsucker), showing relationship to fins and head.

comes just behind the location of the tenth cranial nerve (vagus), because the vagus happens to be the last that is enclosed by any part of the skull. But in the degenerate modern cyclostomes the skull terminates behind the position of the eighth nerve. On the other hand, when we come to reptiles, birds and mammals there are twelve pairs of cranial nerves, and it is doubtful, then, whether the first trunk myomere is the same in all cases.

A special feature of segmental muscles is the zigzag form of the myomeres in most fishes. This seems to be related to speed in swimming, for the swiftest kinds, as tuna, swordfish, and mackerel, are those with the most acute angles in their myomeres. Moreover, in any fish the angles become sharper in the more posterior segments (Fig. 207), for the greatest activity is near the base of the tail. Here the myosepta are almost longitudinal and serve as tendons to transmit movement from muscle segments to tail vertebrae. The more nearly the myosepta lie in the direction of the muscle fibers (lengthwise), the more effective is their action. Any young fish shows at first vertical myosepta, but presently they become angled. The more sluggish Amphibia, although a "higher class" of animals than fishes, have only slightly bent myomeres.

Fishes show scarcely any differentiation of the hypaxial muscles into layers, but we find these in Amphibia (Fig. 209) and all higher animals. Ventrally, from pelvis to thorax, runs a ribbonlike paired muscle, the ***rectus abdominis.*** Laterally, but below the horizontal septum, are three layers, differing in the direction of their fibers: ***external oblique*** (runs

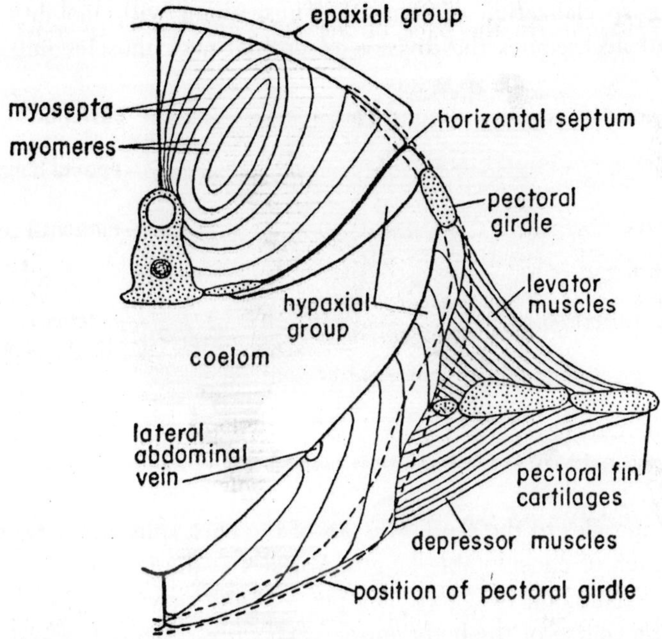

Fig. 208. Diagram of cross section through a shark, showing muscle groups and concentric pattern of myomere cones.

posteroventrad), ***internal oblique*** (anteroventrad), and ***transversus abdominis*** (dorsoventrad). Between successive ribs are segregated the ***external*** and ***internal intercostals,*** in higher tetrapods, which take part in breathing movements.

The ***diaphragm*** of mammals is a membranous sheet between the pleuropericardial and peritoneal cavities, to which has been added a layer of muscle derived from myomeres in the neck; it is supplied by the phrenic nerve, formed by branches of the third, fourth, and fifth cervical nerves. The action of the diaphragm is rhythmically to flatten itself from a dome shape, thereby enlarging the thorax cavity and admitting air to the lungs. This movement synchronizes with that of the intercostal muscles.

Integumental (skin) muscles of the body in reptiles, birds, and mammals are variously derived from segmental muscles directly, or from appendicular muscles. The skin-twitching muscle found in mammals, for instance (***panniculus carnosus***), develops from the ***latissimus dorsi***, a shoulder muscle, and from the edge of the ***pectoralis*** group; it frequently spreads to envelop most of the body.

Development of special activities in certain regions results in a corresponding specialization of muscles. Thus, along with mobility of the head and neck comes the division of epaxial neck muscles into several

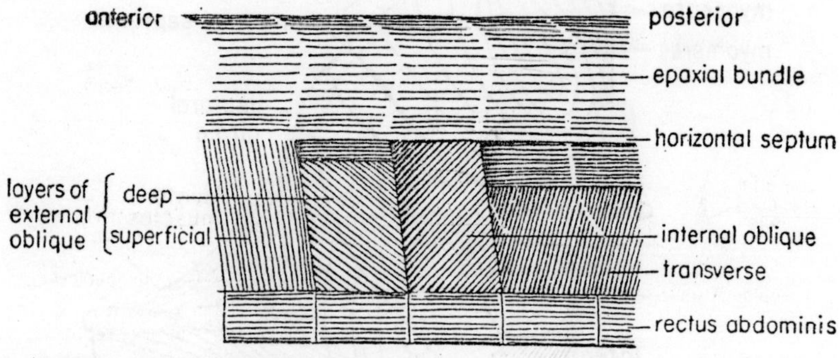

Fig. 209. Muscles of body wall in a salamander, *Dicamptodon ensatus*, partly dissected, left side.

occipital muscles to the skull. Muscles of the back split into three or four groups, each of considerable length. Muscles at the base of the tail may be specialized as tail waggers. The pectoral and pelvic girdles partly interrupt the course of the body muscles. This is especially true of the pelvis in birds and mammals because its bones are enlarged for increased attachment of limb muscles. The pectoral girdle, because it forms the posterior wall of the gill chamber in fishes, acts as a complete barrier to all muscles running forward, *except* those dorsal to it, and a small ventral group which extends to the lower jaw. In tetrapods, where the weight of the body must be borne by the limbs, we find portions of the body muscles attaching to the shoulder girdle internally or on its margins. For instance the ***serratus*** muscles act as a sling supporting the body between the scapulae, especially in mammals, and the ***rhomboideus*** takes a part in the same action, at a different angle.

The ventral group of throat muscles, the ***hypobranchial*** group, is actually a forward continuation of the hypaxial trunk muscles. It consists of paired muscles originating on the pectoral girdle (as in elasmobranchs) or sternum, and inserting on the mandible and hyoid (***coracomandibu-***

laris and *coracohyoideus* of sharks; *sternohyoid* plus *geniohyoid*, etc., of tetrapods). These are the ventral divisions of myomeres which have their dorsal divisions *above* the gill chambers; consequently the nerves which supply them do so by taking a detour back behind the gill region and then forward ventrally. In tetrapods the tongue muscles develop largely from this group.

Eye Muscles. No muscles have departed more widely from their presumed original segmental form than the muscles that move the eyes.

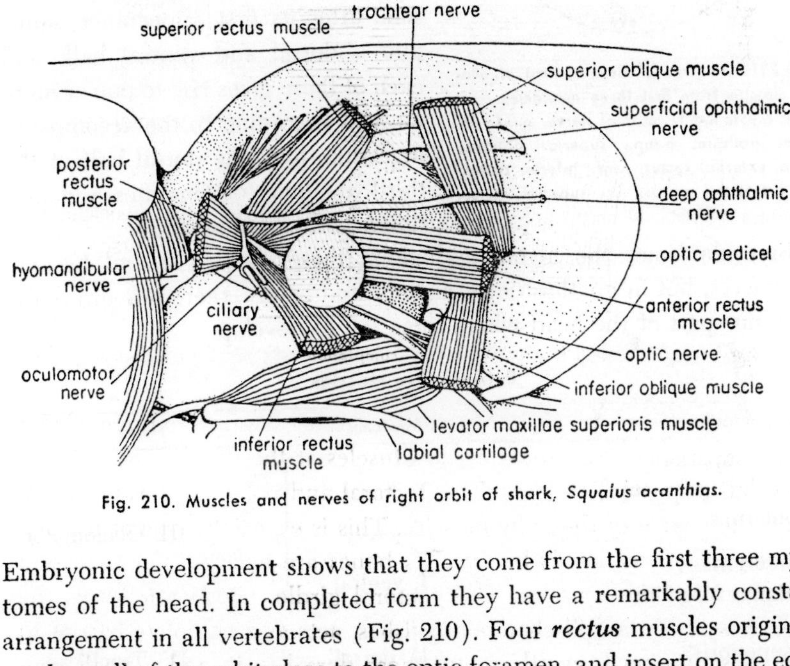

Fig. 210. Muscles and nerves of right orbit of shark, *Squalus acanthias.*

Embryonic development shows that they come from the first three myotomes of the head. In completed form they have a remarkably constant arrangement in all vertebrates (Fig. 210). Four *rectus* muscles originate on the wall of the orbit close to the optic foramen, and insert on the edge of the eyeball at places indicated by their names: ***rectus superioris, r. inferioris, r. anterioris*** (or ***internus*** in man), ***r. posterioris*** (or ***externus*** in man). Two ***obliquus*** muscles originate at a point on the anterior wall of the orbit (in man the medial wall), and insert on the upper and lower edges of the eye; hence ***o. superioris*** and ***o. inferioris.*** These six, so placed, are found from cyclostomes to birds and mammals.

There is in Amphibia, reptiles, and mammals also a ***retractor bulbi,*** derived from the external rectus, which helps withdraw the protruding eyes. Reptiles and mammals have a ***levator palpebrae superioris*** to the upper eyelid, to raise it; this evidently is derived from the superior rec-

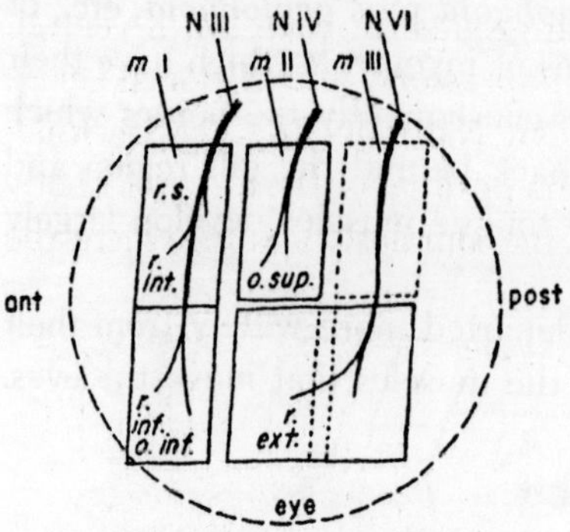

Fig. 211. Diagram of development of ocular muscles from first three myotomes.

m, myotome; N, cranial nerve; *o.inf.*, inferior oblique; *o.sup.*, superior oblique; *r.ext.*, external rectus; *r.inf.*, inferior rectus; *r.int.*, internal rectus; *r.s.*, superior rectus.

tus. A *quadratus* and *pyramidalis*, to the nictitating membrane (third eyelid) of birds, are derived from the third myotome, like the external rectus. In mammals the superior oblique takes a detour through a pulley-like loop in the wall of the orbit before reaching the eye.

Each of the first three myotomes, soon after its first appearance, splits into a dorsal and ventral half, and each of these gives rise to one or more muscles according to the accompanying table; only the dorsal half of the third myotome degenerates and produces nothing (Fig. 211).

In cyclostomes the myotomes behind these all persist and become myomeres, but in gnathostomes, as indicated before, there is a gap in the posterior part of the head where some fail to develop.

Extrinsic Eye Muscles

Muscle	Myotome	Nerve
Rectus superioris (+ lev. palp. sup.) R. internus	I, dorsal	III. Oculomotor
R. inferioris Obliquus inferioris	I, ventral	
O. superioris	II, dorsal	IV. Trochlear
R. externus (+ retr. bulbi, + quadr. + pyram.)	II, ventral + III, ventral	VI. Abducens

Relation of Body Muscles to the Fins in Fishes. The supporting rays of the dorsal and anal fins usually have separate small muscles to move them (Fig. 207). Typically a fin ray has on each side an *erector*, a *depressor*, and an *inclinator* (to pull it sideways); often the fin as a whole will have a *protractor* and a *retractor*.

Muscles of the paired fins, like those of the unpaired, arise directly from the myotomes in all fishes in which their development has been studied. They lie characteristically in two groups for each fin: a ventral, more anterior group, the ***abductors*** or ***depressors,*** and a dorsal, more posterior group, the ***adductors*** or ***levators.*** Naturally the precise functions of these groups differ according to the position of the fin on the body in various fishes. A shark gives perhaps the simplest picture, where the

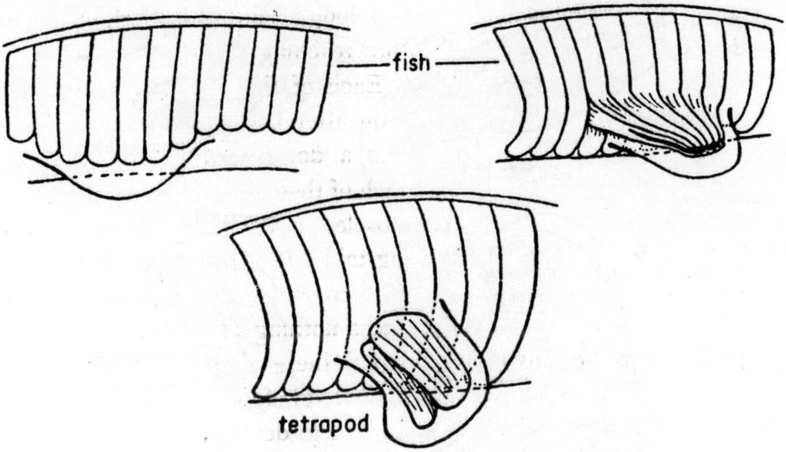

Fig. 212. Relationship of myotomes to limb bud in embryo fish (above) and tetrapod (below).

muscles act as depressors and levators, and are but little divided into separate parts (Fig. 215).

As the limb bud appears in the embryo fish (Fig. 212) the descending tips of the myotomes in the body wall reach the base of the limb bud, and muscle buds grow into it from each myotome in the area concerned. These projecting muscle buds split into dorsal and ventral halves, and each of these again into an anterior and posterior slip. Radial fin cartilages then appear, lying parallel to and between the dorsal and ventral portions. Deeper cartilages (basals) may also arise in the body wall proper; indeed it is likely that the entire pelvis and the scapulocoracoid part of the pectoral girdle originated in this way. In any case, the fundamental dorsal-ventral division of fin muscles is thus accomplished, and the same grouping has been shown to persist, more obscurely, in tetrapods.

Muscles of the Limbs in Tetrapods. During development of tetrapod limb muscles there is no direct relationship between them and the body myotomes (Fig. 212). Instead the limb muscles arise from undifferentiated mesenchyme cells in the limb itself. But this is an embryonic

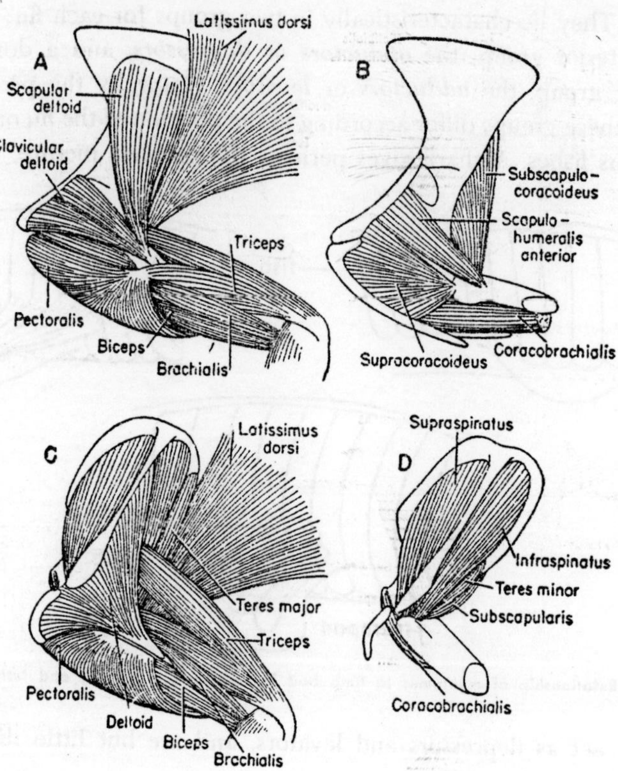

Fig. 213. Shoulder and upper arm muscles in (A, B) lizard and (C, D) opossum, lateral views. B and D are deeper layers. (From Romer.)

specialization, and is not a reason to doubt the derivation of tetrapod limb muscles from those of fishes. Spinal nerves supply the limbs, and the number of nerves to each limb indicates the number of body segments to which that limb is related.

We do not know from direct evidence the intermediate stages between the limb muscle plan of fishes and that of tetrapods, since the soft parts of the appropriate early amphibians and lobefins are not preserved as fossils. But of some points we may be reasonably sure. Unlike most bony fishes the early crossopterygians had the forward edges of the paired fins tilted down; muscles of the dorsal group in these fins were therefore probably abductors and levators, while the ventral muscles were adductors and depressors. They extended out on the fin far enough to cause it to undulate along its axis, as is evident from the axial series of freely movable bones. It is likewise highly probable that the muscles nearest

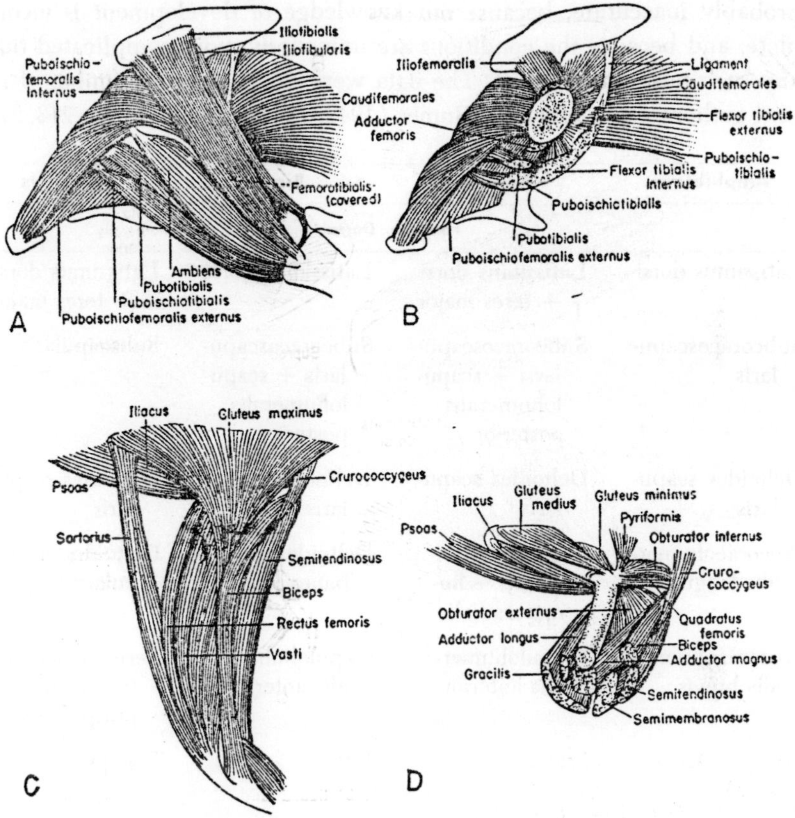

Fig. 214. Muscles of pelvis and thigh in (A, B) lizard and (C, D) opossum, lateral views. A and C, superficial; B and D, deeper layers. (From Romer.)

the base of the fin were derived from more anterior myotomes than those distally, and that the whole set of fin muscles arose from five or six body segments.

On the ventral side of the fore limb the more basal muscles became the *pectoral* group primarily, of tetrapods, while on the dorsal side the basal ones produced several of the upper shoulder muscles, including ***latissimus dorsi*** and the ***deltoids.*** Farther out, the ventral muscles of the fin formed the ***flexors*** of arm and hand, while the more dorsal ones became ***extensors,*** on account of the manner in which the elbow and wrist joints developed as the fin was converted to a walking limb. (See Chapter IV.) In the hind limb similar arrangements took place.

The following table summarizes the dorsal and ventral groups of limb muscles among the classes of tetrapods. It is incomplete, and in places probably inaccurate, because our knowledge of development is incomplete, and because the conditions are immensely more complicated than any such table can suggest. The data were drawn from a number of recent authors, especially from Romer (1942, 1944). See also Figs. 213, 214.

Amphibians	Reptiles	Birds	Mammals
	Forelimb, Dorsal		
Latissimus dorsi	Latissimus dorsi + teres major	Latissimus dorsi	Latissimus dorsi + teres major
Subcoracoscapularis	Subcoracoscapularis + scapulohumeralis posterior	Subcoracoscapularis + scapulohumeralis posterior	Subscapularis
Deltoides scapularis	Deltoides scapularis	Deltoides scapularis	Deltoides scapularis
Procoracohumeralis longus	Deltoides clavicularis + humeroradialis	Deltoides propatagialis	Deltoides clavicularis
Procoracohumeralis brevis	Scapulohumeralis anterior	Scapulohumeralis anterior	Teres minor, infraspinatus, supraspinatus
Triceps (Anconeus)	Triceps	Triceps	Triceps
Also forearm extensors and supinators			
	Forelimb, Ventral		
Pectoralis	Pectoralis	Pectoralis	Pectoralis
Supracoracoideus	Supracoracoideus	Supracoracoideus	Supracoracoideus
Coracobrachialis brevis	Coracobrachialis brevis	Coracobrachialis externus + internus	Coracobrachialis (part)
Coracobrachialis longus	Coracobrachialis longus (Absent in Crocodilia and Birds)		Coracobrachialis (part)
Brachialis	Brachialis	Brachialis	Brachialis
Biceps	Biceps	Biceps	Biceps
Also forearm flexors and pronators			

Hind Limb, Dorsal			
Iliotibialis	Triceps femoris: Iliotibialis	Triceps femoris: Iliotibialis + "Sartorius"	Rectus femoris
	Ambiens	Ambiens	Sartorius
Femorotibialis	Femorotibialis	Femorotibialis	Vastus group
Iliofibularis	Iliofibularis	Iliofibularis	Tenuissimus
Ilioextensorius	Iliofemoralis	Iliofemoralis externus + Iliotrochantericus	Glutei + Tensor fasciae
Puboischiofemoralis internus	Puboischiofemoralis internus	Iliofemoralis internus	Iliacus, Psoas, Pectineus
Also extensors of lower leg and foot			
Hind Limb, Ventral			
Puboischiotibialis	Flexor cruris: Puboischiotibialis		Gracilis
	Flexor tibialis internus	Ilioflexorius	Semimembranosus, Semitendinosus, Biceps femoris
	Flexor tibialis externus	Caudilioflexorius	
Pubotibialis	Pubotibialis	Puboischiofemoralis	
Adductor femoris	Adductor femoris		Adductores femoris
Puboischiofemoralis externus	Puboischiofemoralis externus	Obturator	Obturator externus, Quadratus femoris
	Ischiotrochantericus	Ischiofemoralis	Obturator internus
Caudofemoralis	Caudofemoralis brevis + longus	Coccygeofemoralis	Caudofemoralis
Also flexors of lower leg and foot			

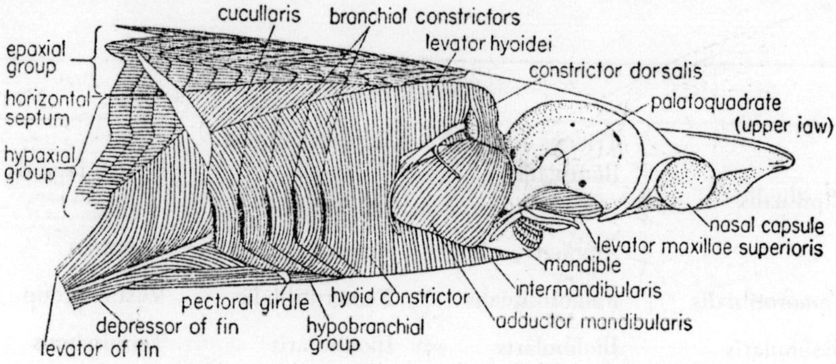

Fig. 215. Head muscles of shark, *Squalus acanthias.*

COMPARATIVE ANATOMY OF THE BRANCHIOMERIC MUSCLES

In embryos of the lower vertebrates the branchiomeric or visceral muscles appear first as slender bands running parallel to the jaw and gill bars, that is, from the cranium down to the mid-ventral line of the throat. At least this was the generalized plan; we have noted already that jaws were developed from the anterior gill arch. (See Fig. 206.)

In cyclostomes (lamprey, hagfish) the cartilaginous gill basket is external to the muscles that move it. Being elastic but not jointed, this gill basket yields to the contraction of the muscles as they squeeze the gill pouches within, and then it springs back, outward, as the muscles relax; thus the cartilages serve as antagonists to the constricting action of the muscles.

But in gnathostomes the cartilages (or bones) of the visceral arches are always jointed and lie internally to the muscles that move them. No one questions that the gill pouches, gills, blood supply, and nerves in these two major divisions correspond. To the author it seems probable that the muscles also correspond, but that the skeletal parts in the gill region do not; in other words, that the internal, jointed visceral arches of gnathostomes were not derived from the external, nonjointed

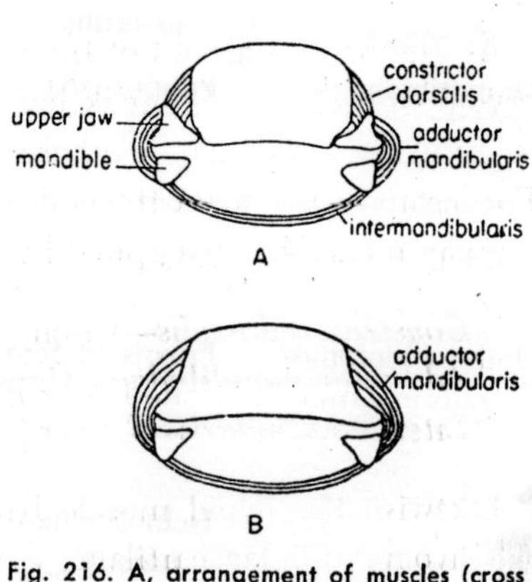

Fig. 216. A, arrangement of muscles (cross section) when upper jaw is movably articulated to cranium; B, when upper jaw is fused to cranium, showing spread of area of origin of adductor mandibularis, and loss of constrictor dorsalis.

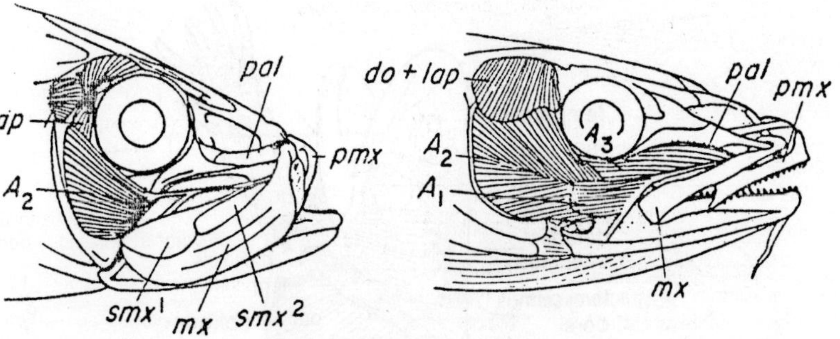

Fig. 217. Modifications of jaw muscles in herring (left) and cod (right), in connection with mobility of upper jaws. (From Eaton, 1935. *J. Morphology*, vol. 58 (1): 171, 172.)

A_1, A_2, A_3, divisions of adductor; *do*, dilator of operculum; *lap.*, levator of primary upper jaw; *mx*, maxillary; *pal*, palatine; *pmx*, premaxillary; *smx*, supramaxillary.

arches of cyclostomes, but that both arose in ostracoderms.

Branchiomeric muscles seem to be derived embryonically from the hypomere. They probably have nothing to do with myotomes. They are supplied by cranial nerves, as follows:

Jaws	V. Trigeminal
Hyoid	VII. Facial
First branchial	IX. Glossopharyngeal
Second and succeeding branchials	X. Vagus

As the visceral arches of the embryo become movable, the simple embryonic muscle bands break into corresponding muscles, so arranged as to raise, flex, constrict, or otherwise move the parts to which they attach. For instance in a shark the embryonic jaw muscle becomes, on each side, a group of three, all supplied by the trigeminal nerve (Figs. 215, 216A):

Constrictor dorsalis—Cranium to upper jaw
Adductor mandibulae—Upper to lower jaw
Intermandibularis—Lower jaw to mid-ventral line

Likewise the hyoid muscle band forms several muscles, a ***levator*** of the hyomandibular cartilage, a ventral ***interhyoid,*** a superficial ***hyoid constrictor,*** etc., all supplied by the facial nerve. For each of the branchial arches there is a similar differentiation of muscles from its embryonic muscle band.

Essentially the plan just described occurs in fishes wherever the upper

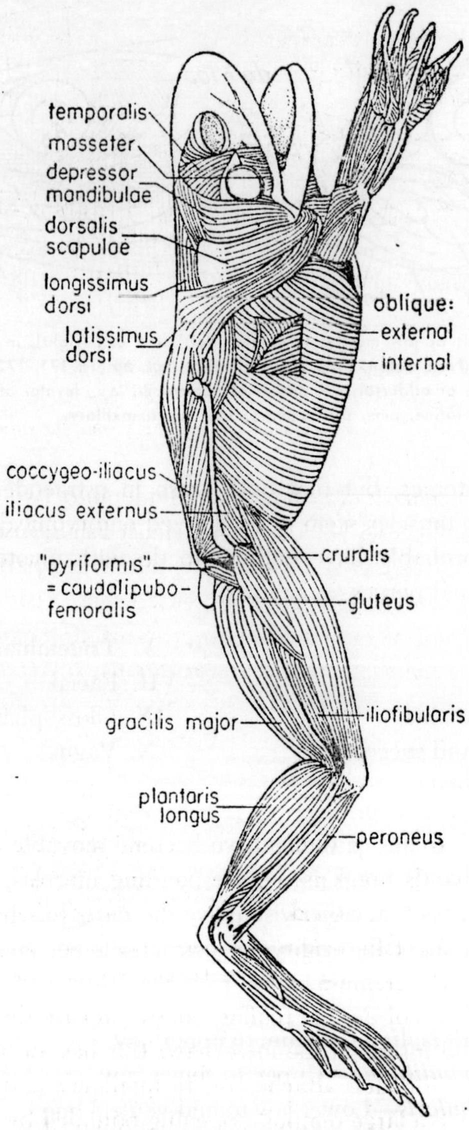

Fig. 218. Muscles of bullfrog, *Rana catesbeiana.*

jaw (palatoquadrate) is not immovably fused with the cranium, that is, in the majority of bony fishes and in elasmobranchs except for Holocephali. Wherever there is a movable operculum a ***dilator operculi*** (from the jaw muscle group) and an ***adductor operculi*** (from the hyoid group) are present to move it.

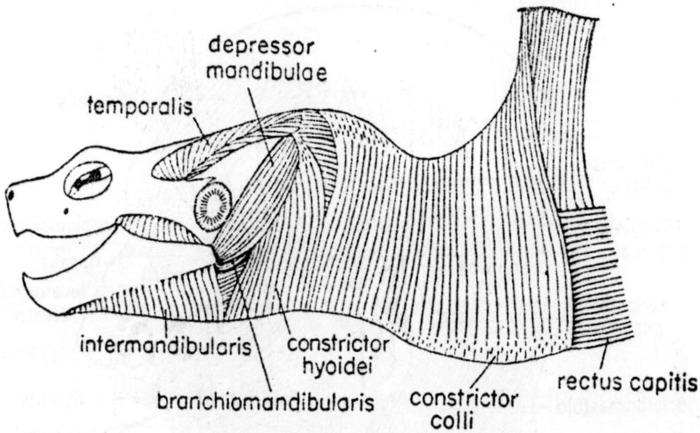

Fig. 219. Muscles of head and neck of snapping turtle, *Chelydra serpentina.*

The *adductor mandibulae* tends to divide into two, three, or more parts, one of which (Fig. 217), in most teleost fishes, inserts on the loose maxillary bone instead of the lower jaw, and takes part in the familiar protrusion of the mouth practiced by most fishes. Fishes such as Holocephali and Dipnoi, which do not have a movable palatoquadrate, have lost at the same time the *constrictor dorsalis* usually found between upper jaw and cranium. This encourages the adductor mandibulae to extend its area of origin upon the cranium itself, just as we find in most tetrapods (Fig. 216B).

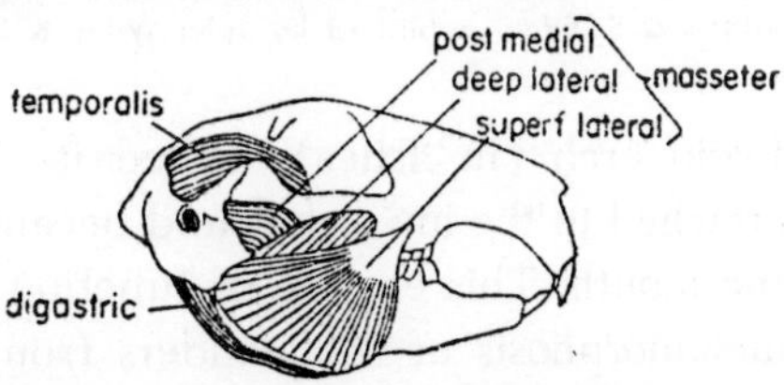

Fig. 220. Jaw muscles of rabbit. (Modified from Parker and Haswell, after Tullberg.)

The adductors of the jaw, by which we chew or bite, are usually more complex in tetrapods than in fishes. They take origin from the sides of the cranium, or from the inner faces of dermal roofing bones (in case the skull is covered by a "deck"), and insert on the lower jaw; this has an upright crest, the coronoid process, for their attachment. In mammals and in several other groups which have a large temporal opening bounded by a bridge of bone laterally, there is a deep adductor, the *temporalis,* originating on the cranium and largely filling the temporal fossa, while a more superficial *masseter* originates on the arch laterally; there are also one or more *pterygoid* muscles, coming from the floor of the cranium and inserting diagonally on the inner face of the lower jaw.

At the time of transformation from fish to amphibian the levator of the

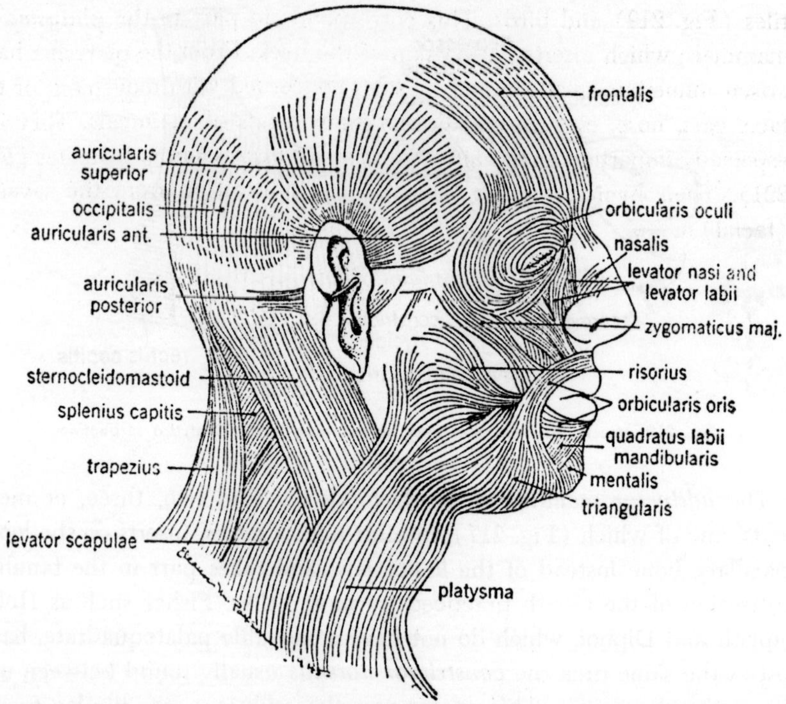

Fig. 221. Human facial muscles. (Reproduced by permission from *Comparative Anatomy* by L. A. Adams & S. Eddy, published by John Wiley & Sons, Inc., 1949.)

hyoid arch (in fishes) shifted its insertion forward slightly so that it attached to the lower jaw and became the ***depressor mandibulae,*** to open the mouth. This change of function and position can be seen also in the metamorphosis of salamanders from larvae to adults. The depressor so formed is present in Amphibia, reptiles, and birds (Figs. 218, 219). But in mammals the muscle to open the mouth is the ***digastric*** (Fig. 220), which develops in part from the embryonic hyoid group and partly from the intermandibularis or mylohyoid muscle of the lower jaw; it is therefore compound, and is innervated by both the fifth and seventh nerves.

We have seen before that the hyomandibular bone of fishes became the stapes, a sound-transmitting bone of the middle ear of tetrapods. With it there went also a small vestige of the levator hyoidei of fishes, to become the ***stapedius,*** which is the smallest skeletal muscle we have; it is served, of course, by a branch of the seventh nerve.

From the ventral portion of the hyoid muscle series there develops a broad thin sheet, the ***constrictor colli***, enclosing most of the neck in reptiles (Fig. 219) and birds. This corresponds in part to the ***platysma*** of mammals, which inserts on the skin of the neck. From the platysma have arisen numerous small dermal muscles concerned with movement of the face, ears, nose, eyelids, and lips in many kinds of mammals. They are especially important as the ***muscles of facial expression*** in Primates (Fig. 221). Their hyoid origin is betrayed by innervation from the seventh (facial) nerve.

CHAPTER VII

Digestive and Respiratory Systems, and Coelom

INTRODUCTORY

The covering of the body turns in at the mouth, to run as a tube through to the anus. The lining of this tube is everywhere continuous, and therefore anything taken into the passage is still, in a sense, outside the body, still a part of the environment (Fig. 222). It is the business of the digestive canal to absorb, through its unbroken lining, some of the materials carried in it, and to convert those which cannot be absorbed into substances which can. The latter process is *digestion.* Food becomes usable only if it can pass through the semipermeable membrane into the blood or other body fluids, and the main problem comes in the relatively large size of many molecules, which must therefore be broken, by hydrolysis or other reactions, into smaller ones. After digestion is completed, there is selective absorption of its products.

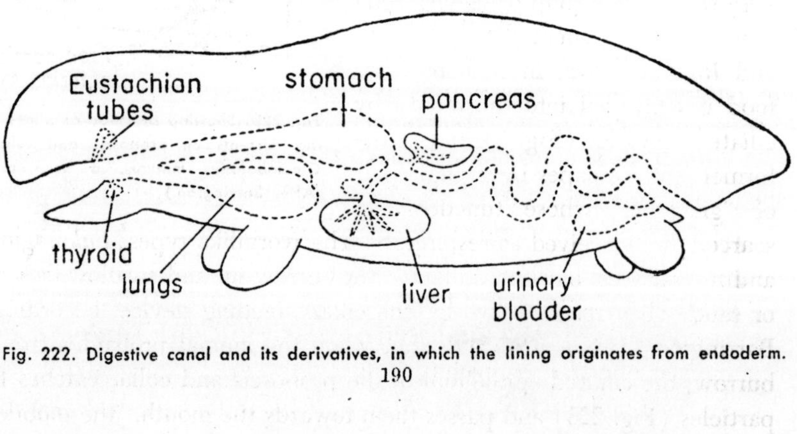

Fig. 222. Digestive canal and its derivatives, in which the lining originates from endoderm.

The respiratory and digestive organs are usually specialized parts of the same canal, and in the lower chordates it would be difficult to consider their anatomy separately. An organ of respiration is one in which blood with low oxygen concentration is given as close access as possible, through membranes, to oxygen of higher concentration dissolved in water outside the body. Such organs, of course, include the lungs of air-breathing vertebrates, because oxygen from the air must dissolve in the water of alveolar inner surfaces before it can actually enter the body.

After dealing with the digestive and respiratory organs it should be easier to visualize the cavities in which they lie, that is, the coelom and its divisions, for these are most readily described by reference to stomach, liver, intestine, lungs, etc.

DIGESTIVE AND RESPIRATORY ORGANS OF PROTOCHORDATES

In each group of protochordates and in the larvae of cyclostomes there is a feeding mechanism fundamentally different from that of the familiar vertebrates. It has these characteristics: sticky mucus secreted by cells in or near the beginning of the digestive canal, entanglement of microscopic food by this mucus, transportation of the mucus and food into the digestive tract by continuously beating cilia on the mucous surface, and openings in the wall of the pharynx through which a current of water produced by the cilia can escape. Since this mechanism and the microphagous (microscopic feeding) habit are present in Hemichorda, Urochorda, Cephalochorda, and up into the lowest vertebrates, they must enter any discussion of the origin of Chordata, and we shall refer to them again in that connection.

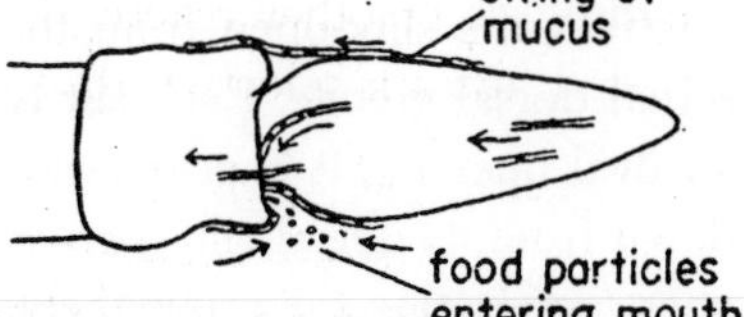

Fig. 223. Showing direction of ciliary feeding currents on proboscis and collar of *Glossobalanus minutus*, a hemichordate. (After Barrington.)

In the sessile group of Hemichorda there are two genera, *Cephalodiscus* and *Rhabdopleura;* in both of these food is conveyed into the mouth by ciliated tentacles (Fig. 5) and in the former water escapes by a single pair of "gill slits" whose function can scarcely be conceived as respiratory. The wormlike types, *Balanoglossus* and its relatives, have no tentacles but burrow in, and swallow, sea mud or sand. They retain, however, the ciliary feeding device, according to Barrington (1940), who finds that when the animal protrudes from its burrow, the ciliated epithelium of the proboscis and collar catches food particles (Fig. 223) and passes them towards the mouth, "the mobile lip of the collar providing a rejection mechanism." Inside the pharynx there is a partial division into upper and lower passages, the former running between the two series of gill slits where water runs out, and the latter towards the opening of the simple straight intestine. But Barrington believes that there is no distinction between these passages so far as food is concerned, and that the gill bars help trap food just as in Amphioxus.

Amphioxus (Cephalochorda) and the tunicates (Urochorda) are much alike in their pharyngeal basket. The ***pharynx*** of these (Figs. 6, 7), and of lower vertebrates as well, might be defined as the anterior region of the alimentary canal where food is separated from the water by a sieve or basket arrangement in its walls. When this sort of pharynx had become established, the persistent water current through the sieve made it possible to use the bars for gills. In Amphioxus most of the food-catching mucus originates in a trough on the floor of the pharynx, the ***endostyle,*** from which cilia carry it forward, then laterad along ***peripharyngeal bands*** to the dorsal side, where an ***epipharyngeal groove*** transmits it back to the esophagus. Nothing is changed from this in the simple tunicates (Ascidia) except the orientation, for a tunicate grows on a surface so that the mouth is farthest up or away from that surface, and the intestine is bent around to bring the anus likewise away from the substratum. In both Cephalochorda and Urochorda the water escapes from the pharynx through a large number of minute slits separated by very delicate bars, and is then in an outer chamber, the ***atrium,*** from which it leaves by the ***atriopore.***

As to the gills in protochordates: in *Balanoglossus* (Figs. 3, 4) and its relatives the slits open from the dorsal part of the pharynx into an external dorsal groove. Each slit is U-shaped because, having developed as an oval opening, it is nearly cut in two by a ***secondary gill bar*** growing down from its dorsal rim. Between these secondary bars and the ***primary*** ones (separating the original slits) there usually extend little bridges, the ***synapticulae.*** The number of gill slits is indeterminate, increasing with age.

A similar arrangement of primary and secondary bars is seen in Amphioxus, except that here the secondaries completely cut the original openings into two. Synapticulae are numerous. The whole basket is more complex and delicate. In tunicates the pharyngeal basket appears to stand upright, as mentioned above, and this means the original slits are now horizontal, while the synapticulae, far more numerous than before, are vertical, and the basket seems to contain many rows of very small slits.

"AMMOCOETES" AND THE THYROID GLAND

The larva of a lamprey is radically different from the adult and was described as another genus, *Ammocoetes,* before its relationship was known; this name is still used as a convenience in speaking of it. The *Ammocoetes* looks and behaves much like Amphioxus, burrowing in the sand of rivers and creeks, but it has no atrium, and fewer (seven pairs of) pouched gill clefts. It feeds, however, by a ciliary and mucous mechanism essentially the same as in Amphioxus; there is an epipharyngeal and a hypopharyngeal ciliated groove, connected anteriorly by two peripharyngeal bands (Fig. 224). But midway in the hypopharyngeal groove is a

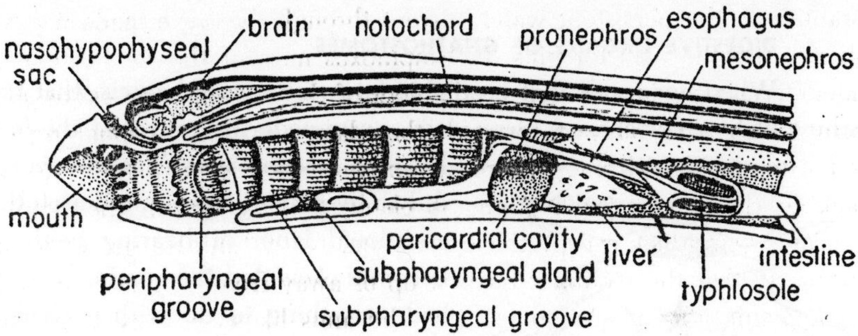

Fig. 224. Sagittal section through anterior half of larval lamprey, Ammocoetes. (After Goodrich.)

ventral slit leading down into a complex glandular sac (Leach, 1944). This has been called the endostyle; it develops as a pocket from the ciliated epithelium of the groove. Evidently the hypopharyngeal groove as a whole corresponds to the endostyle of Amphioxus, and the sac may be called the ***subpharyngeal gland.***

During transformation from the microphagous larva to the parasitic adult lamprey, the sac closes and in part becomes the ***thyroid gland.*** In sharks the thyroid originates likewise as an open pouch in the floor of the pharynx of the embryo, and it also develops from ventral pharyngeal epithelium in higher vertebrates.

The transforming lamprey loses the ciliated bands, enlarges the anterior part of the head into a sucking disc with horny teeth (Fig. 101; compare Fig. 225), and develops a hydrostatic pumping apparatus (Reynolds, 1931), of which the major elements are: a complete separation between the esophagus (dorsally) and the respiratory part of the pharynx (ventrally), a valve called the ***velum*** to block the latter passage while blood is being sucked into the mouth, and a toothed, pistonlike "tongue" for cutting the skin of the fish on which the lamprey clings. The esophagus leads to the straight, slender intestine, whose inner surface is increased by a somewhat spiral fold, the *typhlosole* (Fig. 224). There is a liver, but the gall bladder, present in *Ammocoetes*, is lost; there is no distinct pancreas.

It is important to add that the gills are in pouches, each opening by a small passage inward to the pharynx and by another pore outward to the surface (Fig. 236). Using the muscular outer wall, the lamprey pumps water in and out of the pouches through the external pores. This is the only way to supply a breathing current while the lamprey feeds or clings to stones with its mouth.

DIGESTIVE ORGANS OF GNATHOSTOMES

When we recall that protochordates are microphagous, that the earliest known vertebrates were jawless ostracoderms, and that the first fishes to develop jaws (acanthodians) did so by placing upon a gill arch the additional labor of bearing teeth, it becomes clear that the coming of jaws had something to do with a change from microscopic to larger units of food. These larger units, usually whole animals, take an appreciable time to digest, and so must be held in an elastic pouch, the ***stomach,*** until this is nearly accomplished. Most chemical phases of digestion and the greater part of absorption then take place in the remainder of the canal, the ***intestine.*** Indigestible remains are extruded from the ***anus.*** In most classes of vertebrates the excretory and reproductive systems also employ the same opening for their products; that is, a portion of the lower end of the digestive tract, called the ***cloaca,*** receives the ducts from these systems.

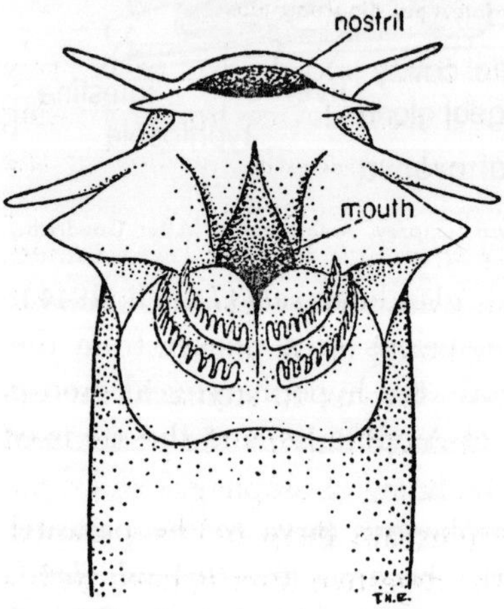

Fig. 225. Hagfish, ventral side of head, showing nostril, tentacles, mouth, and horny teeth. (Sketch from model in American Museum of Natural History.)

Fishes

Mouths of fishes are usually terminal (at the end of the head), but in several groups, as sharks and rays, sturgeons, some catfishes, etc., they are ventral. In a few (e.g., halfbeaks) the mouth is dorsal. It varies

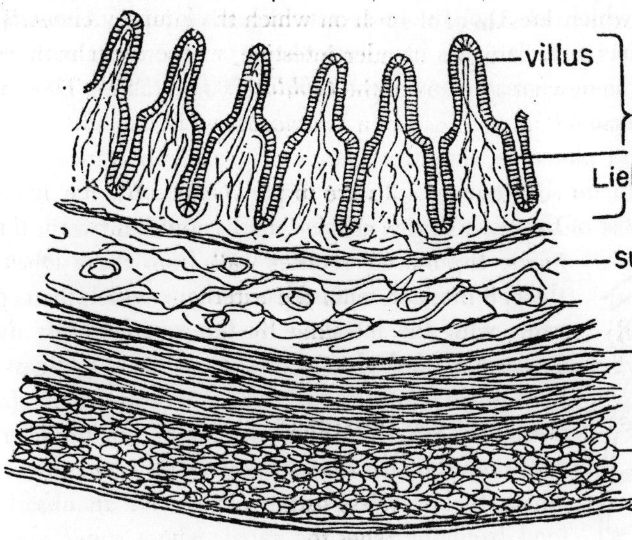

Fig. 226. Layers of wall of small intestine, diagrammatic.

in size from the immense gape of angler fishes and gulpers to the tiny tube of a sea horse or pipefish. Several different devices for protruding the rim of the mouth have been developed, especially by moving the bones of the upper jaw. All these and other modifications are correlated with the nature of the food and manner in which it must be obtained. There is no movable tongue in fishes and no salivary glands, but taste buds and mucous glands are widely distributed in the mouth.

The *esophagus* is a short passage from pharynx to stomach; the entrance to the latter is indefinite as a rule, marked only by the presence of gastric glands and increased diameter. The lining of esophagus and stomach may both be folded. Gastric glands are usually deep and flask shaped. As with higher vertebrates (Fig. 226), the digestive tract is lined with a *mucosa* derived from embryonic endoderm, and to this are added *submucosa* (containing capillaries), *muscularis* (circular and then longitudinal layers of smooth muscle), and *serosa* (peritoneum), all from the embryonic mesoderm. This fundamental arrangement continues from the esophagus to rectum or cloaca. The stomach in fishes is most often J-shaped, but may be a blind sac with both openings anteriorly. Some deep-sea fishes have a stomach so distensible as to hold prey larger than themselves. It is thick and gizzardlike in mullets (Mugilidae) and the gizzard-shad (*Dorosoma*). The exit from the stomach is the *pylorus;* usually but not always there is a *pyloric valve* (sphincter muscle). The majority of teleost fishes have blind pouches or tubes at this point called *pyloric ceca,* which are thought to secrete digestive juice. *Ammodytes* (sand-lance) has one; flatfishes have two, three, or several; salmon, cod, and mackerel have a great many, up to 100 or 200; catfishes, pike, and pipefish have none.

The *liver* and *pancreas,* digestive glands of great importance, empty their ducts into the beginning of the intestine. Ordinarily the liver is large, filling most of the body cavity in some sharks, and has two or three lobes; *hepatic ducts* carry bile from these lobes to the gall bladder, embedded in them, which then connects with the intestine by the *common bile duct* (*ductus choledochus*). Bile is, in part, a digestive fluid, since one of its functions is to lower the surface tension of fats and so emulsify them, but the liver is much more than a digestive gland. It is the organ to which the hepatic portal vein carries all absorbed food from the digestive canal, where sugar can be removed and stored as glycogen, and where broken-down hemoglobin from the red blood cells is cast off in the form of bile pigments. The pancreas is a distinct, two-lobed gland in sharks, but may be very hard to find in a teleost, where it is sometimes hidden in the liver and may use the same duct for its secretion.

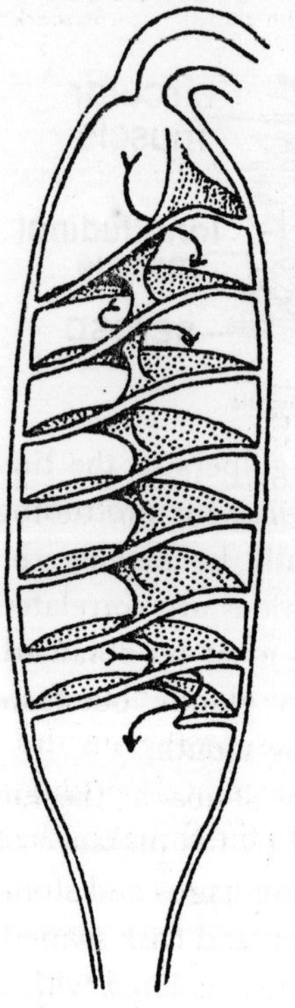

Fig. 227. Spiral valve in *Raja.*

The *intestine* of primitive fishes has a feature of great interest, the *spiral valve* (Fig. 227). It is a fold of the inner wall, usually running spirally down like a winding stairway. The turns may so far overlap as to resemble a scroll. Presumably the valve is homologous with the typhlosole of cyclostomes. There is evidence that it occurred in a placoderm, *Bothriolepis,* of the Devonian period (Denison, 1941). It is best developed in Elasmobranchii but continues in lungfishes, *Polypterus,* sturgeons, and *Amia,* finally vanishing in the most primitive teleosts. It is a device to increase the absorptive area of the intestine. The same object is served among plant-feeding or mud-grubbing teleosts by lengthening and looping the intestine; in the mud-eating stone-roller minnow, *Campostoma,* the intestine coils around the air bladder. As a general rule among vertebrates the herbivorous types have a long intestine, the predatory or carnivorous ones short and simple.

There is little point in trying to distinguish a small and large intestine

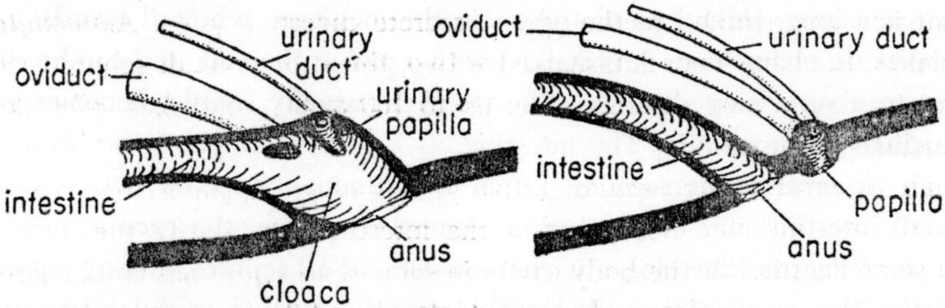

Fig. 228. Relationship of urinary and genital ducts to intestine in fish with cloaca, such as shark (left), and fish without cloaca, as teleost (right).

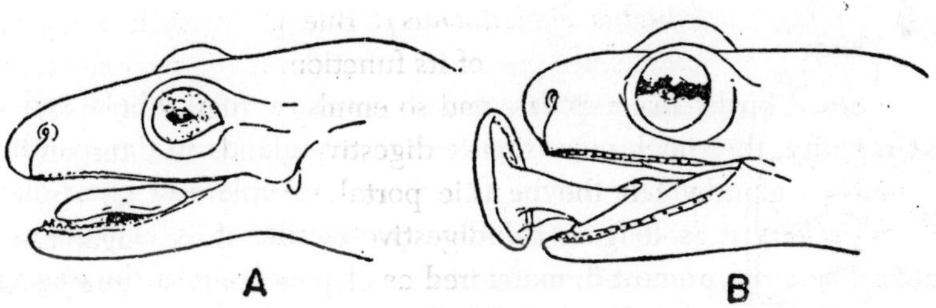

Fig. 229. Tongues of two plethodontid salamanders: A, *Desmognathus fuscus*, tongue attached in front; B, *Eurycea bislineata*, tongue free and stalked. (From *Biology of the Amphibia* by Noble.)

in fishes, but there is often a ***rectum***, from which in elasmobranchs a dorsal blind pouch, the ***rectal gland***, is given off. The elasmobranchs and Dipnoi have a ***cloaca*** (Fig. 228), as no doubt the crossopterygians did also, but in Actinopteri the anus lies in front of the urinogenital opening. The anus in a few teleosts is brought far forward, even to the throat in the electric eel.

Amphibians and Reptiles

The digestive organs of these animals are usually simple and ordinary. A movable ***tongue*** appears for the first time in Amphibia. Although absent in primitive frogs (Aglossa: *Xenopus* and *Pipa*) it is normally present and free behind in this group, so that it can be flipped out of the mouth to catch an insect. Salamanders of the lungless family Plethodontidae have a stalked, mushroomlike tongue (Fig. 229), which they shoot out for the same reason. With life on land came multicellular glands in the mouth, secreting a sticky mucus. The intestine of tadpoles, feeding on plants, is long and coiled, but when they transform to insect-eating frogs or toads it becomes relatively much shorter.

The tongue remains short in turtles, Crocodilia, and the iguanid lizards, but is a long, forked tactile organ in the monitors, whiptail lizards, and snakes. In chameleons it is stalked with a pad at the end, and can be shot out to a surprising distance. The stomach, usually simple, becomes gizzardlike in Crocodilia. The intestine, as before, is longer in herbivorous than in carnivorous reptiles. Often a distinction appears between the small intestine and colon, with an intestinal pouch, the *cecum*, present in some lizards. The cloaca is characteristic of all reptiles; among aquatic turtles this is often provided with large thin-walled vascular sacs just inside the anus, in and out of which water is regularly pumped. They serve as accessory breathing organs.

Birds

Since birds, like reptiles, usually swallow food whole and with great rapidity, there is no occasion for salivary glands in the mouth, but only mucous glands. The tongue is ordinarily simple and spear-shaped; in woodpeckers it is long and highly protractile; it is club-shaped in parrots. The *crop*, present in many birds, is a pouch out of the esophagus in which food can be dropped hastily for temporary storage. Pigeons have a crop secretion, "pigeon's milk," as supplementary food for the young.

The stomach in flesh-eating or fish-eating birds is of the simple reptilian type (Fig. 230), but among seed eaters it becomes divided into an upper, thin-walled, glandular *proventriculus* and a lower, muscular *gizzard*, with a horny lining. Here, with the help of grit, the resistant seeds and grain can be milled.

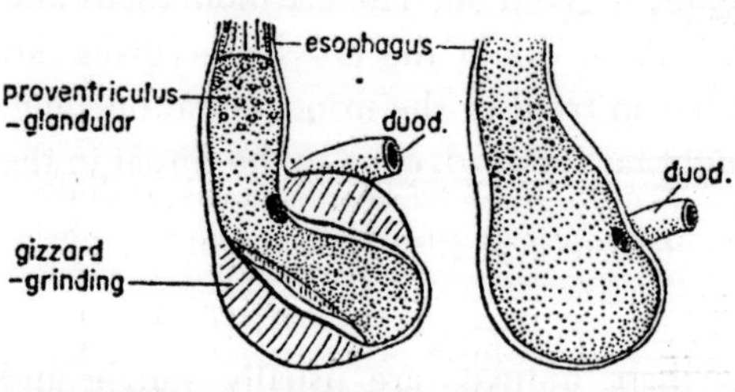

Fig. 230. Stomach of grain-eating bird, turkey (left), and carrion eater, buzzard (right). (Redrawn from Kingsley, *Outlines of Comparative Anatomy of Vertebrates*, 3rd ed., Blakiston.)

The intestine, sometimes divided into ***duodenum, ileum,*** and ***rectum,*** leaves the gizzard near the proventriculus, then receives one or two bile ducts (the gall bladder is often lacking), curves back under the slender, lobed pancreas, and then forms several close loops before reaching the cloaca. It frequently has a pair of ceca near the lower end. Commonly the cloaca is in three parts: ***coprodeum,*** receiving the rectum, ***urodeum,*** into which empty the excretory and reproductive ducts, and ***proctodeum,*** next to the anus.

Mammals

Among the most distinctive characters of mammals are those of the mouth. ***Heterodont dentition,*** a ***hard palate*** (Fig. 231) separating the buccal and nasal cavities, and the presence of fleshy ***lips*** and ***cheeks*** are probably all correlated with the habit of holding food in the mouth for gnawing or chewing before it is swallowed. (The hard palate of Crocodilia has a different meaning.) And in connection with this, the oral mucous glands are supplemented by three or more pairs of enlarged ***salivary glands*** (parotid, submandibular, sublingual, molar, infraorbital, etc.) producing a digestive enzyme, ***ptyalin.*** This starch-breaking enzyme has been demonstrated in many mammals and a few frogs. The tongue,

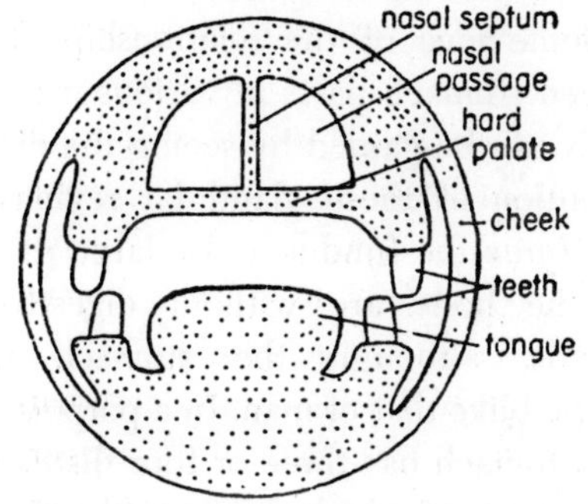

Fig. 231. Diagram of relationship between teeth, tongue, cheeks, and hard palate in mammals.

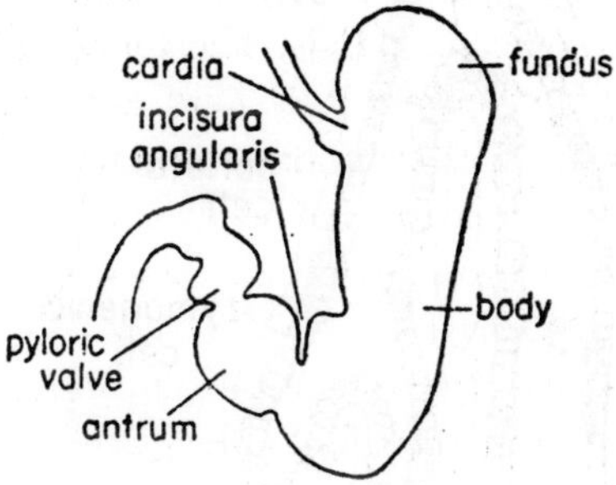

Fig. 232. Outline of human stomach.

equipped with ***papillae*** and ***taste buds,*** is a mobile organ to manipulate food in chewing. Among some herbivores it is prehensile, helping pull grass or leaves into the mouth. It is commonly used for lapping, and in the case of anteaters for licking up ants and termites.

The stomach is simple, with a folded lining, in most mammals. That of man is a good example (Fig. 232). It has no fixed shape, since in life it is active and mobile, but its most distinct landmarks are these: ***cardiac opening,*** from the esophagus; ***fundus,*** the swollen blind pocket on the left, located above the level of the cardiac opening; ***body,*** the main portion of the stomach on the left side, below this; ***pyloric portion,*** the narrow, lower, right side, leading to the ***pyloric opening*** into the small intestine. A constriction, the ***incisura angularis,*** frequently appears below the body, but it is not an anatomical feature, since it moves along towards the pylorus and a new one presently appears above it; it is simply a result of the activity of the stomach wall. Multicellular glands are scattered throughout the epithelial lining, to the number of about 35 million. Each gland (Fig. 233) has a short ***neck*** and a ***body,*** which is often branched. In the neck are ***mucous neck cells;*** the body contains ***zymogenic cells*** and ***parietal cells,*** the former producing pepsin and the latter hydrochloric acid. In the fundus and body of the stomach, then, the glands supply mucus, pepsin, and hydrochloric acid. In the pylorus, however, only mucous neck cells occur in the glands, so that their secretion has no pepsin and is alkaline instead of acid. The same is true of glands adjacent to the cardiac opening. The esophagus is mostly nonglandular but has mucous glands near its lower end.

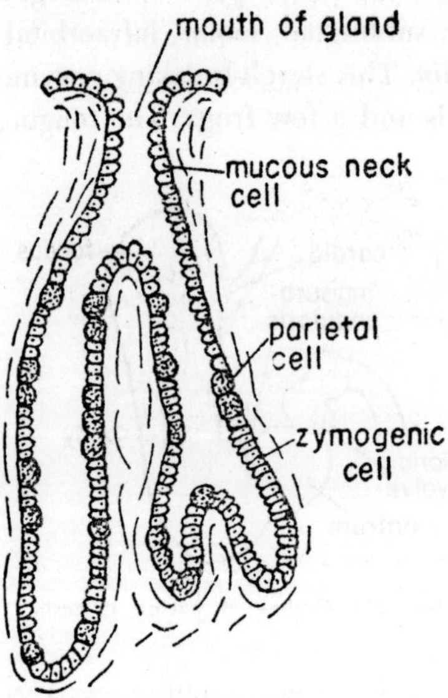

Fig. 233. A gastric gland.

Some idea of the relationship between stomach parts in various mammals can be gained by seeing the distribution of these glandular regions. In *Hyrax* the fundus and a large part of the body are without digestive glands, although the stomach is shaped like that of man. In a porpoise, the stomach has three or four distinct chambers, of which the first, likewise, resembles the esophagus in its nonglandular lining. The well-known four-chambered stomach of a cow (Fig. 234) contains three parts in which glands are absent: ***rumen,*** a large smooth-walled cavity; ***reticulum,*** a smaller space lined with honeycomb pouches (tripe); and ***omasum*** or ***psalterium,*** lined with folds like the leaves of a book. Only the fourth part, the ***abomasum,*** contains glands corresponding to those described in the human stomach. The rumen and reticulum are simply reservoirs from which swallowed vegetation, formed into a ball, can be regurgitated as a cud up the esophagus, for repeated chewing. Camels have numerous small sacs emerging from the walls of their stomachs, where they store water after a long drink, and hold it there by sphincter muscles.

The liver usually has two large divisions, each split again into two or more lobes, and the gall bladder is present in all but a few mammals (perissodactyls, cetaceans, hyracoids, and such rodents as rats). The pancreas is a slender forked or bilobed gland with one to three ducts opening into the duodenum close to the bile duct (Fig. 235).

Herbivorous mammals have not only a relatively long intestine but often an immense cecum at the beginning of the colon, while carnivores tend to reduce or lose the cecum and have a short intestine. The small wormlike *appendix* in man and apes is considered to be a vestige of a more extended cecum. The *Hyrax,* peculiar in so many ways, has two extra ceca along the colon. The small intestine of mammals can be divided into ***duodenum*** (loop below stomach and pancreas), ***jejunum,*** and ***ileum*** (the greater part). Most of its lining is well supplied with folds and microscopic ***villi,*** fingerlike projections of the mucosa, each contain-

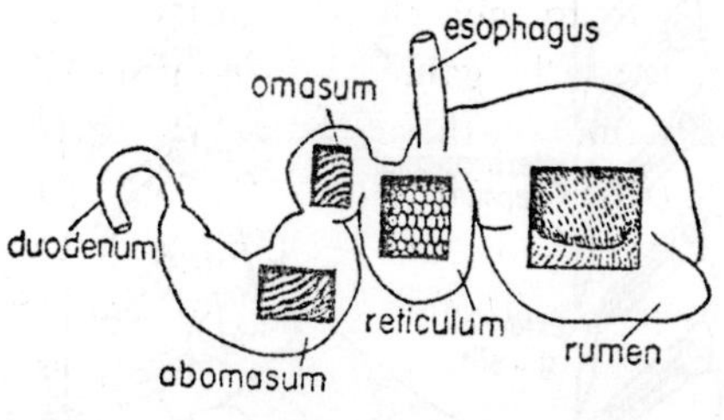

Fig. 234. A ruminant stomach. (Modified from Parker and Haswell, after Flower and Lydekker.)

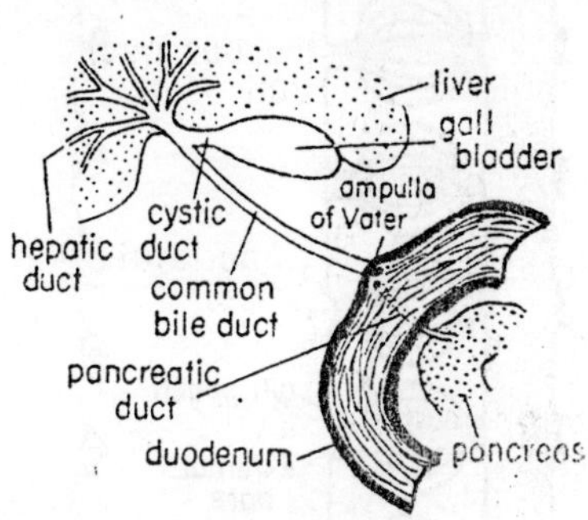

Fig. 235. Connections of liver, pancreas, and intestine (human), simplified.

ing a capillary loop and a lacteal vessel. The latter is a minute branch of the lymphatic system, serving to absorb fat from the contents of the intestine, while the blood capillaries absorb digested amino acids, carbohydrates, salts, etc. Between the longitudinal and circular muscle layers lie plexuses of autonomic nerves, responsible for the slow wavelike contraction of the intestinal walls, called ***peristalsis,*** which urges the food along. Such movement, with local modifications, takes place throughout the tract from esophagus to rectum. The ***colon*** is mainly a place for water absorption. A ***cloaca*** persists in the monotremes and in female marsupials; otherwise mammals have the anus separate from the urinary and reproductive openings.

GILLS IN FISHES AND AMPHIBIANS

The arrangement of gills in pouches, with small internal and external openings (Fig. 236), is characteristic of all cyclostomes and gives them the alternative name Marsipobranchii (pouch gills). The number of pouches is from 5 to 15 pairs (7 in modern lampreys, up to 14 in hagfishes, 15 in some anaspid ostracoderms). The gills proper consist of shelf-like folds of epithelium in series on the anterior and posterior walls of the pouches, each fold supplied by a capillary loop.

In a shark the plan at first seems utterly different: the gills are supported by cartilaginous bars inwardly, and consist of paired sets (*holo-*

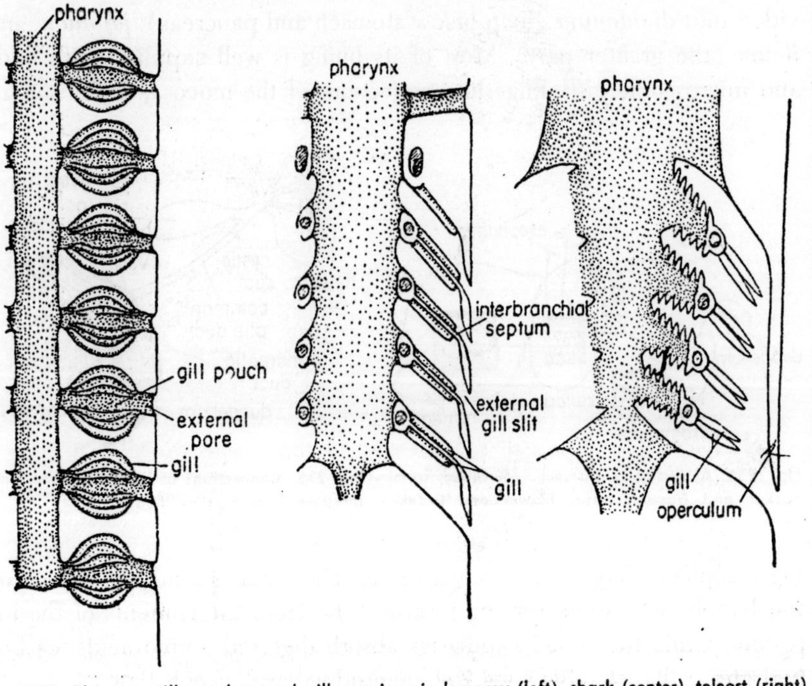

Fig. 236. Pharynx, gill pouches, and gill openings in lamprey (left), shark (center), teleost (right).

branchs) of filaments attached to the two surfaces of an ***interbranchial septum*** (each set is a ***demibranch***); the septum continues to the outer wall of the body, thus separating the exterior gill openings. Actually, however, a gill slit, from pharynx to exterior, corresponds fully to a gill pouch with inner and outer pores; the interbranchial septum therefore is the much reduced wall between pouches.

Bony fishes (Dipnoi, Actinopteri, etc.) as well as the peculiar chimaeras (Holocephali) no longer have separate external slits, but the gill chamber is covered by a flap, the ***operculum;*** consequently the interbranchial septa do not reach to the outside, nor even beyond the tips of the gill filaments; they undergo progressive shortening until virtually absent in teleosts.

Common to the lungfishes and primitive Actinopteri (e.g. *Polypterus*) are temporary bushy ***external gills*** (Fig. 237) in the young fish, developed from the outer epithelium over the gill arches. These occur also in young salamanders, persisting throughout life in the permanently aquatic

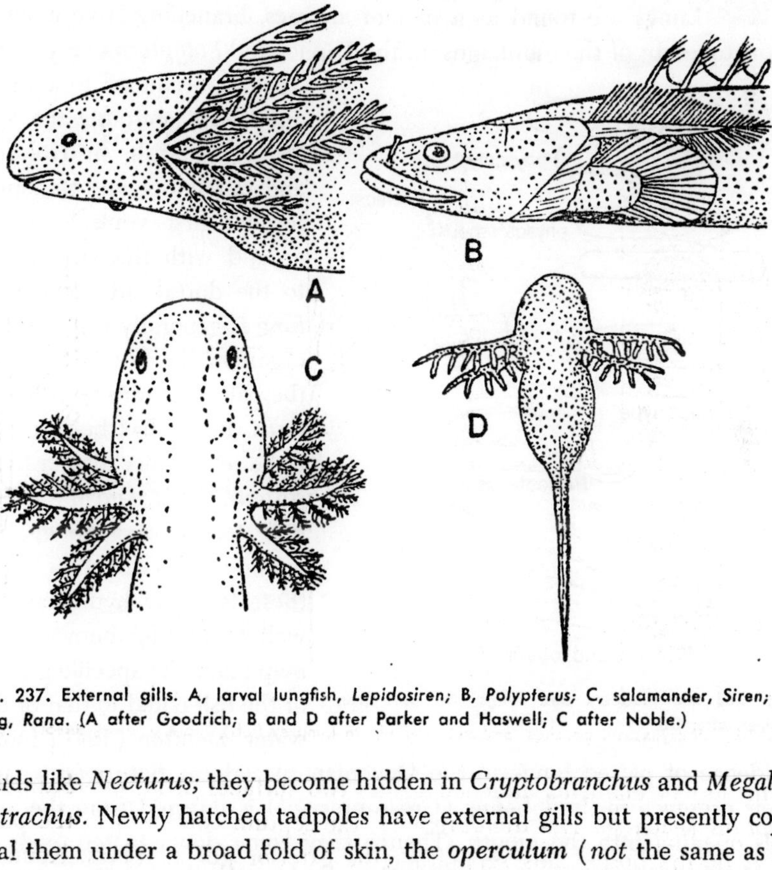

Fig. 237. External gills. A, larval lungfish, *Lepidosiren;* B, *Polypterus;* C, salamander, *Siren;* D, frog, *Rana.* (A after Goodrich; B and D after Parker and Haswell; C after Noble.)

kinds like *Necturus;* they become hidden in *Cryptobranchus* and *Megalobatrachus.* Newly hatched tadpoles have external gills but presently conceal them under a broad fold of skin, the ***operculum*** (*not* the same as in a fish), and then develop true internal gills on their gill bars. Water escapes through a small pore, usually left-sided, the ***spiracle*** (not to be confused with the spiracles of many fishes).

Other devices for breathing should be mentioned here before we touch on the lungs and air bladder: The African "hairy frog" has vascular filaments on its thighs and abdomen; the skin as a whole is a respiratory organ in many Amphibia, especially those like the lungless salamanders that live in damp situations where the skin is always moist; many adult salamanders and frogs breathe through the lining of the mouth and throat, its floor moving in regular respiratory rhythm. This is probably true in some aquatic turtles as well; their rectal breathing sacs have already been mentioned.

LUNGS AND AIR BLADDER

Lungs are found as a pair of air sacs, branching from a single short tube out of the esophagus, in the Dipnoi and *Polypterus* (Fig. 238). They are reduced to a single one in *Epiceratodus*, the Australian lungfish, in Holostei, and in teleosts, and the opening, at first ventral, swings around with the air bladder to the dorsal side. The term *lung* is generally used for this organ in the lungfishes or in the paired condition in *Polypterus*, but it has been shown to be respiratory also in *Lepisosteus, Amia*, and the primitive teleost *Arapaima*. Undoubtedly its function in all these is partly hydrostatic as well, that is, for buoyancy, to help make the specific gravity of the fish equal to that of the water. Denison (1941) found evidence of paired lungs in the Devonian placoderm *Bothriolepis*, and their presence in the lobefins (Crossopterygii) is inferred from the nostrils opening into the mouth. Thence they come down to tetrapods as highly developed organs of respiration. Embryonically they originate as paired internal pouches just behind the last pair of gill slits in the pharynx; they may, then, have started as gill pouches.

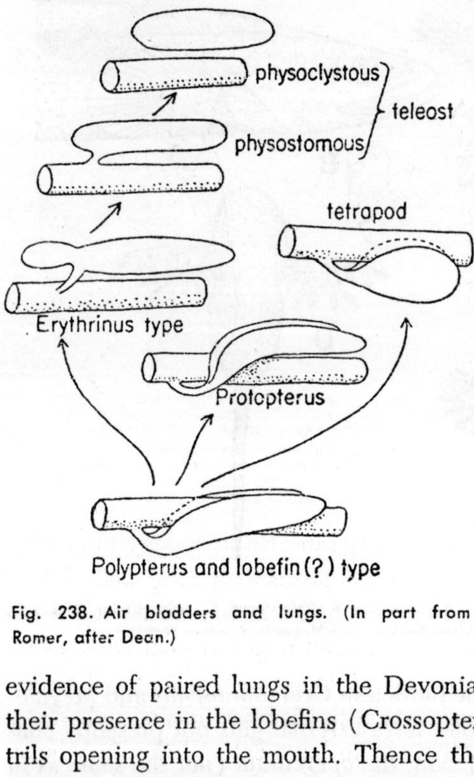

Fig. 238. Air bladders and lungs. (In part from Romer, after Dean.)

In certain salamanders, the Plethodontidae, and the Olympic salamander, *Rhyacotriton*, lungs are reduced to vestiges or lacking; their place is taken by cutaneous and buccal respiration. Otherwise modern Amphibia have simple, thin, saclike lungs with varying degrees of pocketing internally, to increase their absorptive surfaces (Fig. 239). This is essentially true still in reptiles. The tube carrying air to the lungs is the *trachea.* It divides into a pair of *bronchi.* The voices of male frogs and toads are produced by resonance of air entering single or paired *vocal sacs* in the throat, but these have no direct relationship to the lungs. Air may be passed rapidly back and forth from lungs to sacs, however, producing a trill or other multiple call. Snakes and Amphisbaenia have only the right lung (evidently two is a crowd). In chameleons several blind pouches emerge into the body cavity from each lung, by which the body can be swelled up.

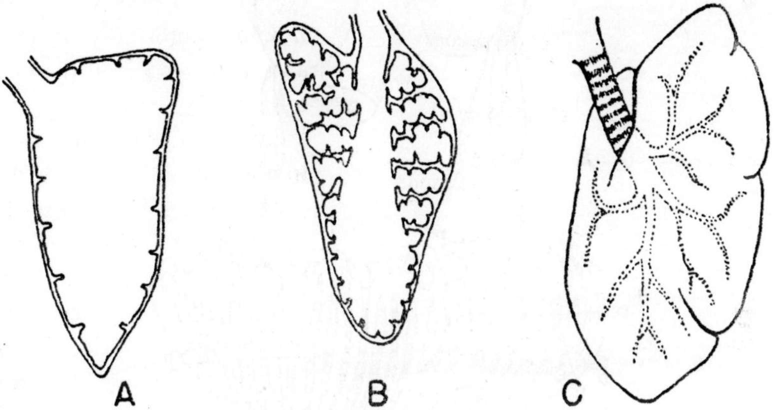

Fig. 239. Lungs. A, toad; B, lizard (after Goodrich, *Studies on the Structure and Development of the Vertebrates,* Macmillan); C, mammal.

The lungs of birds are much more elaborate, being spongy internally. Their most distinctive feature is the series of *air sacs* (Fig. 240) which extend from them into many parts of the body, as between viscera, between muscles, and even into some of the limb bones. Recent work shows that many flying birds, such as grebes and loons, as well as the flightless ones, lack the spaces within bones, so that this is by no means a character of the whole class. Air sacs are thought to have entered the limb bones in pterosaurs, the flying reptiles. Probably air sacs are of greater value in respiration than in merely lowering the specific gravity of the bird, since they make it possible for a more complete change of air in the lungs to take place with every breath, thus increasing the availability of oxygen, as compared with the partial change which occurs in other vertebrates. In bird lungs a series of *parabronchial tubes* with looped *air capillaries* carries on respiration, instead of alveoli. At the point where the bronchi

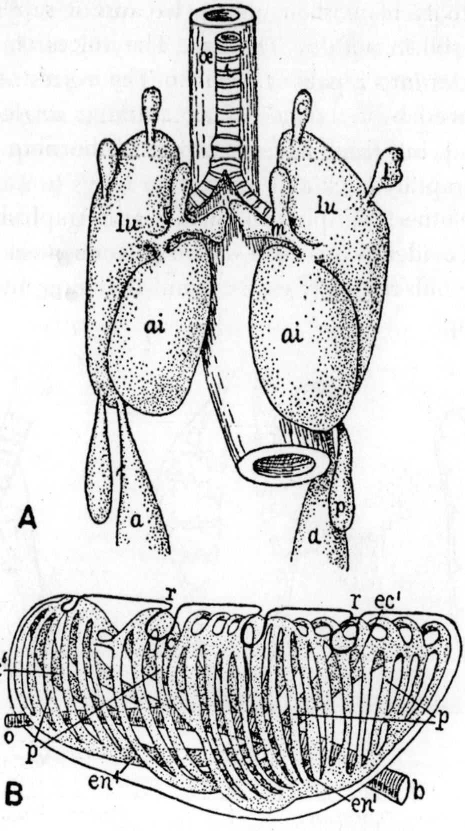

Fig. 240. A, developing air sacs in 12-day chick embryo (from Kingsley, *Outlines of Comparative Anatomy of Vertebrates*, 3rd ed., Blakiston).

a, abdominal sac; *ai*, anterior intermediate sac; *c*, cervical sac; *l*, lateral moiety of interclavicular sac; *lu*, lung; *m*, mesial moiety of interclavicular sac; *oe*, esophagus; *p*, posterior sac; *t*, trachea.

B, diagram of bird lung (from Kingsley, *Outlines of Comparative Anatomy of Vertebrates*, 3rd ed., Blakiston).

b, bronchus; *ec*, ectobronchi; *en*, entobronchi; *o*, connection with abdominal air sac; *p*, parabronchi; *r*, notches occupied by ribs.

meet the trachea in most birds there is an enlargement, the ***syrinx***, in which a ***semilunar membrane*** (or ***pessulus***), vibrating as air passes it, produces the voice (Fig. 241).

In mammals, moreover, the lungs have innumerable microscopic pockets, the ***alveoli***, reached by ***bronchioles*** (branches from the bronchi). The development of a muscular ***diaphragm*** between the thorax and abdomen was undoubtedly a major contribution to the rapid metabolism and high activity of most mammals, because it supplements the movements of the body wall in breathing. The voices of mammals are produced in the *larynx* (Fig. 242), at the beginning of the trachea, by vibration of paired folds, the ***vocal cords.*** Cartilages supporting the larynx in mammals are greatly modified derivatives of the more posterior branchial cartilages of amphibians and reptiles.

Having followed the development of lungs from primitive fishes through tetrapods, we must return to the Actinopteri and see what happens to the air bladder. Through most of the lower teleosts there remains

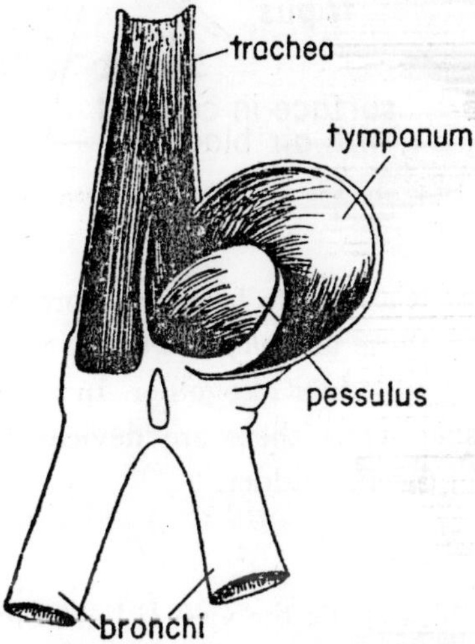

Fig. 241. Syrinx of canvasback duck. (Redrawn from Kingsley, *Outlines of Comparative Anatomy of Vertebrates*, 3rd ed., Blakiston.)

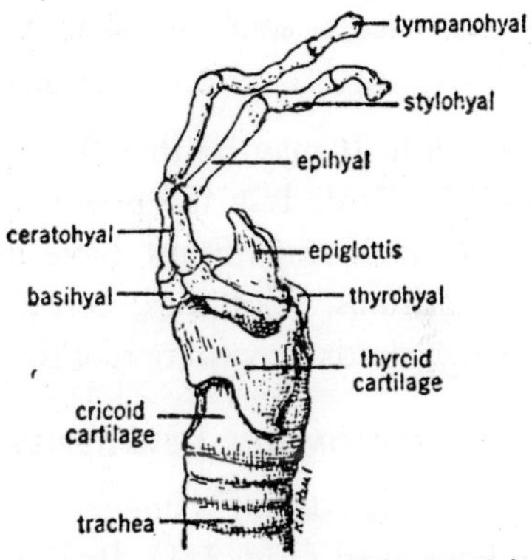

Fig. 242. Larynx of dog. (Reproduced by permission from *Comparative Anatomy* by L. A. Adams & S. Eddy, published by John Wiley & Sons, Inc., 1949.)

a ***pneumatic duct*** connecting the air bladder with the esophagus; such fishes are ***physostomous*** (Fig. 238). Among them some, such as suckers and carp, have an anterior and posterior division with a narrow neck between, and in the very large order Ostariophysi (characins, catfishes, carp, minnows, and suckers) the air bladder connects with the internal ear of the fish by means of a curious series of ***Weberian ossicles*** (Fig. 243), developed from the anterior vertebrae. Presumably this apparatus makes the fish respond to changes in depth as relayed by the increasing or decreasing size of the air bladder. By a reflex mechanism the fish can secrete or absorb into the blood some of the gas (oxygen, primarily) in the bladder, so helping to maintain a constant specific gravity of the

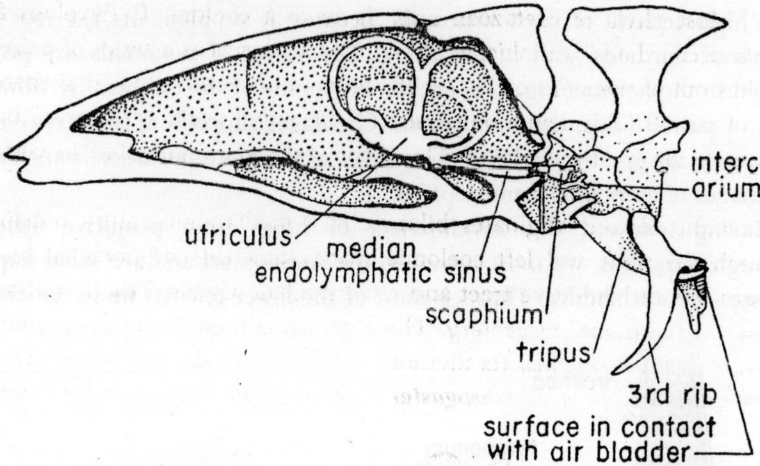

Fig. 243. Weberian ossicles of carp in relation to inner ear and skull. (Modified from Goodrich.)

whole fish. It may be that the apparatus is an aid to hearing. More advanced teleosts lose the pneumatic duct; these are ***physoclystous.*** Some groups, such as flatfishes, have lost the air bladder altogether. In a few cases (drums, toadfishes, certain catfishes, etc.) there are devices for making sounds by vibration of the walls of the air bladder.

COELOM AND MESENTERIES

The coelom is one or more cavities outside the viscera but inside the body wall (Fig. 244). Its lining, the ***peritoneum,*** therefore covers the viscera (***splanchnic peritoneum*** or ***serosa***) and the inner surface of the body wall (***somatic*** or ***parietal peritoneum***). It also is extended as thin sheets of connective tissue, the ***mesenteries,*** which support the viscera and form partitions between various cavities of the coelom. In general no actual space remains vacant, but the organs are able to move or change their shape in cavities which they almost completely fill. According to Arey (1942) the coelom "serves as a large bursa to permit frictionless movement of the heart, lungs and abdominal vis-

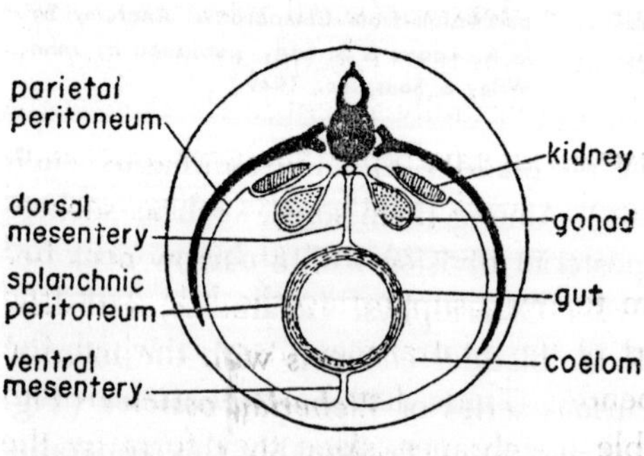

Fig. 244. Diagram of coelom in relation to other parts of vertebrate.

cera." Most phyla of metazoan animals have a coelom. It develops in primitive chordates (Amphioxus) and in echinoderms as a series of paired pouches out of the sides of the archenteron (embryonic gut). It is therefore segmented. Segmentation persists in the upper parts (epimeres) but not in the lower (hypomeres). The latter are never segmented in vertebrates.

Since the coelom originates bilaterally, a median mesentery remains between the right and left coelomic spaces, and this supports, at least temporarily, the digestive tract and other median organs. The part above the gut is the ***dorsal mesentery.*** This remains, although much modified, in reptiles and mammals. Its divisions, according to the part of the tract where they attach, are: ***mesogaster*** (stomach; includes ***greater omen-***

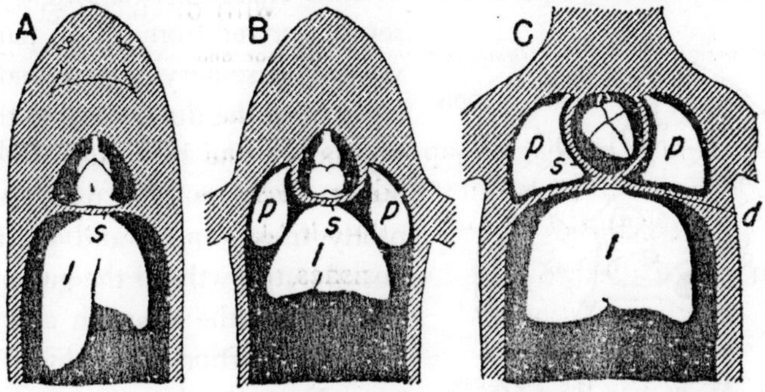

Fig. 245. Coelom (black). A, fish; B, amphibians, reptiles, and birds; C, mammals. (From Kingsley, *Outlines of Comparative Anatomy of Vertebrates*, 3rd ed., Blakiston.)

tum), ***mesoduodenum*** (duodenum), ***mesentery*** (small intestine), ***mesocolon*** (large intestine), ***mesorectum*** (rectum). The dorsal mesentery is incomplete in Amphibia and birds, and virtually absent in elasmobranchs and Actinopteri. The ***ventral mesentery*** (part below the gut) persists in very few vertebrates (lungfishes, eels, garpike), otherwise being reduced to the ***falciform ligament,*** from base of liver to ventral body wall, and ***lesser omentum,*** from duodenum to liver, supporting the bile duct; the ***mesocardium*** is a temporary part of the ventral mesentery in which the developing heart is, at first, suspended. These dorsal and ventral median mesenteries are, of course, double membranes, since they form by the meeting of two coelomic cavities in the middle line, each having its own peritoneum.

Most characteristic of the several special cavities (Fig. 245) developed in the coelom of vertebrates is the ***pericardial cavity*** (enclosed by ***pericardium***), a space in which the heart can beat without friction against other organs. In fishes this lies far forward under the esophagus, just behind the branchial arches; its posterior wall is the ***transverse septum,*** a nearly upright membrane that accompanies the large veins (ducts of Cuvier) which run in from both sides to the heart. Above these veins, however, small openings persist from the pericardial to peritoneal cavities in hagfishes, larvae of lampreys, and in elasmobranchs and Holostei. Otherwise, in all vertebrates, the pericardium is complete. Among tetrapods, with the development of a mobile neck, the heart and pericardium retreat into the thorax, and the transverse septum draws back into an oblique position as a result.

The liver develops in the embryo as a median ventral outgrowth of the gut just behind the stomach and transverse septum. It bulges into the transverse septum, and of course remains enclosed in a single layer of serosa derived from this. A part of the ventral mesentery then remains as a strap from the duodenum to the liver, supporting the bile duct; this is the ***lesser omentum.*** The pancreas arises partly from ventral and partly from dorsal outgrowths of the gut, and so is attached to the stomach and duodenum by a portion of the dorsal mesentery. Now the lesser omentum and bile duct anchor the duodenum, so to speak, to the liver, and as the growing stomach bends to the left, the dorsal mesogaster bends with it, forming a cavity (***bursa omentalis***) with an opening (***foramen of Winslow***). This pouch becomes a large thin sac in mammals, the ***greater omentum,*** and the foramen of Winslow lies in the angle, or lesser curvature, of the stomach.

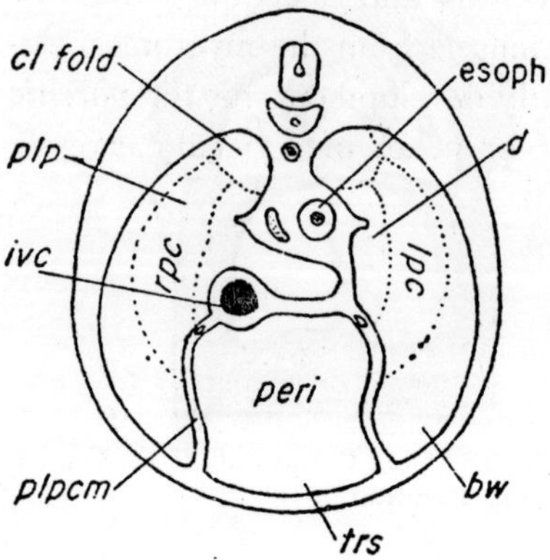

Fig. 246. Face of diaphragm of mammal, showing sources from which its parts are derived. (After Goodrich.)

bw, from body wall; *cl fold,* from posterior closing fold; *d,* from median mesentery and tissue dorsal to liver; *esoph,* esophagus; *ivc,* inferior vena cava; *lpc,* left pleural cavity; *peri,* pericardium; *plp,* from pleuroperitoneal membrane; *plpcm,* pleuropericardial membrane; *rpc,* right pleural cavity; *trs,* transverse septum.

The lungs as they grow out from the foregut push ahead of them a covering derived from the splanchnic peritoneum of the gut, and usually remain attached by ***pulmonary folds*** to the esophagus, dorsal mesentery, and transverse septum. These connections may be partly or wholly lost (the left pulmonary fold is lost in mammals). The Müllerian ducts (oviducts) and kidney ducts may be supported by ***nephric folds*** attached to the dorsal body wall.

In reptiles there is partial but rarely (Crocodilia, etc.) complete separation of the *pleural cavities*, containing the lungs, from the rest of the coelom. This is completed in birds, however, by a ***pulmonary aponeurosis***, which extends more or less horizontally from the transverse septum to the sides of the thorax and dorsal wall in front of the kidneys. This, plus the median mesentery, gives a pair of pleural cavities.

The ***diaphragm*** of mammals (Fig. 246) accomplishes much the same thing in a different way. It results from the spreading of a ***pleuroperitoneal membrane*** across the body cavity to the transverse septum, while the liver withdraws somewhat from the latter, remaining attached to it only by the coronary and falciform ligaments. The sides of the diaphragm spread more posteriorly down the body wall, and it acquires thus the shape of a dome. It receives a sheet of muscle from the myotomes, evidently in the neck region during very early development, for the ***phrenic nerve*** which supplies it is composed of branches of cervical nerves 3, 4, and 5.

CHAPTER VIII

Circulatory System

INTRODUCTORY

One requirement for developing a many-celled body is that the individual cells have access to a watery medium which will transport solutes to and from them, and indeed provide the liquid basis of their own protoplasm. In all the higher Metazoa the body carries such a fluid within itself. It is further necessary, in all animals beyond a certain minimum size, that the fluid be in motion so that dissolved food and oxygen can get to every part and not be wasted on the few tissues lying near the surface where they were absorbed, and that excretory products may not destroy the cells that make them, but be carried off.

Speaking of vertebrates especially, there are three body fluids distinguished simply by their location: ***tissue fluid,*** which occupies spaces between cells or tissues, including the coelom, but is not within any vessels; ***lymph,*** which is the same fluid carried within the lymphatic system of vessels; ***blood,*** which is the same fluid in the blood vessels, with the addition of red corpuscles and some difference in the proportion of proteins. Minor chemical differences exist between these fluids on account of their relationship to various tissues, but it is important to understand that they are fundamentally the same, and that a nearly free interchange goes on between them all the time. This we illustrate in Fig. 247, which diagrams the relationship between a capillary, tissue spaces, and a lymphatic vessel.

Leucocytes, the white phagocytic blood cells, are present in blood, tissue fluid, and lymph because, by diapedesis (squeezing through) they can penetrate a capillary or lymphatic wall. ***Erythrocytes,*** the far more abundant red cells which carry oxygen for respiration, are limited to the blood proper. Except for brief references, we shall be concerned here only with the blood vessels and heart, for, as a well-defined system of tubes and a pump, these display the evolutionary progress of all the vertebrate groups.

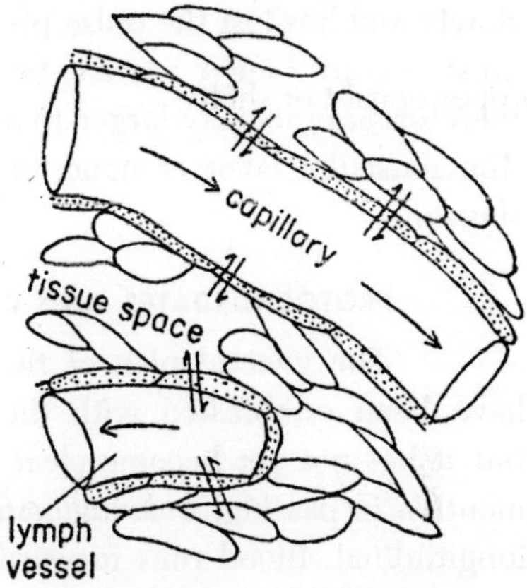

Fig. 247. Diagram of relationship between blood capillary, tissue spaces, and lymph vessels. Arrows show flow of blood and lymph; double arrows show diffusion through membranes.

Capillaries are the microscopic vessels through whose single-layered walls (Fig. 248) nearly all the work of the blood goes on; virtually everything that enters or leaves the blood does so here. It is for these capillaries that the veins, arteries, and heart exist. In an animal the size of man there are hundreds of millions of capillaries, averaging less than a millimeter in length, and permeating all the tissues. Blood must be kept moving in them, but barely so, and the heart, therefore, must deliver it into arteries with a pressure sufficient that the very narrow lumen of the capillaries shall not stop it altogether.

Arteries and their finer branches, ***arterioles,*** therefore have thick, elastic walls with a lining, the ***intima*** or ***endothelium*** (continuous ultimately with the capillary walls), a ***muscularis,*** of circular muscle fibers, and an ***adventitia,*** or connective tissue sheath. Arteries aways take blood *away from the heart,* regardless of its oxygen or carbon dioxide content.

Veins and their finer tributaries, the ***venules,*** however, collect blood from the capillaries and return it *toward the heart.* Since it is moving

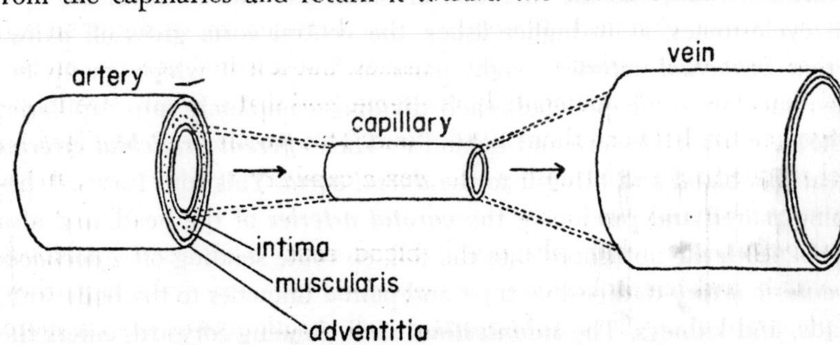

Fig. 248. Differences between artery, capillary, and vein, and continuity of lining.

slowly and has lost the pulse pressure it had in the arteries, venous blood must occupy bigger vessels, but their walls need not be elastic. Veins, therefore, are usually larger than corresponding arteries but have a very thin muscular layer or none; often they expand into large chambers or ***sinuses.***

PROTOCHORDATES AND CYCLOSTOMES; THE VERTEBRATE PLAN

The general plan of the vertebrate circulatory system seems to have been established with the appearance of pharyngeal respiration, but it has not yet become clear in the Hemichorda, which we may just mention in passing; *Balanoglossus* has a dorsal and a ventral vessel, both longitudinal. Blood runs forward above and back below, the connection being made by peripharyngeal vessels in the collar. The "heart" is a contractile chamber at the anterior end of the dorsal vessel, corresponding in development to one in echinoderms, but obviously not to the heart found in other chordates.

Amphioxus, however, has a plan comparable (Fig. 249) to that of vertebrates. Blood flows forward beneath the branchial arches in a pulsating vessel, the ***ventral aorta*** (no actual heart), then up the primary arches as ***branchial arteries*** to enter paired (right and left) ***radices aortae,*** which carry it back to the united ***dorsal aorta.*** It continues in this to the posterior part of the body, including the intestine, where it enters capillaries. Then going forward in a ***subintestinal vein,*** the blood has to pass through capillaries again in the liver (partly ***hepatic portal system***) before entering the ventral aorta. On either side at this point there enters also a ***common cardinal vein*** or ***duct of Cuvier,*** formed laterally by junction of an ***anterior*** and ***posterior cardinal.*** In tunicates the same arrangement is present, with a distinct heart, but no liver and no cardinal veins, and the direction of blood flow can be reversed.

In cyclostomes, as in higher fishes, the ventral aorta gives off paired ***afferent branchial arteries*** (eight pairs in lampreys) which go up between successive gill pouches, each dividing completely into capillaries to the demibranch before and behind it. Then ***efferent branchial arteries*** collect this blood and bring it to the ***dorsal aorta.*** This runs forward, becoming paired and producing the ***carotid arteries*** of the head, and also back beneath the notochord into the tail, en route sending off a ***coeliaco-mesenteric artery*** to digestive tract and paired branches to the body wall, gonads, and kidneys. The ***subintestinal vein,*** running forward, enters the liver as a ***hepatic portal vein*** (that is, breaking up into capillaries there, which unite again), and thence a single *hepatic vein* enters the heart. From the tail the *caudal vein,* embryonically an extension of the subintestinal, splits and sends blood into two *posterior cardinal* (*postcardinal*) *veins* which go forward in the roof of the body cavity beside the notochord. At the level of the heart they meet *anterior cardinals* from the head and unite to form a *common cardinal* entering the right side (in lamprey) of the heart (left in hagfishes). (Other fishes have two common cardinals, left and right.)

This pattern, with the two common cardinals, is the *primitive* one for vertebrates (Fig. 249), and should be learned as something different

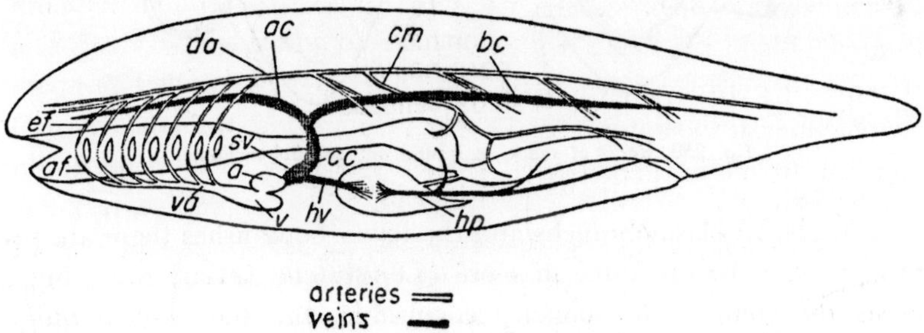

Fig. 249. Diagram of circulation in a generalized vertebrate.
a, auricle; *ac,* anterior cardinal vein; *af,* afferent branchial artery; *cc,* common cardinal vein; *cm,* coeliacomesenteric artery; *da,* dorsal aorta; *ef,* efferent branchial artery; *hp,* hepatic portal vein; *hv,* hepatic vein; *pc,* posterior cardinal vein; *sv,* sinus venosus; *v,* ventricle; *va,* ventral aorts.

from the *embryonic* arrangement of vessels, seen in the early life of an individual fish, bird, or mammal. The embryonic pattern contains some special features, like the vitelline arteries and veins, never present in any adult ancestor, and also eliminates a few details which were present in adults of primitive forms.

HEART

In fishes the heart has two chambers (Fig. 250), a thin-walled, more dorsal *atrium* (*auricle*), receiving blood from behind, and a thick-walled *ventricle,* which propels it forward. Two valves in the opening between permit the flow of blood into, but not back out of, the ventricle. In addition there is a funnel-shaped *sinus venosus,* to which all blood from veins is brought and thence poured into the auricle between *sinauricular valves,* and a *conus arteriosus,* which receives blood forward out of the ventricle; this is the contractile, enlarged base of the ventral aorta, and contains *semilunar valves,* to prevent the return of blood into

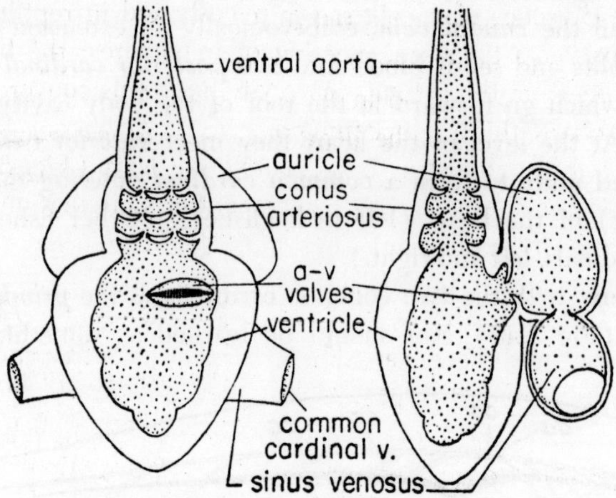

Fig. 250. Heart of shark, in frontal and sagittal sections.

the ventricle. In elasmobranchs and the lower bony fishes there are several of these valves in three or more (*Lepisosteus,* seven) rows, but in teleosts the conus is no longer recognizable, and the swollen ***bulbus cordis*** at the base of the ventral aorta contains usually only one set of valves.

Among these lower vertebrates the heart begins its development as a single median tube, formed from mesenchyme between the right and left coelomic cavities (Fig. 251). It is thus supported, very briefly, by the mesocardium, a portion of the ventral mesentery, but as the pericardial cavity becomes shut off behind by the transverse septum, the mesocardium disappears, leaving the heart as an S-shaped vessel crossing this

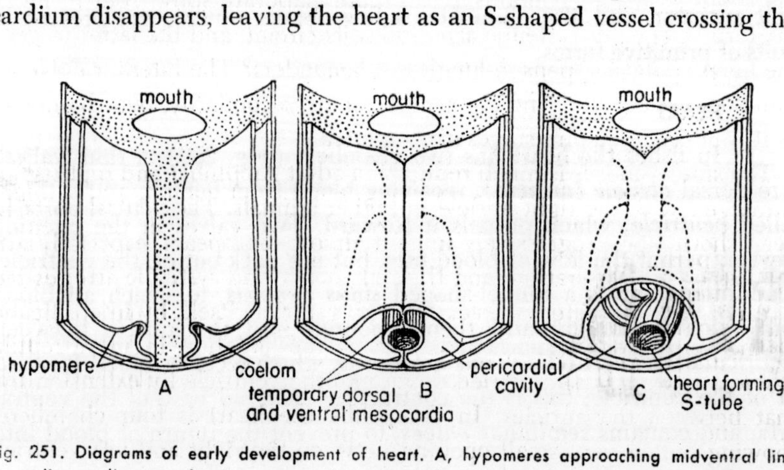

Fig. 251. Diagrams of early development of heart. A, hypomeres approaching midventral line; B, median walls meet, forming mesocardia and tubular heart, with pericardial cavity formed as pocket of coelom on each side; C, mesocardia disappear, leaving heart free to bend in S shape.

cavity. The presence of a yolk sac in teleosts and in reptiles, birds, and mammals prevents the early meeting of hypomeres; the heart in these groups results from the fusion of two vitelline veins converging from the yolk, but of course this is not primitive.

Lungfishes have almost a three-chambered heart (Fig. 252) like that of Amphibia, because with the development of a circulation to the lungs, blood returns by pulmonary veins which enter as one into the left side of the atrium, while the sinus venosus enters the right, and there is an almost complete interauricular septum between. Then by means of a spiral ridge within the conus arteriosus, blood from the right auricle tends largely to veer towards the last (fourth to sixth) aortic arches, from which some of it can get to the lungs, while blood from the left auricle is deflected towards the openings of the arteries leading to the head.

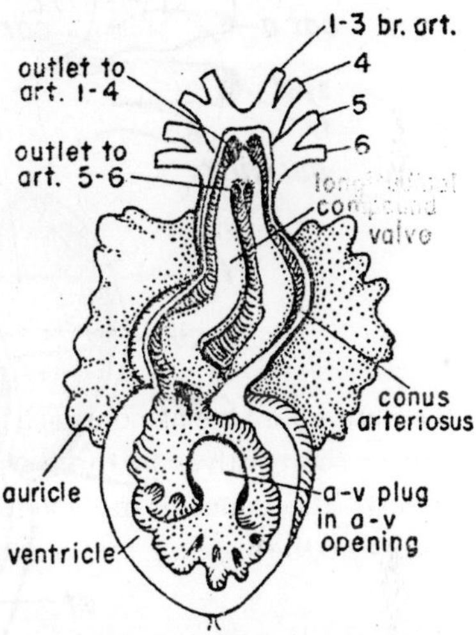

Fig. 252. Heart of lungfish, *Protopterus*, open from ventral side, showing plug valve in a-v opening. (After Goodrich.)

This specialization goes still farther in Amphibia (Fig. 253), where the adults may lose their gills. The auricles are entirely separate, pulmonary ("arterial") blood entering the left, systemic ("venous") the right, and the conus arteriosus contains a still more elaborate spiral fold to induce the former to enter systemic arteries on its circuit, and the latter to get to the lungs. What happens in lungless salamanders? The interauricular septum does not develop, and the heart remains, secondarily, two chambered, as it is in the immature stages of Amphibia.

The sinus venosus is much reduced in adult Amphibia and reptiles, and is merged with the right auricle in higher animals. The ventral aorta has been shortening progressively until it almost disappears, especially after the loss of gill respiration, and the pulmonary and systemic arteries tend to carry their separateness back into the ventricle itself, by intensification of the fold already mentioned. This is the situation in most reptiles, where a septum is also carried down into the ventricle by extension from that between the auricles. In Crocodilia the heart is four chambered, showing only a very small opening between the right and left ventricles (Fig. 254).

Independently of each other, since their common ancestry goes far back among primitive reptiles, the birds (Fig. 254) and mammals attained a four-chambered heart, in which the right auricle brings systemic blood to the right ventricle and it can then go *only* to the pulmonary artery, while the left auricle receives pulmonary blood, sends it to the left ven-

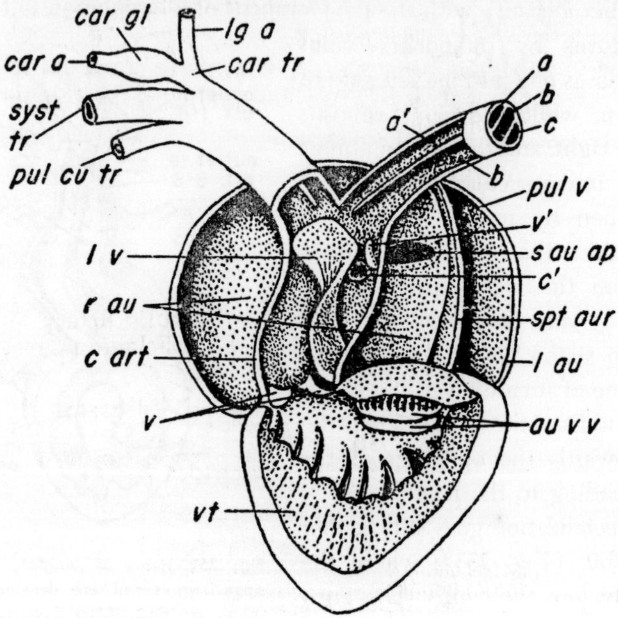

Fig. 253. Heart of frog, *Rana temporaria*, dissected, from ventral side. (Redrawn from Parker and Haswell, *Textbook of Zoology*, 6th ed., Macmillan.)

***a, a'*, left carotid trunk; *au v v*, auriculoventricular valves; *b, b'*, left systemic trunk; *c, c'*, left pulmocutaneous trunk; *car a*, carotid artery; *car gl*, carotid "gland"; *c art*, conus arteriosus; *car tr*, carotid trunk; *l au*, left auricle; *lg a*, lingual artery; *l v*, longitudinal valve; *pul cu tr*, pulmocutaneous trunk; *pul v*, opening of pulmonary veins; *r au*, right auricle; *s au ap*, sinuauricular aperture; *spt aur*, auricular septum; *v, v'*, valves; *vt*, ventricle.**

tricle, and thus *only* to the systemic arteries. Here is another instance of structure perfected in the service of higher physiological activity. No blood has to go twice around the systemic circulation without being brought to the lungs. In mammals before birth (and for a while after, in the case of blue babies) there is a *foramen ovale* between right and left auricles, allowing a small amount of mixing to take place.

We may now summarize the main features of the heart of mammals (Fig. 255). The auricles are thin, and flaccid when empty, while the ventricles have a heavy muscular wall, that of the left being much thicker

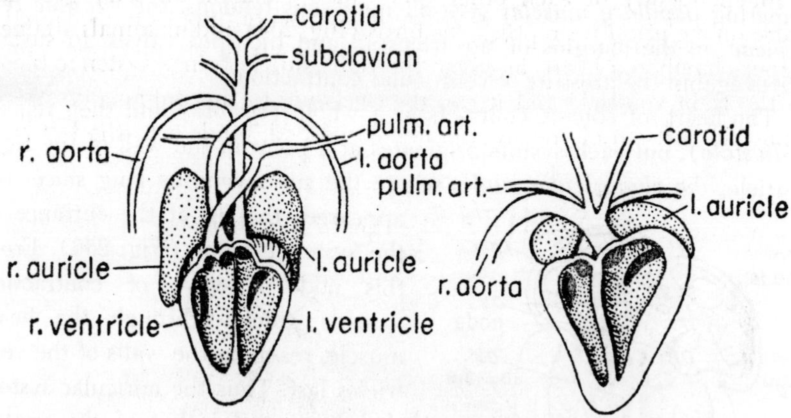

Fig. 254. Heart of crocodile (left, after Parker and Haswell) and chicken, right, diagrammatic.

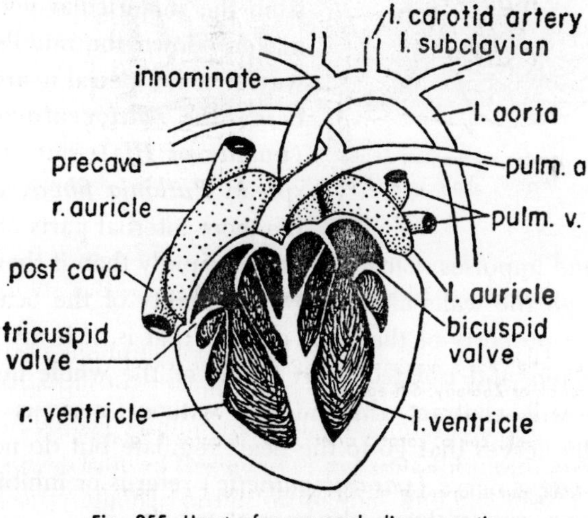

Fig. 255. Heart of a mammal, diagrammatic.

than the right. This is because more pressure is required to circulate blood through the systemic vessels than on the shorter course to the lungs. The auriculoventricular valves are, on the right, the ***tricuspid,*** on the left, ***bicuspid*** or ***mitral.*** The openings into the pulmonary artery and aorta are provided each with three ***semilunar*** valves. In the muscular walls of the heart, especially the ventricles, is a system of coronary circulation; the ***coronary artery*** emerges from the aorta near its base, while the ***coronary vein*** returns blood to the ***coronary sinus*** (a vestige of the right duct of Cuvier), and thence to the right auricle. The inner walls of the ventricles are strongly ridged with muscular ***trabeculae,*** from which tapering ***papillary muscles*** give off numerous tendons, the ***chordae tendineae***, to the margins of the tricuspid and bicuspid valves, to sustain them against the pressure of ventricular contraction.

The heart, of course, contracts as a whole (***systole***) and then relaxes (***diastole***), but each systole originates at a point in the wall of the right auricle, the ***sinauricular node***, where the sinus venosus long since disappeared, that is, at the entrance of the systemic veins (Fig. 256). From this node a wave of contraction spreads quickly through the heart muscle, reaching the walls of the ventricles last. Thus the auricular systole slightly precedes that of the ventricles. The ***impulse*** to contract runs from the sinauricular node, or "pacemaker," down the middle septum by way of some special neuromuscular fibers, the ***atrioventricular bundle*** (***bundle of His***); out of these there spread ***Purkinje fibers***, especially to the more internal parts of the ventricular walls, and impulses reach there more rapidly than if they were transmitted through the walls alone. Now the ability of the heart to initiate each beat is a property of the heart muscle; that is, it is myogenic rather than neurogenic, and isolated heart tissue, or the whole heart in lower vertebrates, will contract rhythmically without nervous connections. Therefore the nerves that go to the heart regulate but do not initiate its action. The ***vagus*** nerve (***parasympathetic***) retards or inhibits, while the ***sympathetic*** nerves accelerate the beat.

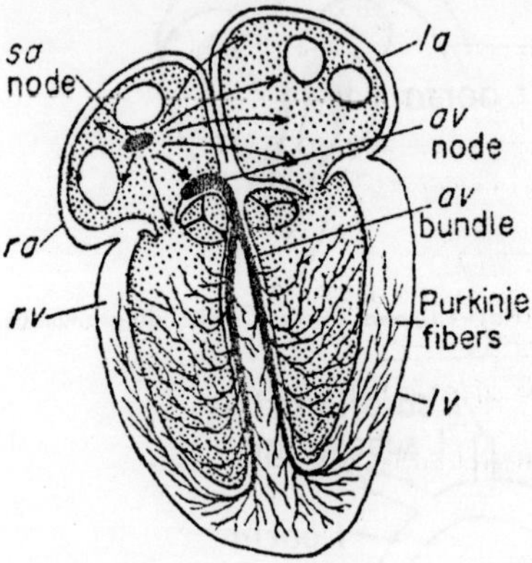

Fig. 256. Neuromuscular mechanism of the mammalian heart. Arrows show spread of contraction wave.

ARTERIES

The history of the arteries in vertebrates centers about the ***aortic arches*** which supply the gills in fishes and Amphibia, and their later derivatives in Amniota, because evolutionary changes in these vessels are diagrammatically clear. They form a classic illustration of the method of comparative anatomy.

Primitively, the ***ventral aorta*** led forward from the heart in the mid-ventral line of the gill region, giving off a right and left ***afferent branchial artery*** to the outer side of every visceral arch including the jaws, for all these bore gills and were essentially identical. The arteries are numbered (Fig. 257) like the arches: I (to jaws), II (hyoid), III-VI (branchial arches). In cyclostomes, of course, the number is higher, including a premandibular arch in some ostracoderms. Capillaries return blood from the gill filaments to *efferent arteries,* although early in development each branchial artery is continuous. In Elasmobranchii (Fig. 258) and

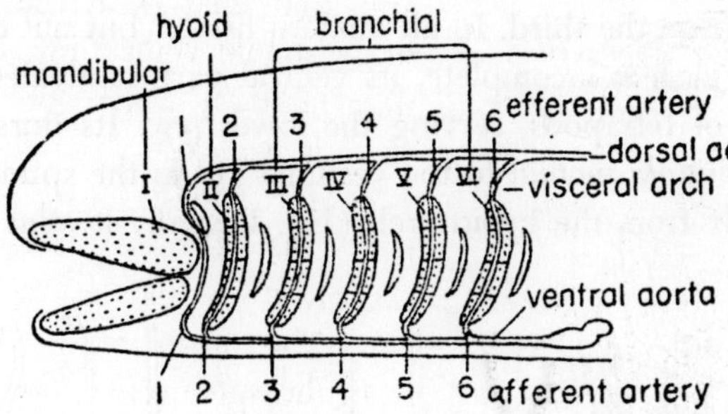

Fig. 257. Customary system of numbering visceral arches and aortic arches.

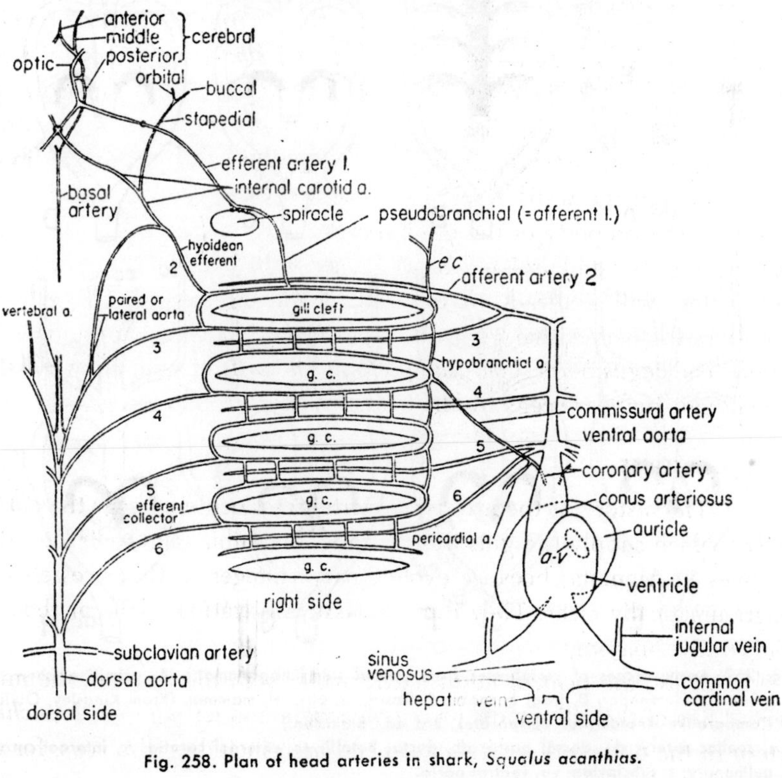

Fig. 258. Plan of head arteries in shark, *Squalus acanthias.*

Dipnoi the efferent arteries connecting with each complete gill (holobranch) are double; one receiving blood from each demibranch, while in teleosts a single efferent serves each holobranch. The afferents are always one per arch.

As seen from beneath, the ventral aorta tends to split back in the more advanced groups (Fig. 259), so that the first afferent comes off the second, and that from the third. In all modern fishes (but not cyclostomes) the first aortic arch is incomplete, its ventral part, equivalent to the *external carotid* of tetrapods, serving the lower jaw. Its dorsal part may persist as a ***pseudobranchial*** to the vestigial gill in the spiracle, deriving its blood supply from the hyoid arch (II). Dorsally it joins the forward

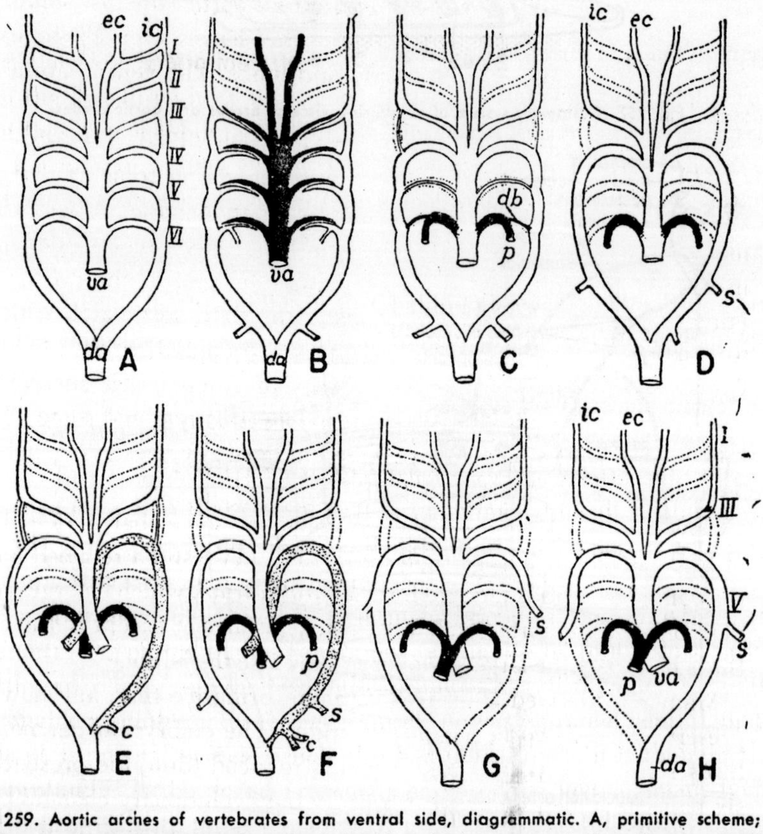

Fig. 259. Aortic arches of vertebrates from ventral side, diagrammatic. A, primitive scheme; B, lungfish; C, salamander; D, frog; E, snake; F, lizard; G, bird; H, mammal. (From Kingsley, *Outlines of Comparative Anatomy of Vertebrates,* 3rd ed., Blakiston.)

c, coeliac artery; *da,* dorsal aorta; *db,* ductus Botalli; *ec,* external carotid; *ic,* internal carotid; *p,* pulmonary; *s,* subclavian; *va,* ventral aorta.

extension of the dorsal aorta to form the ***internal carotid*** and associated branches to the brain and skull.

In larval lungfishes and again in larval salamanders the external gills are supplied by capillary loops from the branchial arches (IV-VI in lungfish, III-V in salamanders), while in these and other fishes with a respiratory lung or air bladder the sixth arch gives off a ***pulmonary artery*** on either side. From adult Amphibia on, we find no arches I and II, but the external carotid persists to the lower jaw, while arch III forms the connection from the ventral aorta to the dorsal internal carotid. The portion of the dorsal aorta between III and IV (***ductus caroticus***) disappears. Blood to all parts of the body behind the neck goes through arches IV, V and VI in adult salamanders; in Salientia arch V vanishes and VI loses the connection with the dorsal aorta (***ductus Botalli***), thus becoming solely a pulmonary artery. This leaves arch IV to carry all systemic arterial blood to the body and limbs. The ventral aorta is now split back to the heart.

Here we quote a paragraph from Goodrich (1930, 524–525): "Since . . . the ductus caroticus still persists in Apoda, in some adult Urodela, such as *Triton*, and some adult Reptilia (*Sphenodon, Alligator?*), and many Lacertilia, its obliteration has probably taken place independently in Amphibia and Amniota. The same may be said of the ductus Botalli (ductus arteriosus), that dorsal part of the sixth aortic arch which joins the pulmonary artery to the dorsal aorta. Closed and reduced to a fibrous strand in adult Anura and the majority of Amniotes, it survives as an open vessel in Apoda, Urodela, *Sphenodon, Alligator,* and some Chelonia. *Sphenodon,* indeed, is the only Amniote normally preserving both these ducts in the fully developed adult."

Now in modern reptiles the arrangement somewhat anticipates that of birds, in having a large ***right aortic*** arch (IV) and a reduced ***left*** arch; the common carotids unite to form a ***carotid trunk*** which emerges near the base of the right aorta. Arch V, of course, is absent except in the embryo. Birds have only the large ***right*** aorta; they lose the left one.

Mammals, derived from reptiles more primitive than any now living, preserve the ***left aorta*** but not the right. The embryos of mammals and birds show, very rapidly and in much reduced form, the presence of all six pairs of arches, and repeat broadly the evolutionary changes outlined here. The ductus Botalli (Fig. 260) closes at the time of birth, in mammals, and of hatching, in birds.

The arteries to the body wall and limbs, as well as those to the gonads

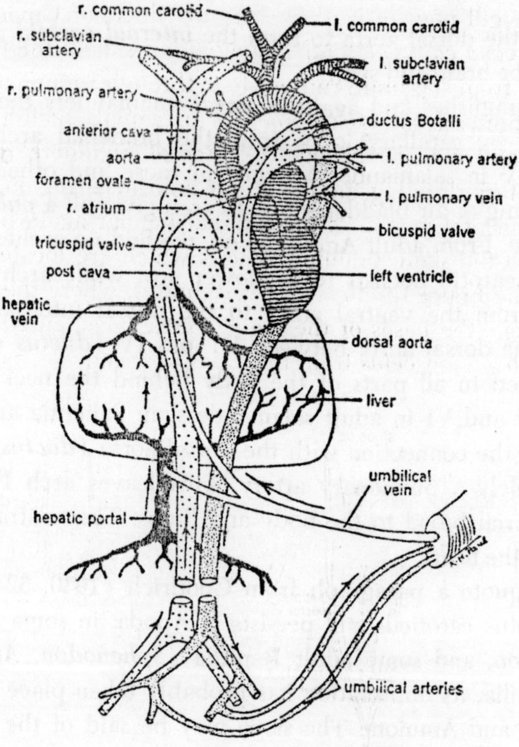

Fig. 260. Embryonic circulation in a mammal. (Reproduced by permission from *Comparative Anatomy* by L. A. Adams & S. Eddy, published by John Wiley & Sons, Inc., 1949.)

and kidneys, are paired segmental branches of the aorta, variously enlarged or modified. On the other hand, the main arteries to the viscera (*celiac, superior* and *inferior mesenteric*) are single, and there is no reason to suppose that they ever were paired in adult vertebrates. But they develop, in those animals that have an embryonic yolk sac, as irregularly paired branches of the aorta, just as the heart itself in such cases comes from a pair of omphalomesenteric veins. This is another illustration of the divergence in ontogeny from the actual sequence followed in adult evolution. The superior mesenteric, to the intestine, is the survivor of an embryonic pair of *omphalomesenteric arteries* to the yolk sac, while the *umbilical arteries* of the mammal embryo arise near the caudal end of the aorta, as the allantois is being formed.

VEINS

The primitive plan has been outlined already. In elasmobranchs the *anterior cardinal veins* of the embryo, running close beside the notochord over the gill pouches and back to the ducts of Cuvier, are partly replaced by a ***vena capitis lateralis*** along either side of the head, and the ***cerebral veins*** from the brain run out to it through foramina in the skull. In most vertebrates from cyclostomes to marsupials such a lateral head vein persists, and may be called the ***internal jugular*** to distinguish it from the tributary draining the lower jaw and facial region, the ***external jugular.*** In placental mammals the internal jugular again becomes actually internal to the skull, draining sinuses which are located among the meninges of the brain. In all higher vertebrates, where the blood enters the right auricle, the bases of the jugular veins (i.e., anterior cardinals) receive the ***subclavian veins*** from the forelimbs, and are known as ***precavals;*** the left precaval often crosses over (birds, most mammals) to join the right before entering the heart as one.

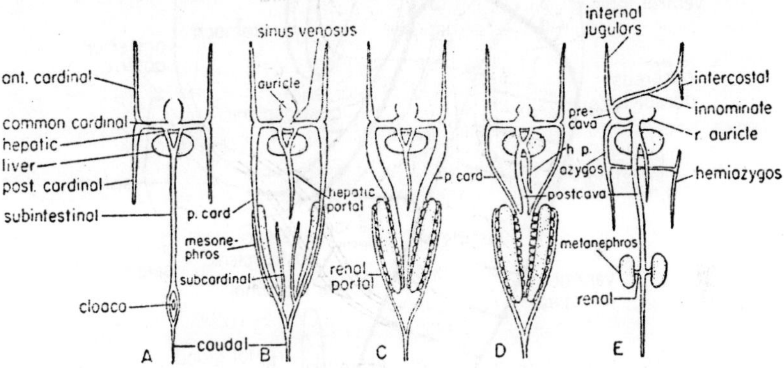

Fig. 261. Development and relationships of principal veins in vertebrates. A, fish, early development; B, fish larva; C, fish adult; D, lungfish, amphibian; E, mammal.

Going back to lower vertebrates again (Fig. 261), we find anterior and posterior cardinals meeting at either side of the heart and there forming ***common cardinals*** or ***ducts of Cuvier,*** which then enter the sinus venosus. Into the sinus venosus from the liver also enter one or two ***hepatic veins.*** The ***hepatic portal system,*** from which these come, is in part a system of capillaries within the liver, fed by venous blood from the digestive tract, by way of the large ***hepatic portal vein*** and its tributaries (Fig. 262). This portal system ensures that all blood from the stomach and intestine is brought into close contact with the liver tissues, where some of the absorbed sugar can be stored. Primitively this system is represented by the ***subintestinal vein*** (Amphioxus). In embryos of all vertebrates that have yolk sacs two ***vitelline veins*** enter the body from the yolk, become embedded in the developing liver as the ***omphalomes-***

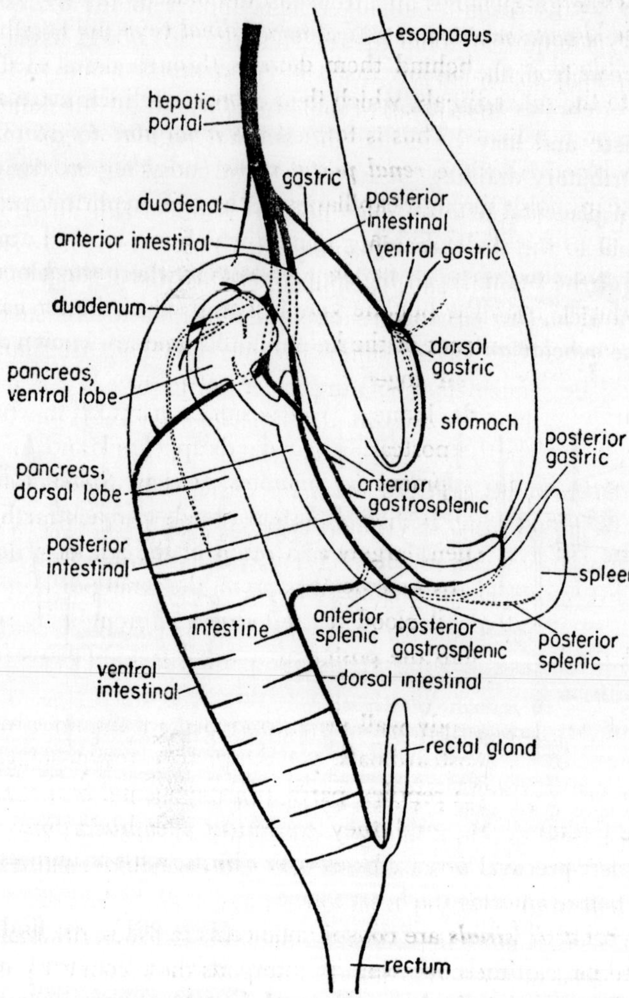

Fig. 262. Hepatic portal system in a shark, *Squalus acanthias.*

enteric veins, and then, as the intestine develops, they produce branches along it which presently combine to form the hepatic portal. This is one of the most constant features in the circulatory system of vertebrates, from fishes to man.

At the earliest stages of development in fishes the subintestinal vein continues into the tail as the *caudal,* making a split around the anus to do so, but when the hepatic portal system begins to function the caudal drops its connection with the subintestinal and drains into the two ***posterior cardinals*** along the dorsal side of the body cavity. These postcardinals receive segmental branches from the muscles, gonads, and kidneys. As the kidneys develop, a pair of ***subcardinal veins*** appears between them, and blood from behind them detours through capillaries in the kidneys into the subcardinals, which then connect with the more anterior part of the postcardinals. Thus is formed the ***renal portal system,*** bringing venous blood from the ***renal portal veins*** (originally posterior ends of the postcardinals) through capillaries in the mesonephric type of kidney, found in fishes and Amphibia.

Beginning in lungfishes a new, important vein, the *postcava,* extends back and dorsad from the hepatic vein just in front of the liver to meet the subcardinal just in front of the kidney; it becomes a bigger vein than the two postcardinals put together, and usurps their function almost entirely, draining gonads, kidneys, and posterior parts of the body. In adult frogs and toads the postcardinals have disappeared, and they occur only during early development in amniotes (except *Sphenodon,* as a part of its assorted hangovers). The postcava tends to reach farther back between the kidneys. Then, largely as a result of the rise of a new type of kidney, the metanephros (see next chapter), the renal portal system as such disappears, so that all blood to the kidneys is brought by renal arteries, and discharged into the postcava by renal veins. This is fully accomplished in mammals.

A part of the dorsal body wall was abandoned by the atrophy of the anterior parts of the postcardinals, yet skipped by the postcava. Here ***vertebral veins*** appear in reptiles, partly from segmental branches of the embryonic postcardinals, and they constitute a ***supracardinal*** system. This, in mammals, survives as the ***azygos vein,*** sometimes paired, which enters the precaval.

The ***lateral abdominals*** are conspicuous veins in fishes, draining lateral and ventral parts of the body wall. In tetrapods these connect posteriorly with the ***iliacs*** (pelvic limbs) and renal portals, but unite forward to form a single ***ventral abdominal*** which meets the hepatic portal. In the embryos of placental mammals the separate lateral abdominals become ***allantoic*** or ***umbilical veins;*** these fuse in the umbilical cord, and presently only the left umbilical survives, bringing blood from the placenta to the embryo. At birth, of course, it is shut off and atrophies.

LYMPHATIC SYSTEM

At the beginning of this chapter we mentioned the relationship between blood, tissue fluid, and lymph, the last being a fluid which drains in thin-walled vessels away from the tissue spaces, ultimately entering the veins. The lymphatic vessels and associated organs are not easily treated in a comparative manner, because transitions from the arrangement in one group of animals to that in another have not been studied adequately.

Lymphatic spaces are present in all groups of vertebrates, especially between the skin and muscles, and among the muscles next to the bones. Some authors treat the coelom as a lymphatic space, and many make no distinction between these and tissue spaces in general. The ***lymphatic vessels*** (Fig. 263) open from such spaces, as well as commencing blindly among the tissues, and drain towards the ***tracheal trunks*** or the ***thoracic ducts.*** Special lymphatics, the ***lacteals,*** arise in the wall of the small intestine, collecting absorbed fats and transmitting them, in minute droplets, along the larger vessels in the mesenteries. All groups of vertebrates appear to have this much in common. A few fishes, probably all amphibians, and also reptiles possess saclike ***lymph hearts,*** which by slow contraction of their smooth-muscle walls help propel the lymph.

Scattered through many parts of the body, primarily in mammals, are ***lymph nodes*** (better not called ***glands***), where lymphatic vessels meet in spongy, vascularized clumps of tissue of any size from bare visibility to a conspicuous organ. These have an important function in trapping bacteria and other invaders of the body fluids. The ***tonsils*** (on sides and posterior wall of the pharynx) and ***thymus*** "gland" (between the two walls of the mediastinum) are important ***lymphoid organs,*** the former because of their liability to infection, and the latter because of a suspicion that it may have an endocrine function. It is well known that the thymus is very large in young animals (and man), decreasing or disappearing at maturity.

The ***spleen,*** an important dark red organ in the abdominal cavity from fish to man, also belongs in this category, but is particularly concerned with the blood. It contains pulpy lymphoid tissue in which are irregular sinuses that hold blood. Apparently a high concentration of red corpuscles can be maintained here, as a reservoir, and by contracting the spleen can release them into the circulation. It takes a part in making red cells during embryonic development.

The lymphatic vessels contain many valves, so arranged that pressure from the movement of muscles and surrounding organs forces the lymph along, and so close to each other that the vessels have a beaded or moniliform appearance. Birds and mammals have no lymph hearts. The

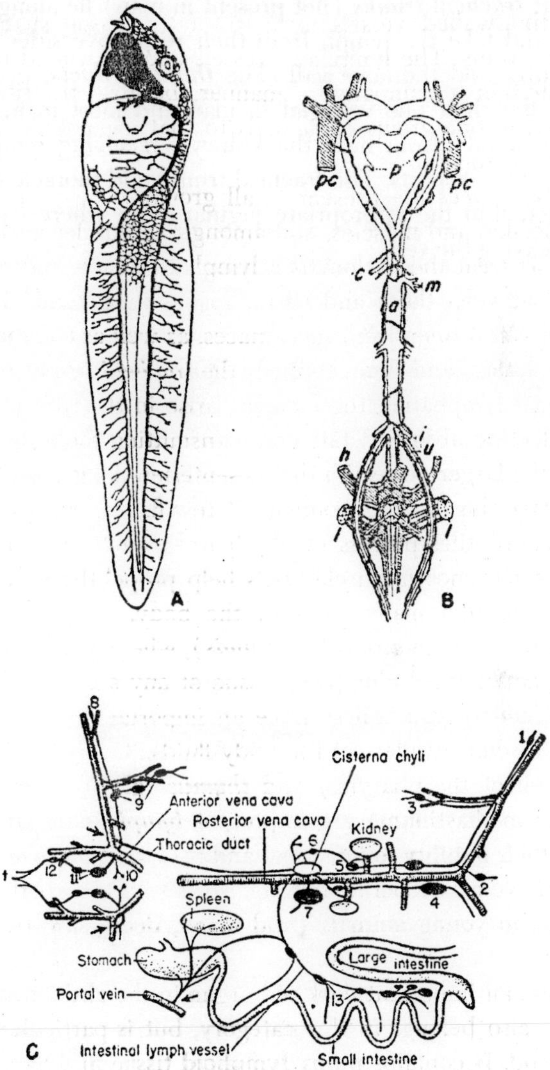

Fig. 263. A, superficial lymph vessels of tadpole; B, thoracic ducts of 17-day chick embryo; C, deep lymphatic vessels of rat. (A and B from Kingsley, *Outlines of Comparative Anatomy of Vertebrates*, 3rd ed., Blakiston; C from Romer.)

In B, arteries white, veins crosshatched, lymph vessels black.

a, dorsal aorta; *c*, coeliac artery; *h*, hypogastric vein; *i*, ischiadic artery; *l*, lymph hearts; *m*, mesenteric artery; *pc*, precava; *u*, umbilical artery.

In C, anterior end is left; lymphatic nodes numbered as follows; 1, knee; 2, tail; 3, inguinal; 4, lumbar; 5, kidney; 6, nodes around cisterna chyli; 7, intestinal; 8, elbow; 9, axilla; 10, thoracic; 11, cervical; 12, submaxillary; 13, mesenteric; t, lymphatic plexus of tongue and lips. Arrows show where lymph vessels enter veins.

right and left ***tracheal trunks*** (not present in man) lie along the sides of the trachea and take the lymph from their respective sides of the head, neck, forelimbs, and thoracic wall. The ***thoracic ducts,*** paired in most animals but the right one vestigial in man and most mammals, extend anteriorly (craniad) from near the kidneys, receiving lymph from the viscera and posterior parts. The tracheal trunks and thoracic ducts empty, close together, into the appropriate jugular vein, where valves prevent blood from leaving the vein.

CHAPTER IX

Urinary and Genital Systems

INTRODUCTION

The kidneys excrete metabolic wastes, especially urea and uric acid. But this is not the only, nor perhaps even the primary, function of these organs, for teleost fishes excrete nearly all their urea through their gills, and elasmobranchs reabsorb urea into their blood from the kidneys, in order to maintain a higher osmotic pressure in their body fluids than that of sea water. Rather, the primary function of kidneys in vertebrates is to regulate the water content of the body, keeping it constant against the tendency (in fresh water) to absorb more, or (in sea water) to lose water by diffusion, or (on land) to lose it by evaporation. The work of Homer W. Smith (1937) in this connection is of great value and has illuminated the whole subject of kidney evolution.

A system of paired ***coelomic pores,*** opening from the coelomic cavities to the outside, is present in the lowest chordates (Hemichorda), and undoubtedly originated far back in the invertebrates, not necessarily for excretion or for reproductive passages, but perhaps merely as a channel to keep internal and external pressures equal. In primitive vertebrates paired ducts from the coelom are arranged segmentally, and clusters of capillaries (***glomeruli***) make it convenient for the blood to use the same openings for release of water by filtration. Therefore these ducts, in more advanced vertebrates, came to be almost the only available outlets for metabolic wastes from the body. Likewise they were available and near to the ***gonads*** (reproductive organs), so that by appropriate changes the tubes could be used for emission of eggs or sperms. Thus it becomes difficult to separate the study of urinary organs from that of reproductive organs, but we shall begin with the former because the kidneys and their ducts show most clearly what the primitive structure was.

EXCRETORY AND GENITAL ORGANS OF PROTOCHORDATES

Among protochordates there is little suggestion, in these systems, of anything found in vertebrates, but for an adequate picture of the Chordata they should be mentioned. In some but not all Hemichorda the proboscis contains a small glandular organ, the *glomerulus* (not related

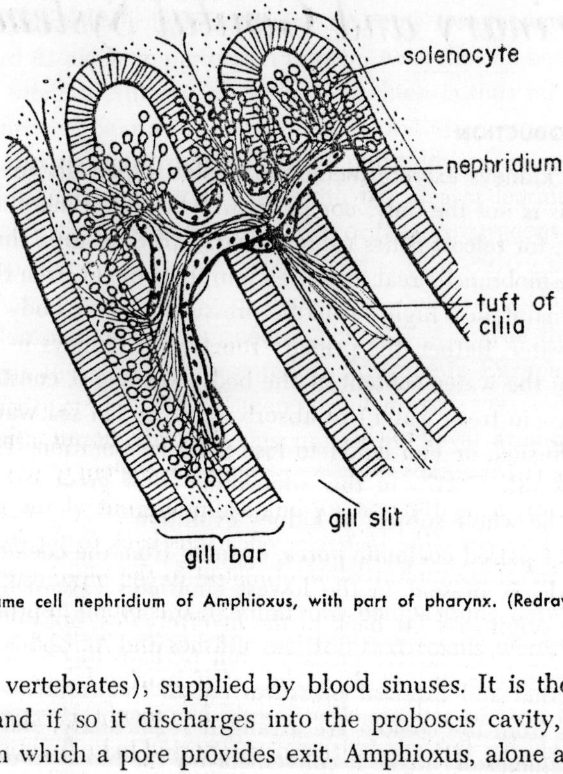

Fig. 264. A flame cell nephridium of Amphioxus, with part of pharynx. (Redrawn from Parker and Haswell.)

to those of vertebrates), supplied by blood sinuses. It is thought to be excretory, and if so it discharges into the proboscis cavity, a coelomic pouch, from which a pore provides exit. Amphioxus, alone among chordates, has an excretory system in which *flame cells* (*solenocytes*) are the essential part (Fig. 264). Although many lower invertebrates have these, the paired tubular *protonephridia* which bear them in Amphioxus most nearly resemble protonephridia of some marine polychaete worms (Annelida). It is highly improbable that any close relationship exists between chordates and polychaetes, however. About 90 pairs of these nephridia are present, located above the gill clefts and opening into the atrium; there seems to be no connection with the coelom. In tunicates it is uncertain what organs are excretory; this function has been attributed to the "*neural gland*," a blind pouch in the dorsal wall of the pharynx, and also to a vesicular organ without any duct, along the intestine.

The reproductive organs in the wormlike Hemichorda (Fig. 4) are very numerous pouches lying in rows along the dorsolateral part of the trunk, including the gill region. They open by pores to the outside. Ovaries and testes are in separate individuals, although some of the Pterobranchia (sessile Hemichorda) are hermaphroditic. The gonads of Amphioxus are segmentally arranged in the inner surface of the body wall, actually within coelomic cavities, but they break open when mature, letting the eggs or sperms fall into the atrium; the sexes are separate. Sexual reproduction among the tunicates is hermaphroditic, the single or paired ovaries and testes having their own ducts to the atrium, but many tunicates add to this an asexual budding, sometimes most complex and remarkable, so that either sessile or drifting colonies may result. Some (*Salpa, Doliolum*) alternate sexual with asexual (budding) generations, as if they had no more ambition than a jellyfish.

THE VERTEBRATE KIDNEY

Pronephros

Among vertebrates it is impossible for the segmental coelomic pores to open separately to the surface of the body, because the myotomes of the body wall have spread down around the coelom to the mid-ventral line. The pores must then become tubes, going out of the coelom to a longitudinal duct on either side, and this runs back to the best available opening, the cloaca. Such a kidney, limited to the more anterior part of the body cavity, is a *pronephros* (Fig. 265). It is found in primitive vertebrates, functioning in the larval stages of fishes and Amphibia but persisting to maturity only in hagfishes and a few marine teleosts (where this results from secondary delay of kidney development). It is present briefly in the embryos of some amniotes, and is reported to be functional in those of turtles and crocodilians. Thus the pronephros appears to be the first stage of the vertebrate kidney, in both evolution and ontogeny, although one might question whether it is a clear-cut entity in any particular case.

Each pronephros has one ***tubule*** per segment, which develops out of the ***mesomere*** or ***nephrotome*** (that portion of the mesoderm between epimere and hypomere in the embryo). The nephrotome is segmented, like the somites, but retains a coelomic cavity in each segment, and this connects with the main coelom below (Figs. 265, 266). A complete

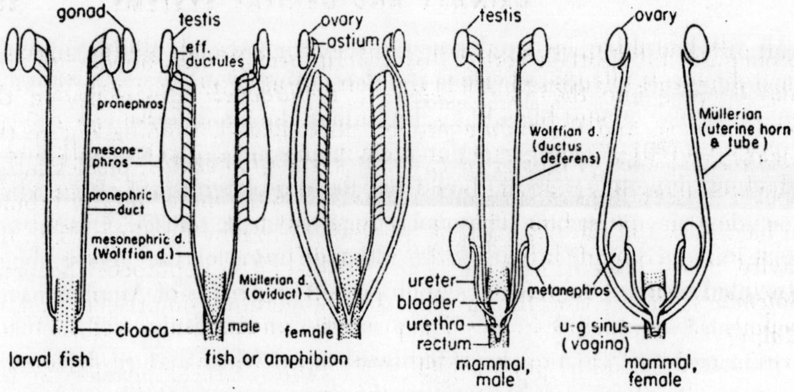

Fig. 265. Relationships of excretory and reproductive ducts among vertebrates.

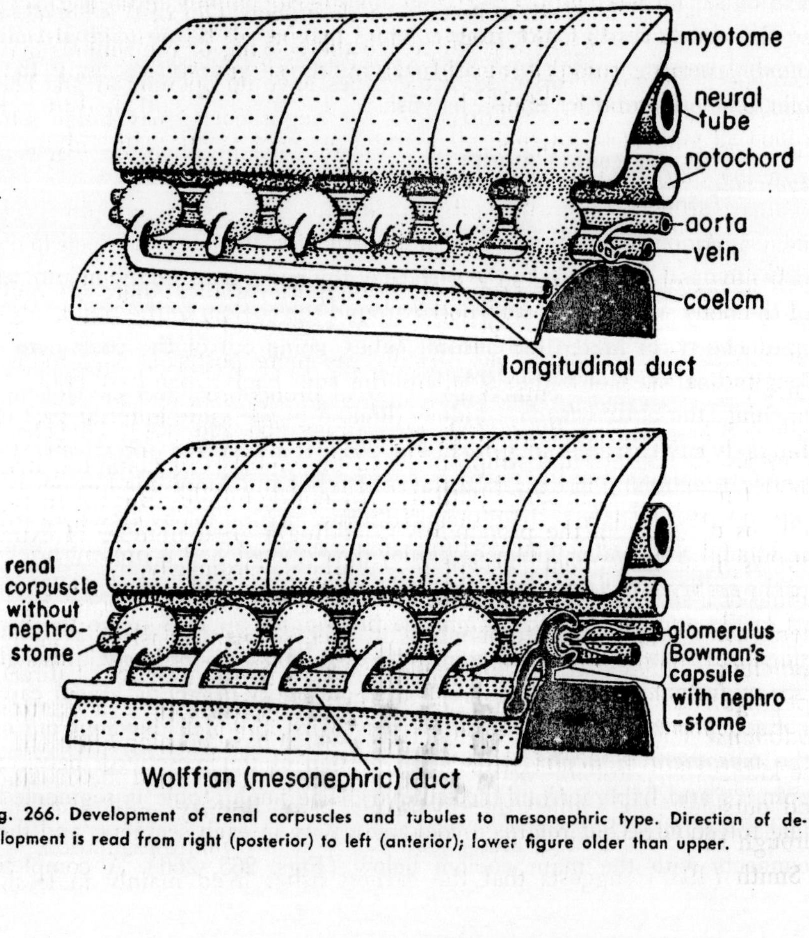

Fig. 266. Development of renal corpuscles and tubules to mesonephric type. Direction of development is read from right (posterior) to left (anterior); lower figure older than upper.

pronephric tubule then consists of a ***peritoneal funnel,*** lined with cilia and opening into the main coelom, a ***nephrocoel*** (the cavity in the nephrotome), and the ***tubule*** proper, also ciliated, running out to the longitudinal duct. A cluster of capillaries, the ***glomerulus,*** pushes into a pocket in the wall of each nephrocoel, except in the more anterior tubules in hagfishes, and the plasma from the blood can now be filtered into the cavity, minus proteins and red cells of course. The nephrocoel becomes ***Bowman's capsule,*** and the capsule plus glomerulus comprise a ***renal,*** or ***Malpighian, corpuscle.***

Commonly several nephrocoels and tubules start to develop in the embryo but a few anterior and posterior to the level of the pronephros degenerate, leaving from one to seven or eight tubules functional. Often the nephrocoels fuse together into a ***pronephric chamber*** and the glomeruli associated with them unite as a single ***glomus.*** Out of this chamber one or a few peritoneal funnels open to the coelom, or the funnels may combine into a single opening; some fishes have no opening at all. The glomeruli, instead of associating with the nephrocoels, may bulge into the coelom proper, so that capsules do not develop; such are ***external glomeruli.***

Mesonephros

Regardless of the fate of the pronephric tubules, there are only the most anterior members of a series of segmental tubules which can develop from the nephrotome. Usually the more posterior ones follow quickly after the brief embryonic or larval pronephros, and produce an elongate trunk kidney, the ***mesonephros*** (Figs. 265, 266, 267). This characterizes adult fishes and Amphibia, and the embryos of amniotes; it is present in hagfishes, in series with the pronephric tubules. But the mesonephros differs from the pronephros in adding a great number of extra tubules in each segment (except in hagfishes). Ordinarily the primary tubule of a segment loses its peritoneal funnel and the others never have them; consequently the mesonephros in each segment consists of many branching tubules that end in renal corpuscles. In the absence of direct openings from the kidney to the coelom, many fishes have a pair of ***abdominal pores*** in or close to the cloaca, which have nothing to do with the kidneys but probably serve to equalize pressure between the coelom and outside. This emphasizes the fact that the kidneys now perform through the blood system alone their function of osmotic regulation.

Smith (1937) suggests that the earliest fishes lived mainly in fresh

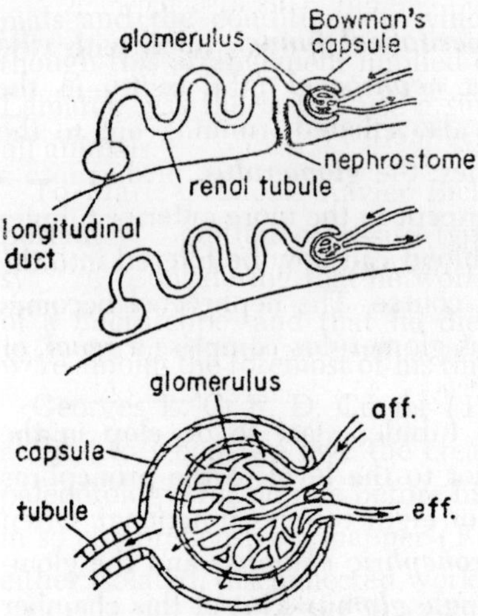

Fig. 267. Upper figures: renal tubule and corpuscle with and without nephrostome and ciliated funnel from coelom.

Lower figure: renal corpuscle, showing filtration from glomerulus into capsule.

water (and this is strongly supported by geological evidence), where the body fluids had a far higher salt concentration than the surrounding water, and the development of an efficient glomerular kidney was associated with the need to excrete large quantities of water, since water tends to enter the body by osmosis at any available place. The glomerulus does not, of course, work by osmosis but by an actual filtration under rather high blood pressure (Fig. 267). But to protect the fish from loss of salts at the same time, the tubules must reabsorb these into the blood, while letting most of the water go. So the kidney is now an organ for filtration and reabsorption. By the Upper Devonian period some of these fishes (sharks; later Actinopteri) had become marine, and were now at a disadvantage because the osmotic pressure of their body fluids was less than that of the sea. The elasmobranchs, without changing the structure of the mesonephros, reabsorb enough urea from the tubules to give the body fluids a high concentration of it, and thus lift their osmotic pressure above that of sea water. But marine teleosts (we can only judge by modern types, of course) show a marked degeneration of the glomeruli and sometimes lose them altogether (most deep-sea fishes, also sea horse, pipefish, toadfish, angler, etc.). As a result they save water and salts in their bodies, excreting only from 1 to 5 per cent as much as fresh-water kinds, and get rid of urea through their gills.

Mesonephric tubules, much coiled to give them added length for reabsorption, open into the longitudinal duct that served the pronephros; it is now the *mesonephric* or *Wolffian duct,* for the pronephric tubules and anterior part of the duct degenerate when the mesonephros becomes active. There is another distinction: the pronephric duct either splits lengthwise, or, independently, a new duct forms parallel with it, and in female vertebrates the more lateral part, not connected with mesonephric tubules, becomes the *Müllerian duct* or *oviduct,* retaining the pronephric funnel as its opening into the coelom. This also happens, for no good reason, in males, but the Müllerian duct in this case degenerates as described above, leaving only the Wolffian duct. These will be mentioned farther on, in dealing with the reproductive organs.

Metanephros

Embryos of reptiles, birds, and mammals have a mesonephros, but behind this, from the most posterior part of the nephrotome, new renal tubules begin to develop (Fig. 268) as the forerunners of the third type of kidney, the ***metanephros*** (Fig. 265). Up towards this cluster of tubules grows a new duct, from the posterior end of the Wolffian duct. This is the ***ureter.*** It expands distally and produces many short branches

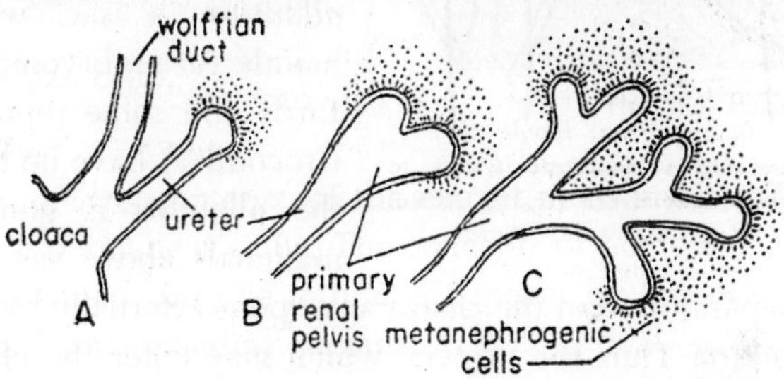

Fig. 268. Embryonic origin of ureter and renal pelvis, human. (Redrawn from *Manual of Embryology* by Frazer.)

which connect with the metanephric tubules, thus forming a kidney separate from, and posterior to, the mesonephros. Now the latter degenerates and disappears, and so does the Wolffian duct with it (except in males, where the latter is already in use as a sperm duct).

The metanephros is distinguished by an enormous number of renal tubules and corpuscles (over a million in the average human kidney), the lack of any peritoneal funnels or segmentation, a common cavity, the renal ***pelvis,*** formed by the expanded end of the ureter, and the fact that the ureter is a new duct, replacing the Wolffian. Also, since amniotes are mainly terrestrial animals, the renal tubules are highly convoluted to increase the available surface for reabsorption of water into the blood. In man more than 99 per cent of the water passing through the glomeruli is taken back before it leaves the tubules. Mammals have also introduced a new feature in the tubule, a hairpin-shaped ***loop of Henle,*** located between the ***proximal*** and ***distal convoluted portions*** of the tubule. The convolutions and the loop are well supplied by capillaries from the *efferent arteriole* which has just left the glomerulus of the same tubule.

The metanephros may be either lobed or oval. With the increasing bulk and discreteness of the kidney, it tends to push into the body cavity, but cannot pierce the peritoneum, so that it and the ureter remain encased in a part of this membrane, which holds them to the dorsal wall.

Associated with life on land is the ***urinary bladder*** (Fig. 269), developing as a ventral pouch from the cloaca in some amphibians, in reptiles, monotremes, and the embryos of higher mammals. It appears in the embryos of all amniotes and enlarges to produce the ***allantois,*** a sac which serves mainly for embryonic respiration. Birds and some reptiles (snakes, Crocodilia) have no bladder after the allantois is gone, while in mammals above the monotremes it becomes separated from the cloaca and opens externally by a duct of its own, the ***urethra.*** Thus the ureters, which may enter the cloaca in reptiles, connect with the bladder directly in mammals (Fig. 274).

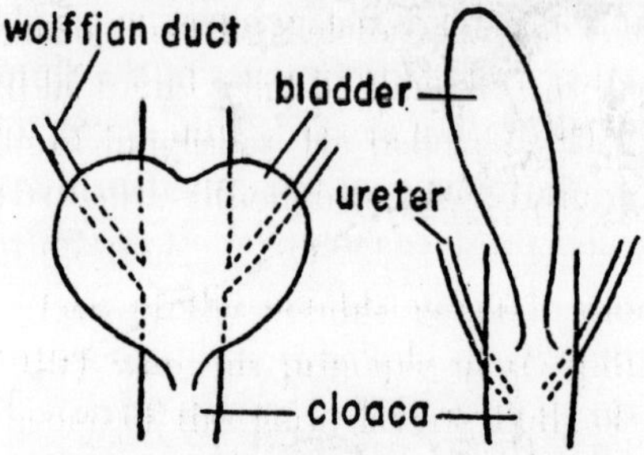

Fig. 269. Relationship of excretory system to cloaca in frog (left) and lizard (right), ventral view.

REPRODUCTIVE ORGANS OF VERTEBRATES

Introductory

All vertebrates reproduce sexually, by the fusion of ***sperm cells*** (male gametes) with ***egg cells*** (female gametes). Not only does this union of gametes bring together in a single fertilized egg the hereditary factors from both parents, but it induces the egg, otherwise inert, to start forming a new individual.

Gametes are made by the ***gonads—testes*** in the male, ***ovaries*** in the female—where ***primordial germ cells*** transform through a series of stages into sperms or eggs (Figs. 270, 271). But the ultimate source of these germ cells appears to be in the endoderm of the embryo, from which they migrate to the developing gonads in the roof of the coelom. (A few of the lower vertebrates are ***hermaphroditic,*** having both male and female gonads maturing at different times in the same individual, but this is not a primitive condition.) The gonads develop from paired but not segmented ***genital ridges*** near the median line, above the peritoneum, which is pushed down as they grow.

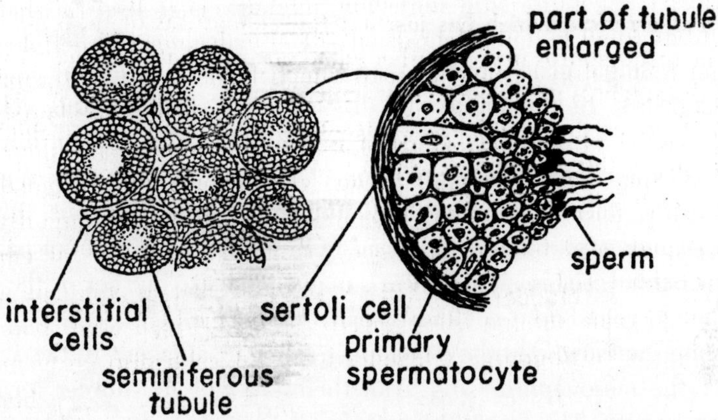

Fig. 270. Structure of testis. On left, section through several seminiferous tubules; on right, detail of part of one tubule, showing maturation of sperm cells.

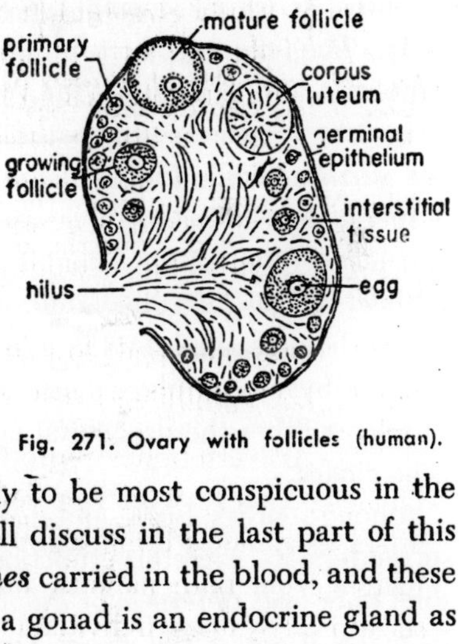

Fig. 271. Ovary with follicles (human).

Many differences in form and function arise between males and females. A part of these have to do with the reproductive organs directly; such ***primary sex characters*** concern the production of eggs and sperms, their passage to the exterior, organs for retaining the embryo in the female when the latter is viviparous, and so on. But there are also a great many ***secondary sex characters,*** which may involve almost any part of the body and are likely to be most conspicuous in the breeding season. These features we shall discuss in the last part of this chapter; they are induced by ***sex hormones*** carried in the blood, and these hormones originate in the gonads. Thus a gonad is an endocrine gland as well as a source of gametes.

The reproductive system proper must transmit gametes to a place where their union is possible, perhaps in the open water of sea, lake, or stream, or perhaps only as far as the oviducts of the female, to which sperm must be brought from the male. Egg-laying animals are ***oviparous,*** their eggs being fertilized externally in most fishes and amphibians (not in all), but internally in terrestrial forms. Animals which retain their eggs until the young can be "born alive" (not hidden in an eggshell) are ***viviparous;*** of course these always have internal fertilization, and the oviducts must be enlarged and specialized to accommodate the embryos.

Male Reproductive Organs

A single gonad, rather than a pair, is the rule among lampreys and hagfishes. In the former, as the sexes are separate, this gonad is either a testis or an ovary, made of two fused in the median line. The gametes escape into the coelom, thence to the outside by a pair of ***genital pores*** in the walls of the ***urogenital sinus*** just behind the anus. But hagfishes are hermaphrodites; the gonad is an ovary in the anterior part and a testis behind, although individuals may emphasize one sex or the other. It is the left gonad only, not a fused pair.

Among gnathostomes the developing testes lie adjacent to the anterior end of the mesonephros, and avail themselves of its tubules for transmitting sperms. The ***seminiferous tubules*** in each testis connect with a longitudinal duct, which then receives ***efferent ductules*** from the mesonephros, across the membrane (mesorchium) supporting the testis (Fig. 265). The efferent ductules are extensions of segmental mesonephric tubules, and thus lead into the Wolffian duct.

In fishes and amphibians the Wolffian or mesonephric duct runs along the surface of the kidney, as an excretory duct, but primarily serves as a sperm duct in males, since anteriorly it receives efferent ductules from the testis (Fig. 272). Then it may enlarge posteriorly into a ***seminal vesicle,*** or may be convoluted, partitioned internally, or otherwise modified to hold seminal fluid in adult males, finally entering the cloaca. This entry is by way of a urogenital sinus in elasmobranchs (Fig. 228), while in teleost fishes the urogenital opening is just behind the anus, and there is no cloaca.

Among reptiles, birds, and mammals (Figs. 273, 274) the Wolffian duct surrenders its excretory function to the ureter, and becomes a sperm duct only (***vas deferens,*** or ***ductus deferens***); it disappears in the female. In males it enters the cloaca (reptiles, birds, monotremes) or meets the urethra just below the bladder (marsupial and placental mammals).

If the eggs of a fish or amphibian are fertilized externally, of course no organ for the transfer of sperms from the male is necessary. Even in salamanders, where fertilization may commonly be internal, no direct transfer occurs, but the male may deposit ***spermatophores*** (small clusters of sperms, Fig. 275) during courtship, to be picked up by the cloaca of the female. One primitive frog, *Ascaphus* (Fig. 276) has a tubelike extension of the cloaca in the male, by which internal fertilization is accomplished, while male caecilians, everting the cloaca, do likewise.

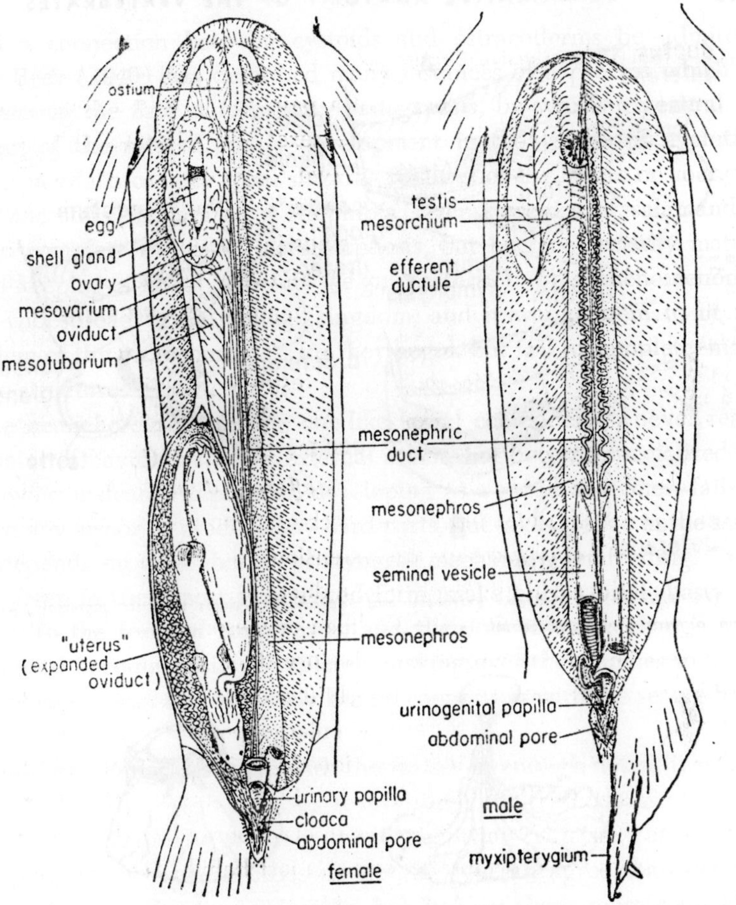

Fig. 272. Urinary and genital organs of shark, *Squalus acanthias.*

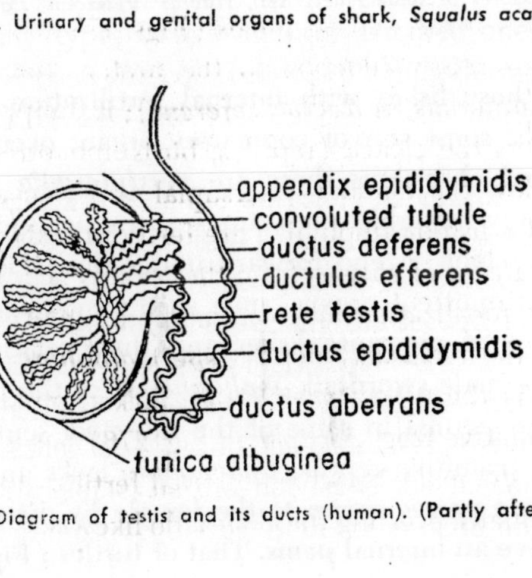

Fig. 273. Diagram of testis and its ducts (human). (Partly after Romer.)

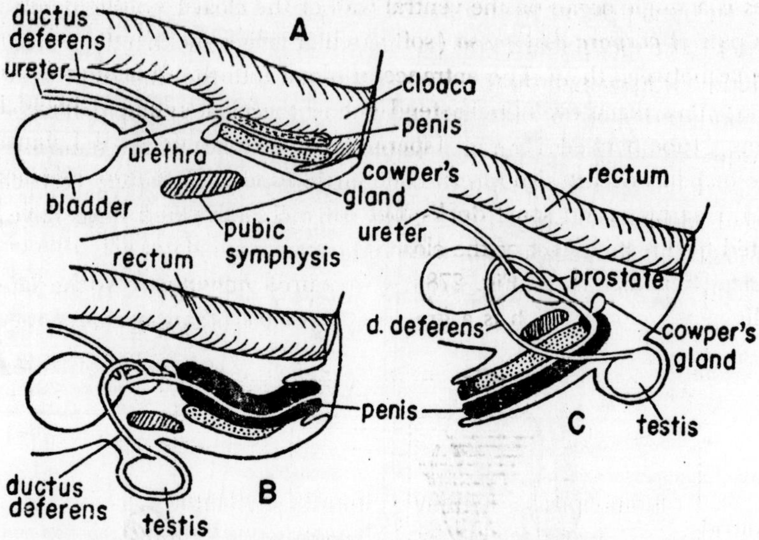

Fig. 274. Relationship between excretory and reproductive ducts in male mammals. A, monotreme; B, marsupial; C, placental.

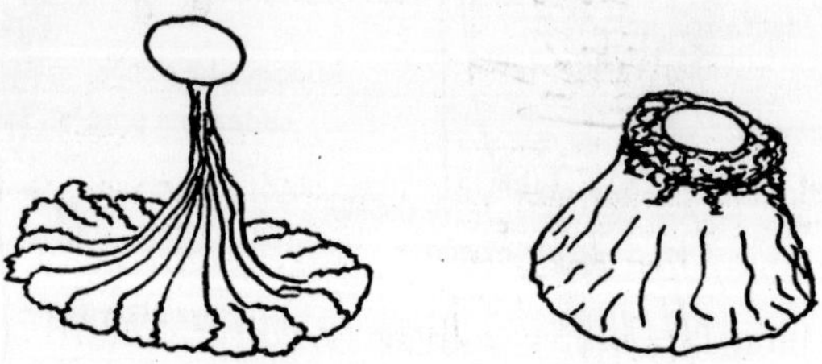

Fig. 275. Spermatophores of salamanders. Left, *Triturus viridescens;* right, *Desmognathus fuscus.* (After Noble.)

But among those fishes with internal fertilization, including all the viviparous kinds, some sort of copulatory organs occur in the males. In sharks, rays, and chimaeras these are ***myxipterygia*** (Fig. 14), special grooved and spiny fingerlike projections of the pelvic fins, supplemented by ***siphon sacs*** which extend forward under the abdominal skin. Many teleosts have intromittent organs, such as the modified anal fins (***gonopodia,*** Fig. 277) of mosquito fishes and their relatives, the extended caudal fin of the male swordtail, *Xiphophorus,* or the penislike extension of the urogenital papilla in some of the tide-pool sculpins.

Although the primitive reptile, *Sphenodon,* lacks any intromittent device and the male passes sperm to the female by direct cloacal contact, most reptiles have an internal ***penis.*** That of turtles (Fig. 278) and crocodiles is a single organ on the ventral wall of the cloaca, consisting mainly of a pair of *corpora cavernosa* (soft vascular ridges) with a deep *urethral groove* between them. The entrance of blood into the sinuses of the corpora makes them swell and extend out of the cloaca, while the groove forms a tube between, to carry sperm into the cloaca of the female. This type of penis occurs also in primitive birds, especially Ratitae (ostriches, etc.), tinamous, and some ducks, but the more advanced birds have reverted to direct contact of the cloacas.

Lizards and snakes (Fig. 278) have paired *hemipenes* in the lateral walls of the cloaca; each is a grooved, fingerlike organ armed at the tip

Fig. 276. Male of tailed frog, *Ascaphus truei*, showing extension of cloaca, a copulatory device, not a tail. (From *Biology of the Amphibia* by Noble.)

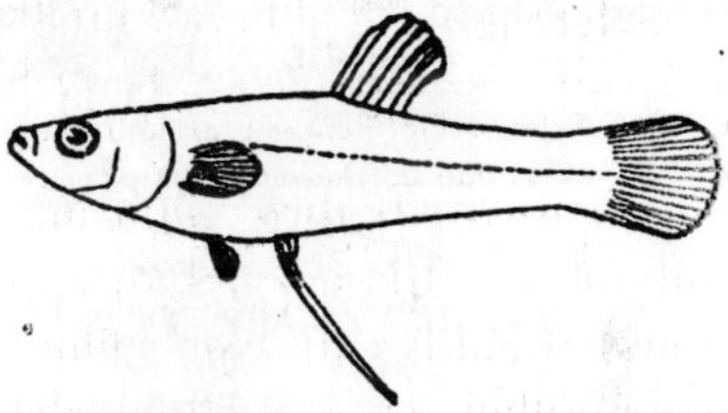

Fig. 277. Top minnow, *Gambusia*, showing male gonopodium. (Redrawn from *General Zoology* by Storer.)

with short spines, when everted, but otherwise it is kept in a pocket, with the outer surface turned inward like the inverted finger of a glove. The ornamentation of the hemipenis is used to some extent in classification.

Monotremes (Fig. 274A) have a partial but not complete separation of the penis from the wall of the cloaca, where it is withdrawn into a ventral pocket; instead of receiving the whole urethra it receives only a ventral division of the latter. The cloaca then disappears in higher mammals with the separation of urogenital and digestive canals. The urethra traverses the penis as a tube rather than a groove; in addition to the corpora cavernosa there is a *corpus spongiosum*, carrying the urethra, and ending usually in a slight expansion, the *glans*. This is forked in the opossum and some other marsupials, as is the vagina of the female. A *sheath* may invest the whole penis, or, in man and a few others, the glans only is covered with a *prepuce*. Many carnivores and other groups have an

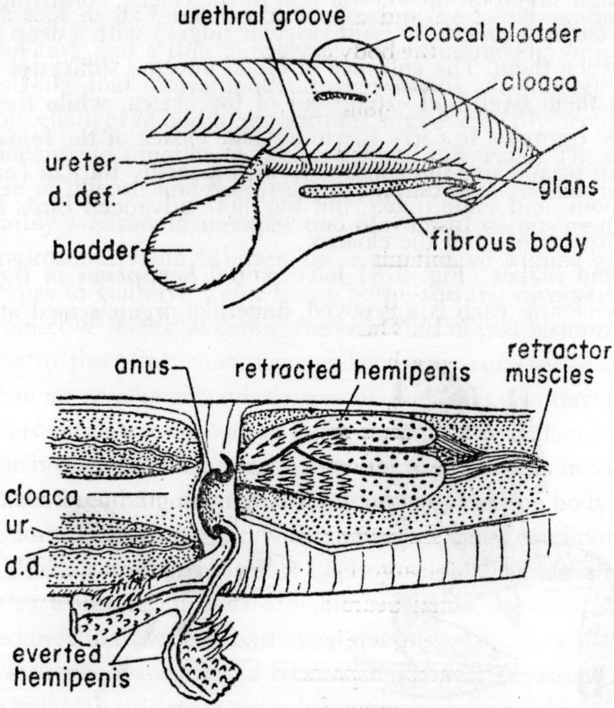

Fig. 278. Upper: cloaca and penis of tortoise. (Redrawn from Romer.) Lower: hemipenes of rattlesnake, *Crotalus*. (Redrawn from Kingsley, *Outlines of Comparative Anatomy of Vertebrates*, 3rd ed., Blakiston.)

os priapi or penis bone as a stiffening rod; it develops in the corpus spongiosum, not from the pelvis.

We should refer here to the descent of the testes into the scrotum in most mammals. The primates, ungulates, most carnivores, and the marsupials have a complete and permanent migration of the testes from their anterior, internal position back into *inguinal sacs* (pockets of the body cavity, lined with peritoneum), which are then enclosed in a common sac of skin, the scrotum. The inguinal canals then close, except that the sperm duct and accompanying artery, vein, and nerve must return to the abdomen from each testis. It has been found that the production and activity of sperms is hindered by the heat of the body, but is normal under the slightly lower temperature of this external position; this justifies but of course does not explain the descent. In some of the rodents, a few carnivores, bats, and the camel family, descent of the testes occurs during the breeding season only, and they are otherwise withdrawn. Monotremes, whales, elephants, and a few others have no scrotum and the testes never migrate out of the body cavity.

Female Reproductive Organs

The *ovaries*, each encased in a fold of peritoneum, lie usually just anterior to the kidneys, and are paired except in birds (left only) and a few fishes (two fused into one in perch; left only in *Mormyrus*). The tissue of the ovary contains no tubules, but the eggs develop within capsules or follicles consisting of a sphere of cells, with or without a fluid-filled cavity (Fig. 271). The ovarian cells which do not become eggs are occupied in producing a hormone, estrone, which helps bring about the secondary female characteristics including behavior.

In the breeding season ovaries may become much enlarged, especially in the animals that lay very numerous eggs. Some marine teleosts produce millions of eggs in a season. Generally the more numerous the eggs, the smaller they are and the more random is their fertilization. But animals producing extremely large eggs (sharks, reptiles, birds) can have very few, and if the animal bears its offspring in an advanced stage of development the number frequently comes down to half a dozen or less.

Since a layer of peritoneum separates the ovary from the coelom, and since, ordinarily, no duct emerges from the ovary itself, the eggs break out into the body cavity and are then obliged to find a way out. This is through the genital pores in cyclostomes and also, no doubt secondarily, in Holostei and a few of the lower teleosts. The problem is solved in most teleosts (Fig. 279C) by the ovaries themselves making ducts, with the help of added peritoneum, so that the eggs can reach the urinogenital opening without falling free in the coelom. This, however, is specialized and peculiar. Elasmobranchs (Fig. 272), Dipnoi (Fig. 279A), and the tetrapods illustrate what we may call the characteristic arrangement in vertebrates. A pair of ***oviducts*** reaches from the neighborhood of the ovaries to the cloaca, or to the exterior if no cloaca is present.

These oviducts originate in the embryo as ***Müllerian ducts,*** which split off from the pronephric ducts in elasmobranchs, carrying with them the peritoneal funnels, and leaving the remainder to become the Wolffian or mesonephric ducts. Thus Müllerian ducts, thereafter, never have anything to do with excretion. They develop from the mesomere proper in higher vertebrates, but there can be no question of the homology.

The peritoneal funnel becomes the ***ostium,*** through which eggs must pass to leave the coelom. Cilia in the coelomic walls and the ostium convey the eggs until they enter the ducts, and in a few mammals the ovary is so completely enclosed that the eggs do not escape into the coelom. Fertilization, when it is internal, takes place near the upper ends of the oviducts; an eggshell may then be added, or gelatinous capsules, if the eggs are to be deposited outside the body.

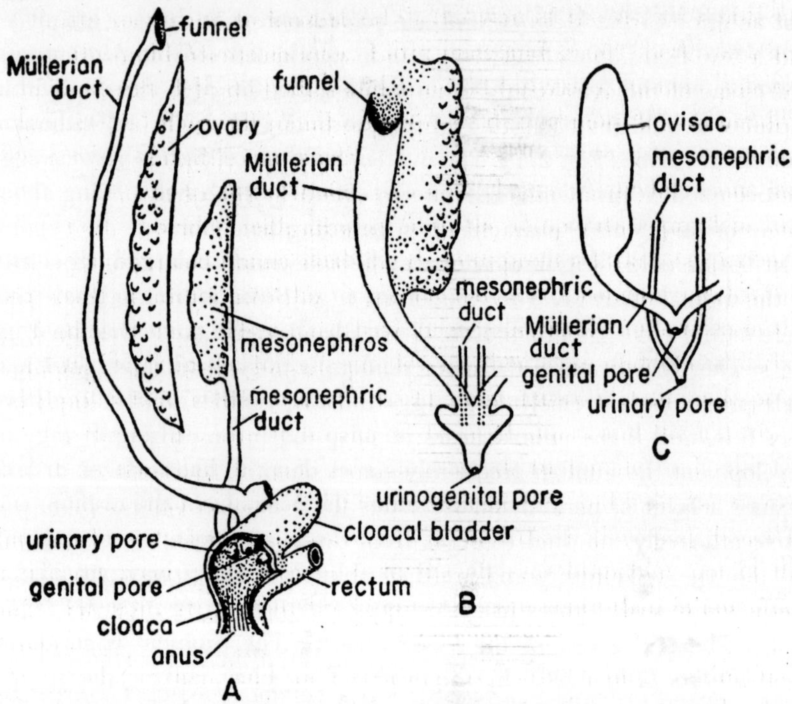

Fig. 279. Female reproductive and excretory systems in various fishes. A, *Protopterus;* B, *Amia;* C, a teleost with closed ovisac and no cloaca. (Redrawn from Goodrich.)

Several kinds of sharks are viviparous, and have therefore only a thin, temporary eggshell, or none; the shell gland persists as a swelling in the oviduct. When the fertilized eggs pass part way down the ducts, the latter become greatly expanded and retain them while the embryos develop far enough to consume the great quantity of yolk with which they are provided, so that they can be born fully equipped for active life. Such an expanded oviduct, usually with vascular folds internally for respiratory exchange with the embryos, is called a "uterus," by analogy with that of mammals (Fig. 272).

Fusion of the lower ends of the oviducts may occur in Amphibia, and does so in mammals, except monotremes. Birds, with only the left ovary, keep the left oviduct but have lost the right. Müllerian ducts are present as useless rudiments in the males of lungfishes, newts, and some frogs.

Now when the urinary and genital systems separate from the rectum, so that a cloaca no longer exists (marsupials, placentals), the lower end of the united oviducts plus a part of the separated cloaca becomes the *vagina,* differentiated by thinner walls from the ***uterus*** above it, and adapted to receive the penis in copulation (Fig. 280). It may be largely divided, as in marsupials, and among most mammals it serves also as an excretory outlet, since the urethra meets it internally. But the urethra has become separate from the vagina in Primates.

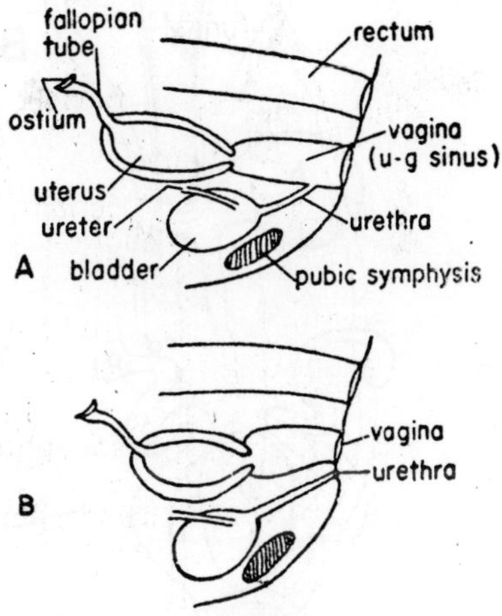

Fig. 280. Relationship between vagina and urethra in female mammals. A, most placental mammals; B, primates, rodents.

The uterus of mammals, then, is the paired or united muscular part of the oviducts, in which the embryos can be carried until birth. In the lower types, with large families, the two horns of the uterus are well separated, each bearing a series of embryos (Fig. 281). If, as in rodents, the horns enter the vagina separately, the uterus is ***duplex;*** if they fuse at the base but are largely separate, it is ***bipartite*** (carnivores); if the fusion is more complete, so that the lower ends can form a common chamber containing embryos more or less in contact, it is ***bicornuate*** (ungulates); and if the uterus appears as a single body, with one cavity holding normally one embryo, it is the ***simplex*** type (apes and man). In all cases the smaller, anterior portions of the ducts, where fertilization takes place but the embryos do not attach, are the ***Fallopian tubes.***

SECONDARY SEX CHARACTERS

Features distinguishing adult males and females from each other, but not a part of the reproductive system, occur commonly in all major groups of vertebrates, but are by no means universal. Their physiological basis is in the hormones originating in the gonads, and therefore such characters must await the development of this hormonal secretion before

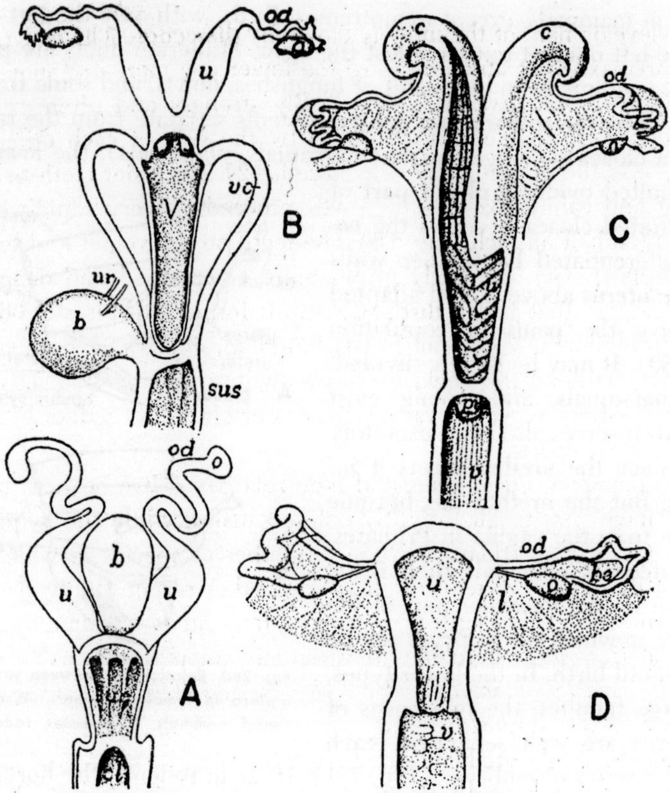

Fig. 281. Uteri of (A) monotreme, *Ornithorhynchus;* (B) marsupial, *Halmaturus;* (C) placental, sheep; (D) *Inuus.* (From Kingsley, *Outlines of Comparative Anatomy of Vertebrates,* 3rd ed., Blakiston.)

they can appear. Often elaborate differences of behavior come at the same time, and under the same stimulus. The adaptive uses of these characters are so diverse and overlapping that the best classification seems to be simply by the natural relationships of the animals that show them.

Among fishes other than teleosts, secondary sex characters are rare. The claspers (myxipterygia) of male elasmobranchs have been mentioned already, as they might justly be considered primary. In Holocephali there is, also, a curious club-shaped hook on the forehead of the male (Fig. 15), armed with toothlike scales, and used as an accessory clasper.

In teleosts a great number of kinds show striking differences between the sexes. These are well known in salmon (Fig. 20, male), for instance. Jordan (1925) writes: "As the season advances the difference between the males and females becomes more and more marked, and keeps pace with the development of the milt, as shown by dissection. The males have (1) the premaxillaries and the tip of the lower jaw more and more prolonged, both of the jaws becoming finally strongly and often extravagantly hooked, so that either they shut by the side of each other like shears, or else the mouth cannot be closed. (2) The front teeth become very long and canine-like, their growth proceeding very rapidly, until they are often half an inch long. (3) The teeth on the vomer and tongue often disappear. (4) The body grows more compressed and deeper at the shoulders, so that a very distinct hump is formed; this is more developed in the humpback salmon, but is found in all. (5) The scales disappear, especially on the back, by the growth of spongy skin. (6) The color changes from silvery to various shades of black and red, or blotchy, according to the species."

The top minnows, Poeciliidae, and their relatives ("live-bearers" to the aquarist) have very small, brightly colored males, while the oviparous killifishes (Cyprinodontidae) include a number of species in which the male is brilliant, a pugnacious fighter, and takes responsibility for nest building and care of the eggs and young. Many kinds of minnows (Cyprinidae) and darters (Percidae) develop gay colors in the males, especially in the breeding season, to which may be added epidermal tubercles on the head, as in *Campostoma*, the stoneroller minnow. Male sticklebacks (Gasterosteidae) have a special mucous gland in the ventral abdominal wall, used in gumming together bits of plants to make a barrel-shaped nest.

It is significant that these features are common in fresh-water fishes but rare in marine kinds, an exception being the brood pouches of male pipefishes and sea horses, which live among seaweeds and eelgrass. These abdominal pouches consist mainly of a pair of lateral folds enclosing a groove, in which the male carries the eggs until they hatch (Fig. 282).

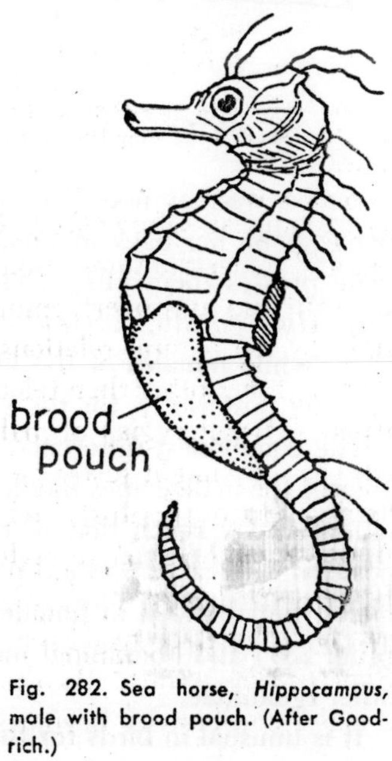

Fig. 282. Sea horse, *Hippocampus*, male with brood pouch. (After Goodrich.)

In tetrapods secondary sex characters are so general as to be rather the rule than the exception. We must judge by modern types, however, for many of the characters do not affect the skeleton, and therefore the sexes can seldom be distinguished in fossil vertebrates. Only a few examples need be used in each large group.

Males of the terrestrial newts, such as *Triturus* in North America and *Triton* (Fig. 283) in Europe, on their return to the water for breeding, develop nuptial colors, enlarged tail fins, stout hind legs and a swollen cloacal region. Male frogs (*Rana*) have enlarged thumbs which aid in clasping the sides of the female. Practically all male frogs and toads, except Ascaphidae, the bell toads, have single or paired vocal sacs for their "spring chorus." While it seems that almost all these characters are limited

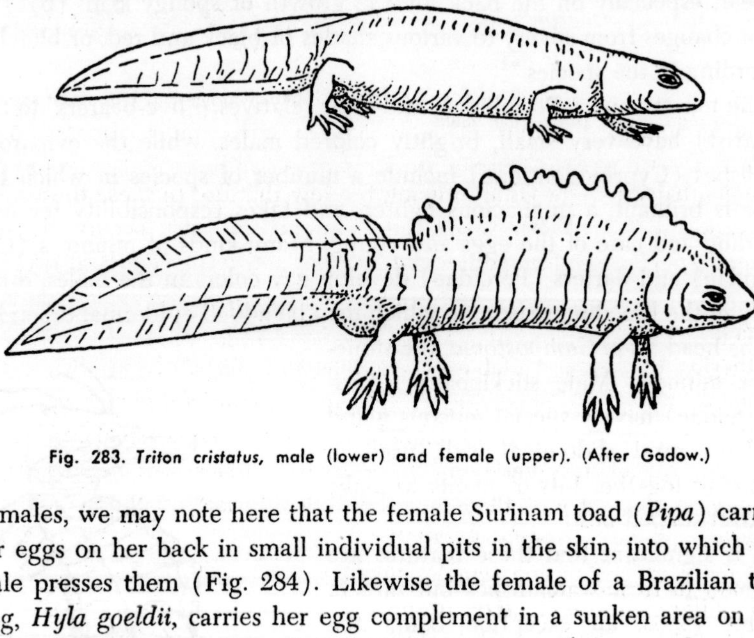

Fig. 283. *Triton cristatus,* male (lower) and female (upper). (After Gadow.)

to males, we may note here that the female Surinam toad (*Pipa*) carries her eggs on her back in small individual pits in the skin, into which the male presses them (Fig. 284). Likewise the female of a Brazilian tree frog, *Hyla goeldii*, carries her egg complement in a sunken area on the back, while females of the Oriental genus *Nototrema* have a dorsal sac opening posteriorly, for the same function.

The commonest difference between male and female turtles is that the anus of the male comes much farther out on the tail, in order that in copulation it may reach that of the female, from its mechanically awkward position. In lizards femoral pores (see Chapter III) are always present in males, but reduced in females, while colorful throat patches, throat fans (Fig. 285) and abdominal markings are common in males, as *Anolis* and other Iguanidae.

It is unusual in birds for the sexes to look alike; penguins, gulls, crows,

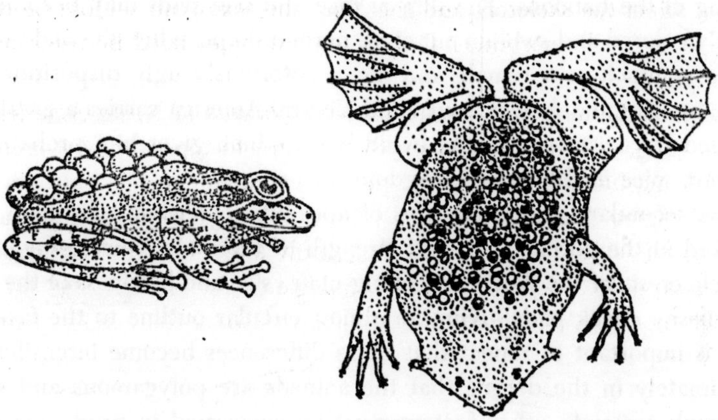

Fig. 284. Females of *Hyla goeldii* (left) and *Pipa pipa* (right), showing modifications for carrying eggs. (*Hyla* after Gadow; *Pipa* after Noble.)

dippers, hawks, and owls are among the few that do. In these the voices, too, are alike, but where a marked difference in voice occurs, there is generally a corresponding difference of color and form. Nearly always the male is larger, more showy, and more vocal. In grouselike birds and many others, he has combs, wattles, crests, ruffs, throat pouches, or ornamental feathering on any part of body, wings, or tail. Birds of Paradise in the Papuan region carry the feather development to incredible extremes. The multitudes of song and perching birds have conservative females in which the ancestral patterns and colors of their families seem reluctant to change, in contrast with the exuberant nonconformity of the males.

With mammals, aside from the mammary glands of the female, the main differences are in size, proportions, voice, and special fighting equipment. Few of these occur in marsupials, insectivores, bats, rodents, or lagomorphs, but they reach an extreme in the polygamous social orders, especially Primates, artiodactyls, and the pinnipeds (sea-lions, walrus, etc.). In elephants and swine, large tusks are a mark of the male; in deer, antlers; in sheep, goats, and cattle, the size of horns, or the hump; in lions, the mane. But in monkeys and apes we find a spectacular assortment of anatomical differences, culmi-

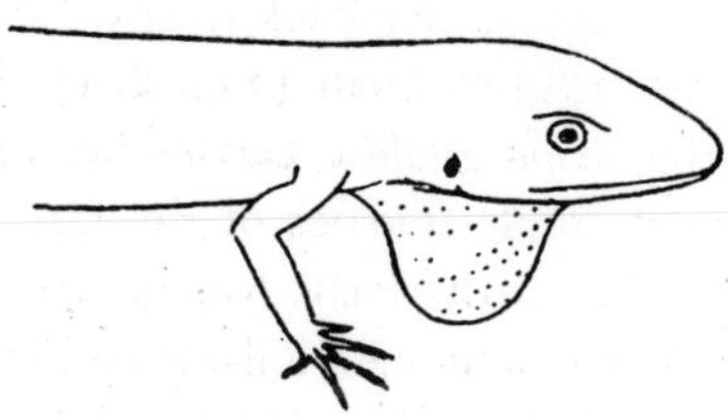

Fig. 285. Throat fan of a male lizard, the "American chameleon," *Anolis*.

nating in the baboons. The skin of face and seat pads may become brilliantly colored and swollen out of all normal shape, while the male weighs twice as much as a female and has a notoriously ugly disposition. The male howling monkey of South and Central America carries a greatly expanded larynx, involving the hyoid bone, which gives him probably the loudest voice in the animal kingdom.

The secondary sexual features of apes are somewhat as in man, most marked in the gorilla (size and strength of the male, voice, exaggerated muscle crests of the skull) and orang-utan (size and hairiness of the male, and fleshy cheek pads giving an almost circular outline to the face).

It is important to repeat that these differences become intensified approximately in the degree that the animals are polygamous and social, although certainly other factors must be concerned in producing them.

CHAPTER X

Sense Organs

It is a familiar idea that animals which are more advanced in evolution make more different responses to their environment, more precisely, more aptly, than those which are primitive. The rule has exceptions, of course, but usually animal behavior becomes increasingly fitted to a special way of life in each twig of the family tree. This may result in some astonishing abilities, while at the same time it imposes certain limitations. So, too, with the history of any system of organs, and so with the development of tissues and structures in early individual life. This is what "specialization" means—a division of labor and correspondingly increased efficiency.

Nowhere is the rule of specialization more evident than in the study of behavior. The physical basis of behavior in vertebrates is a combination of sense organs (receptors), nervous system (transmitting, associating, coördinating), muscles and glands (effectors), to which we should add the regulating power of hormones. In this chapter we need consider only the first element, but without forgetting the others.

A ***sense organ*** is one which responds to a particular stimulus by starting an impulse along a sensory nerve. A ***stimulus*** is simply any influence from the environment or within the animal which will produce such a reaction; without a nerve impulse we usually have no evidence that there has been a stimulus. But this impulse does not affect behavior unless it, in turn, induces a ***response*** by muscle fibers or the secretory cells of glands, equipped to act when the impulse gets to them. A simple ***reflex act***, carried out automatically, is thus an illustration of the coördinated work of different specialized organs.

As a sense organ is made of special tissues which may concentrate or amplify a source of stimulation, so the nervous system is made of special cells, ***neurons***, able to carry impulses at high speed. Neurons have their cytoplasm and plasma membranes drawn out into enormously long ***fibers***, on the surface of which an impulse is thought to travel as a wave of negative electric charges (but not an electric current). All sense organs are associated with sensory nerves, in which there may be thousands of neurons, but there are also free nerve endings and networks, for example in the skin, which may receive stimuli but are not organs. Many kinds of microscopic organs and receptor cells, however, occur in the skin and internal parts of the body, serving sensations of touch, pressure, heat, cold, pain, etc.

But our concern here is with organs in which a series of significant changes can be traced from the primitive to the higher vertebrates. These are: (1) ***organs of chemical sense:*** taste buds, olfactory organs, Jacobson's organs; (2) ***organs detecting pressure change and movements of the medium:*** ear, lateral line organs, pit organs, ampullae of Lorenzini; (3) ***organs of vision:*** the median and paired eyes.

ORGANS OF CHEMICAL SENSE

Essentials for a chemical sense organ in the vertebrates are: epithelial sensory cells, with water from the environment in contact with them, containing dissolved substances which provide stimuli. This water may be that which surrounds any aquatic animal, or it may be on the moist surface of a mucous membrane, in a terrestrial animal.

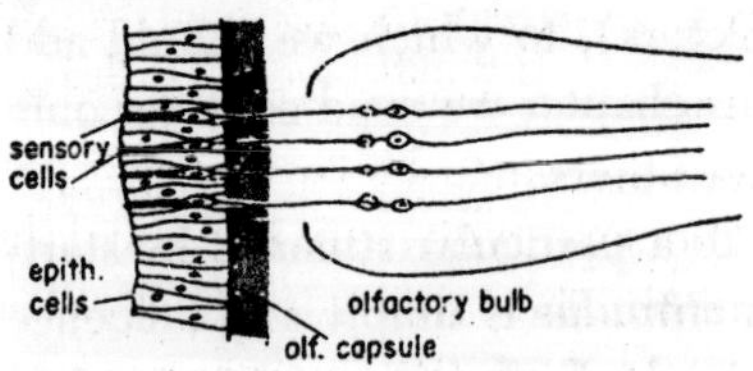

Fig. 286. Diagram of section through olfactory epithelium and olfactory bulb, showing connection of neurons and sensory cells.

The ***olfactory organs*** are pits or open sacs lined with an epithelium of neurosensory cells from the ectoderm, that is, cells which are sensory outwardly, but at their deeper ends produce nerve fibers that synapse with those of the olfactory tract from the brain (Fig. 286). Although the olfactory sacs are paired in gnathostomes, they are largely fused in the cyclostomes, including many early ostracoderms; but it seems, even in the latter, that the fusion is secondary, since the olfactory tracts are always paired. There can be little doubt that the earliest olfactory organs were paired.

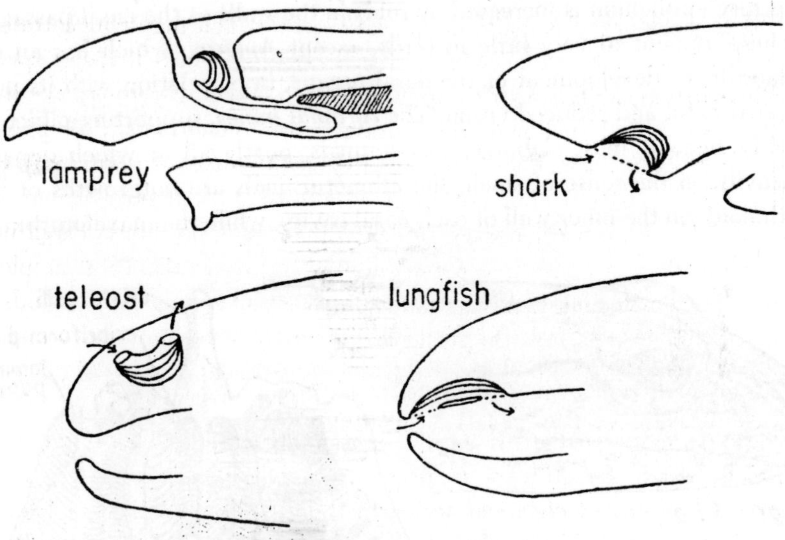

Fig. 287. Diagrams of nostrils and nasal passages in fishes.

The *nostril* of cyclostomes (Fig. 287) is single, opening on top of the head in lampreys and some ostracoderms, but just above the mouth in hagfishes (Fig. 225) and, presumably, the remainder of ostracoderms. At the bottom of its cavity is the *hypophyseal sac*, extending ventrad beneath the brain until it reaches approximately the tip of the notochord; in hagfishes it goes through into the pharynx.

Nostrils of gnathostome fishes generally open into pits lined with upright folds of olfactory epithelium. In Elasmobranchii these openings are incompletely bridged by a flap of skin so that the water flows readily through the anterior part and out the posterior. Among Actinopteri as a rule the bridge is complete, so that there appear to be two pairs of nostrils. The Choanichthyes (Dipnoi and Crossopterygii), are the only fishes in which the nostrils open into the mouth; they show, in the skull, as two notches in the margin of the upper jaw. With the emergence of Amphibia on land these notches withdrew dorsad, closing beneath so that the rim of the jaw became complete and the nasal passages went through the snout. Plethodontid salamanders show a nasolabial groove below each nostril, but it is a special adaptation within the family (Fig. 288). The human abnormality called harelip (usually single, rarely double) results from delayed closure of the gaps in the rim of the embryonic upper lip.

Beginning in reptiles, especially lizards, the surface available for olfactory epithelium is increased by folds in the walls of the nasal passages. These amount to very little in birds, except *Apteryx,* which has an extraordinary development of the nasal organs, in correlation with its nocturnal habit and reduced vision. The ***turbinal bones,*** supporting olfactory folds, become most elaborate in mammals, nearly all of which depend heavily on the sense of smell; the ethmoturbinals are outgrowths of the ethmoid, on the inner wall of each nasal cavity, while the maxilloturbinals

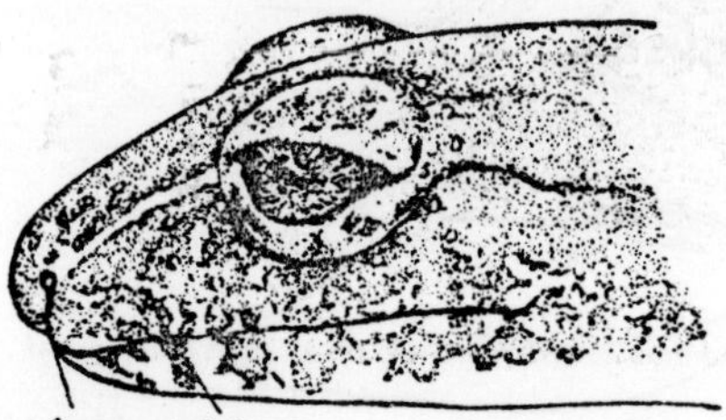

Fig. 288. Head of purple salamander, *Gyrinophilus porphyriticus.* (From *Biology of the Amphibia* by Noble.)

n l gr, nasolabial groove; l l o, lateral line organ.

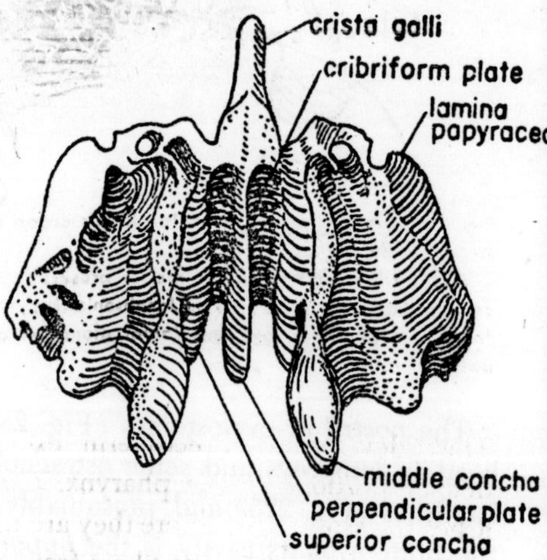

Fig. 289. Human ethmoid bone, from below. (Redrawn from Millard and King.)

are from the maxillaries, on the lateral walls. They are lost secondarily in porpoises and toothed whales. We have alluded already to the separation of oral from nasal cavities in (1) Crocodilia, and (2) mammals and some of the therapsid reptiles, by a secondary hard palate.

Jacobson's organs are a pair of sensory sacs in the posterior part of the nasal cavities of many tetrapods (Fig. 290), whose function is related to, and probably derived from, that of the olfactory organs. Among birds, bats, and primates they are absent, as well as in aquatic reptiles and mammals, while they enter the mouth, rather than nasal cavities, in Squamata. The part of the olfactory tract and lobes innervating these organs is somewhat separated from the rest.

Taste buds (***gustatory organs***) are small clusters of sensory epithelial cells inside, and sometimes outside, the mouth; external taste buds occur in many fishes (but not elasmobranchs) and some aquatic Amphibia.

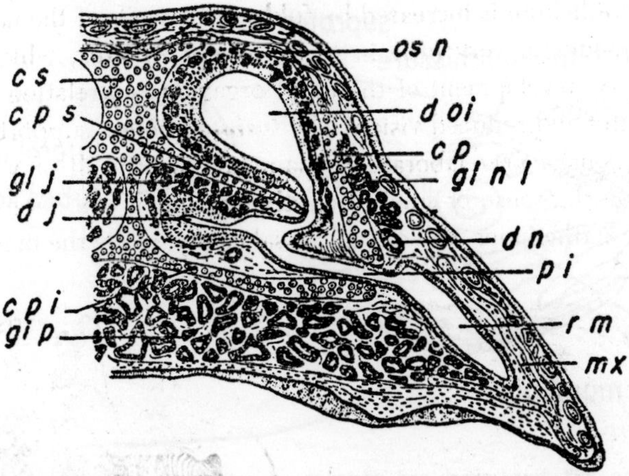

Fig. 290. Transverse section of Jacobson's organ and nasal cavity of tree frog. (From *Biology of Amphibia* by Noble.)

c p, cartilage plate; c p i, interior process from septum; c p s, superior process from septum; c s, cartilaginous septum; d j, Jacobson's duct; d n, nasal duct; d ol, olfactory passage; gl j, Jacobson's gland; gl n l, nasolabial gland; mx, maxillary; os n, olfactory nerve; p i, inferior passage; r m, maxillary recess.

They develop from ectoderm except in the deeper portions of the pharynx, and in some fishes the esophagus, where they are presumably endodermal. To them go fibers from the appropriate cranial nerves: seventh if external or in the buccal cavity, ninth and tenth if in the pharynx. Although more widely dispersed in fishes, taste buds are most specialized in mammals (Fig. 291), which give their food a more careful investigation before swallowing it than is customary in other groups. The tongue and soft palate, epiglottis, and opening of the larynx are the usual locations.

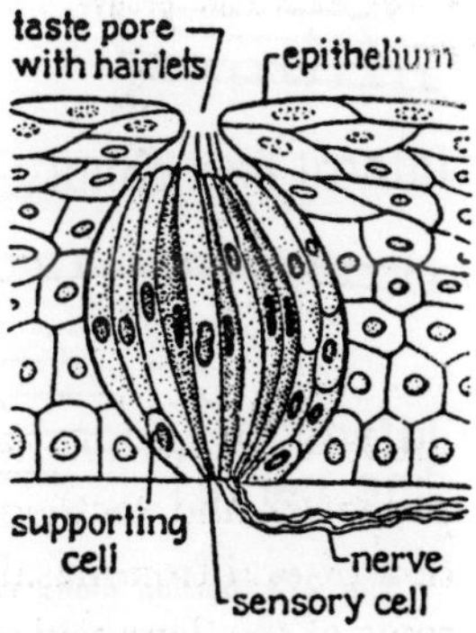

Fig. 291. Section through taste bud in man. (Redrawn from Parker, *Smell, Taste and Allied Senses in the Vertebrates*, Lippincott.)

ORGANS DETECTING PRESSURE CHANGE AND THE MOVEMENT OF THE MEDIUM (NEUROMAST SYSTEM)

This group of organs has in common a structure, the ***neuromast***, of sensory epithelial cells ending outwardly in hairs, which are affected by motion of a liquid medium covering them. Fibers of sensory cranial nerves meet these cells at their deep ends. The motion of the medium

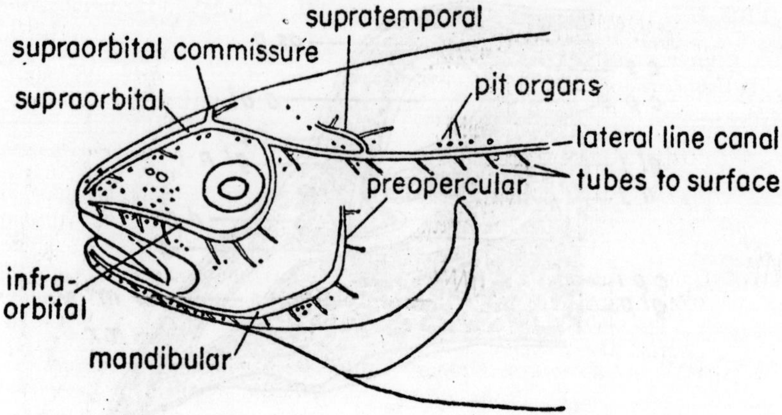

Fig. 292. Head canals and pit organs of pollack. (Redrawn from Bridge.)

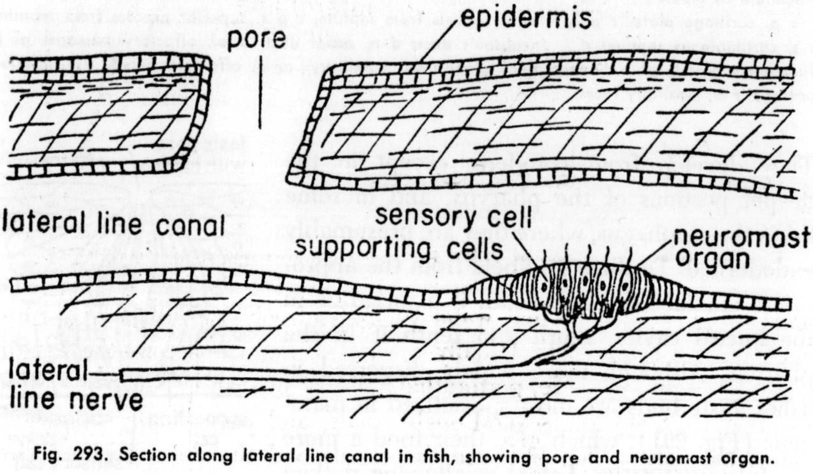

Fig. 293. Section along lateral line canal in fish, showing pore and neuromast organ.

may be either a slow current or vibration caused by pressure waves of high frequency; the medium itself may be water, slime, or endolymph (in the inner ear).

The *lateral line system* is a series of neuromasts arranged over the head and body of fishes and aquatic amphibians in linear patterns (Fig. 292), and supplied by cranial nerves VII, IX and X. In the earliest known fishes, some armored ostracoderms, these organs apparently occupied canals in the skin, the impression of which shows on the dermal bones of the head, or which were buried in the dermal bones, just as in most of the bony fishes today. The canals open to the surface by pores at short

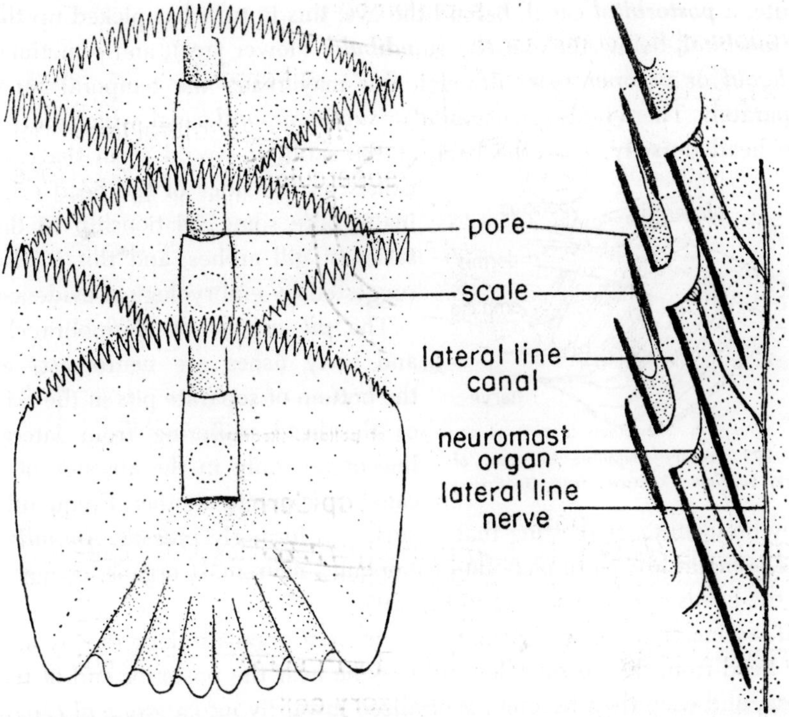

Fig. 294. Lateral line scales of yellow perch, external view and longitudinal section.

intervals (Fig. 293) so that water circulates freely within them in contact with the neuromasts. Usually one lateral line canal runs the length of the body on each side, perforating one row of scales (Fig. 294); in a few teleosts there are several such canals, and occasionally none. Of course the relationship between the canals and the dermal bones (or scales) no longer holds when the bones and scales are absent or too deep beneath the skin. The canals may then, as in most Elasmobranchii, modern Dipnoi, and many smooth-skinned teleosts, lie embedded in the skin alone, or form simply an open groove (Holocephali, Fig. 15, and the frilled shark, *Chlamydoselachus*), or disappear, allowing the neuromasts to lie openly on the surface (modern cyclostomes, some advanced teleosts, and the modern aquatic Amphibia).

The pattern of the lateral line canals is remarkably constant (Fig. 292). A horizontal line runs back from the snout above each eye (***supraorbital***), along the side of the otic region (***temporal***), and continues on the side of the body (***lateral line*** proper). It receives from below, en route, a ***postorbital*** canal, behind the eye; this in turn has picked up the ***infraorbital,*** below the eye, the ***mandibular*** (lower jaw), and sometimes a ***hyoid*** or ***preopercular,*** although this may reach the temporal canal separately. There is also an ***occipital*** or ***supratemporal*** canal, which crosses the head dorsally, near the back of the skull. It would seem that the canals on the sides of the head originally bore some relationship to the series of gill arches, and this is also suggested by embryological evidence.

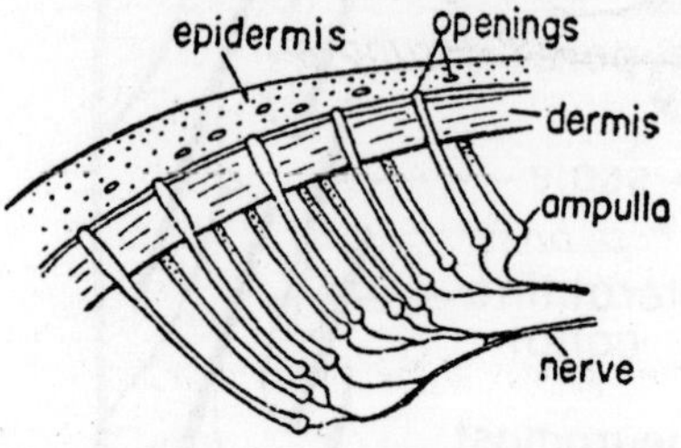

Fig. 295. Sensory ampullae in snout of shark, *Scyllium.* (Redrawn from Goodrich.)

The ***pit organs*** of elasmobranchs and bony fishes are neuromasts at the bottom of separate pits in the skin of the head, differing from lateral line organs only in the absence of a canal from one to another. Frequently they lie in rows, suggesting that a canal was formerly present. ***Ampullae of Lorenzini*** are more elaborate, compound neuromast organs, common on the heads of elasmobranchs (Fig. 295).

If we imagine a pair of primitive neuromast organs sinking deeply into the head from the dorsal side until they lie near the posterior end of the brain, and each then becomes specialized inwardly into a group of canals and hollow vesicles, with three or more neuromasts, we will have the plan of the ***inner ears.*** In sharks and rays the ***endolymphatic ducts*** by which these organs still connect with the surface are visible externally, but they are closed in all others, including extinct and modern cyclostomes. This suggests that the retention of an opening is due to a lag in development and is not primitive.

The inner ear of a cyclostome contains a small ***endolymphatic sac,*** out of which open the ***utriculus*** and ***sacculus,*** each equipped with a ***macula*** (modified neuromast) for detecting movement through stimulation by the endolymph bathing its sensory hairs. There are two ***semicircular canals*** only, anterior and posterior; the horizontal canal of higher vertebrates is absent. This is the case with lampreys and the ancient cephalaspids, as shown by Stensiö (1927), but hagfishes have only one semicircular canal. Each canal has an ***ampulla*** containing a ***crista*** (a modified neuromast), so that it serves to detect changes in equilibrium by the flow of endolymph in the canal as the head tips. The same principle holds in all vertebrates, except that three canals are the rule (Fig. 296).

The semicircular canals and associated vesicles make the ***membranous***

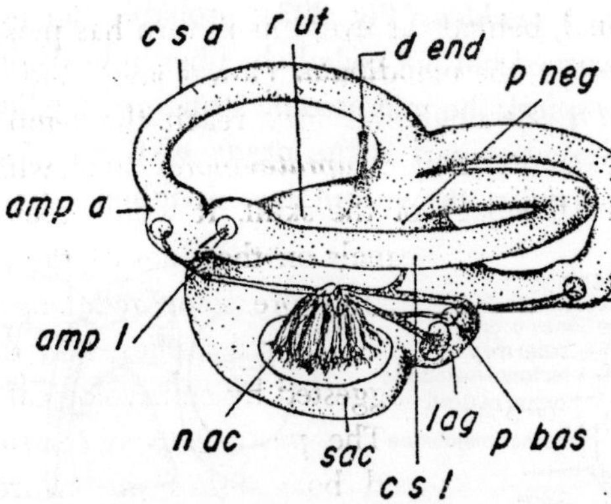

Fig. 296. Inner ear of salamander, *Megalobatrachus*. (From *Biology of the Amphibia* by Noble, after Okajima.)

amp a, ampullae of anterior canal; *amp l*, ampullae of lateral canal; *amp p*, ampullae of posterior canal; *c s a*, anterior semicircular canal; *c s l*, lateral semicircular canal; *c s p*, posterior semicircular canal; *d end*, endolymphatic duct; *lag*, lagena; *n ac*, acoustic nerve; *r ut*, utriculus; *sac*, sacculus.

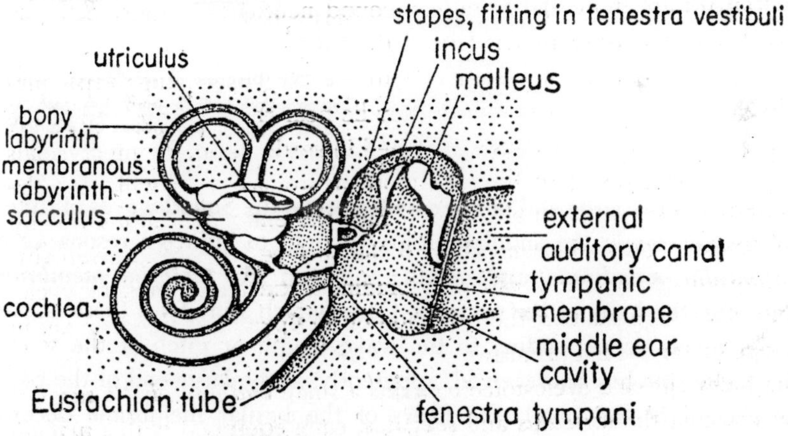

Fig. 297. Diagram of parts of human ear.

labyrinth, within which is the fluid ***endolymph.*** The cartilaginous or bony case enclosing the labyrinth is the ***otic capsule.*** The labyrinth is separated from the surrounding walls of the capsule by a slight gap filled with a similar fluid, ***perilymph*** (Fig. 297).

The principal change in this organ of equilibrium met in higher fishes is the development of a short blind pocket, the ***lagena,*** out of the sacculus. This is still small in Amphibia (Fig. 296, Lag.) and reptiles, but the birds and mammals, with their greater acuteness of hearing, have enlarged the lagena into a *cochlea* (Fig. 297), probably independently of each other. This is a curved but not coiled chamber in birds; in mammals it becomes a snaillike spiral, the microscopic structure of which is amazingly complex. The essential part is the *organ of Corti* (Fig. 298), a highly specialized neuromast drawn out into a ribbon extending along

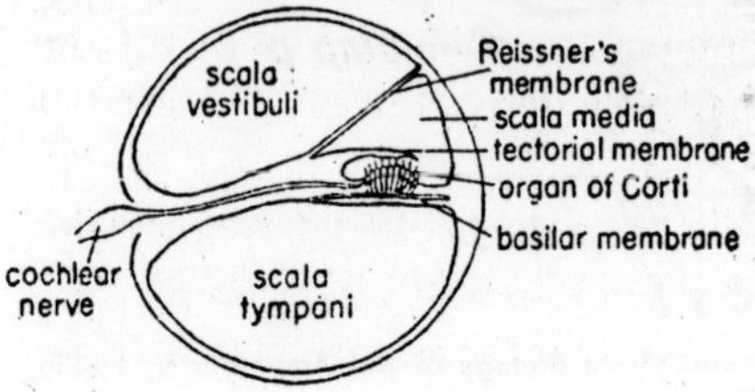

Fig. 298. Cross section of cochlea.

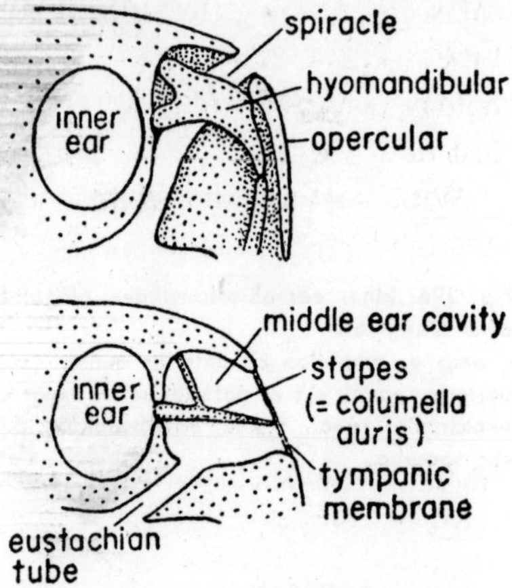

Fig. 299. Diagram of relationship between hyomandibular of a bony fish such as a lobefin (upper), and stapes of a primitive amphibian (lower).

a shelf from the large end of the cochlea to the tip. Sound waves of different frequencies in the endolymph are thought to induce a resonance of the *basilar membrane* underneath, and also the *tectorial membrane* overlying, the hairs of the organ of Corti, which stimulates these at particular levels corresponding to the wave length or pitch of the sound. The higher pitches are received, apparently, at the large end of the cochlea instead of the small; the fibers of the basilar membrane become shorter and finer towards the large end.

Of most significance to us here is the relationship between the inner ear and the eardrum or tympanic membrane, by which the ear becomes an organ for receiving sound waves from the air, as well as an organ of balance. We have spoken already of the transformation of the hyomandibular bone in lobefinned fishes into the *stapes* of Amphibia (Fig. 299). This was accompanied by closure of the spiracle with a *tympanic membrane,* to the inner side of which the stapes (or columella) attached. The inner passage of the spiracle remained as the cavity of the *middle ear* and of the *Eustachian tube* to the pharynx. The stapes therefore bridged the gap from the tympanic membrane to the wall of the inner ear, where presently it came to fit a small opening, the *fenestra ovalis,* and vibrations of the eardrum were thus transmitted to the inner ear. Roughly this is the kind of ear the labyrinthodonts had, and most Salientia still have, but the salamanders and caecilians have lost the middle ear secondarily, at least the cavity, tympanic membrane, and Eustachian tube. (The conditions in modern Amphibia are complex, and their terminology needs revision, but they are evidently derived from the basic type just described.)

With reptiles and birds the middle ear has made no fundamental change, except that the eardrum withdraws into a pit, the ***external auditory canal*** (not so in turtles, where it remains at the surface, nor in snakes, where it is lost). But in mammals two more small bones, the ***quadrate*** of the (reptilian) upper jaw and ***articular*** of the lower jaw, have joined the stapes in transmitting sound (Fig. 159). The quadrate becomes the ***incus,*** the articular becomes the ***malleus,*** and the latter is now in contact with the eardrum. The external auditory canal tends to hide beneath the swelling mass of the cranium, and the ***external ear*** (***pinna***) is added as a sound trap. This reaches its highest efficiency in bats, but becomes reduced or lost in the aquatic orders; the large ungulates often use their external ears as switches to shoo away insects from the canal and neighboring parts.

ORGANS OF VISION, THE MEDIAN AND PAIRED EYES

Paired eyes of vertebrates are cameralike organs of marvelous construction, usually fitted to focus a clear image of the surroundings upon a film of sensory cells, the ***retina,*** and these cells collectively start impulses toward the brain, giving their specific interpretations of intensity, color, or movements in their own parts of the image. But there are also, in some primitive vertebrates, two different median organs which serve as receptors for light, although not to obtain visual images.

According to Walls (1942), "There are indications, from elasmobranch embryology, that the provertebrate possessed a metameric series of paired visual organs on the roof of the head. Most of them rapidly disappeared as the lateral . . . eyes became perfected; but two pairs of dorsal eyes still hung on almost to the cyclostome level. . . ." There is doubt about this hypothesis, but in the lamprey one member of each of these supposed pairs may be seen as a small bulb attached by a stalk to the roof of the brain (diencephalon). The anterior one (***parietal eye***) does not quite reach, but the posterior (*pineal eye*) does reach a semi-transparent spot in the skin of the head. Possibly both bodies were present in primitive Amphibia, for in frogs the pineal is found, nearly reaching the skin, yet in some modern reptiles (*Sphenodon* and a number of lizards) the parietal is present, with the pineal reduced. Among early

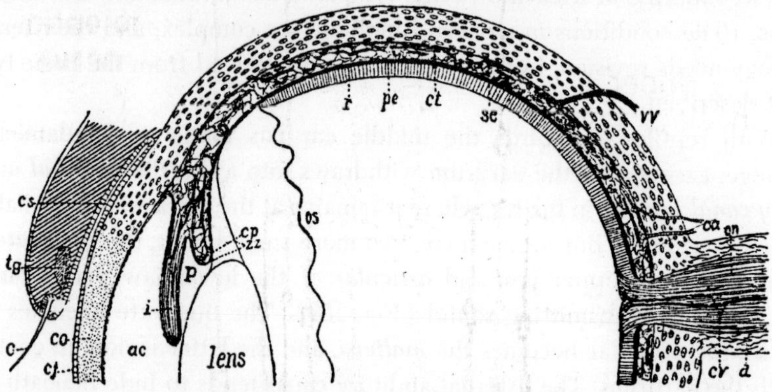

Fig. 300. Partial section through a mammal eye. (After Kingsley, *Outlines of Comparative Anatomy of Vertebrates*, 3rd ed., Blakiston.)

ac, anterior chamber; *c*, eyelash; *ca*, ciliary arteries; *cj*, conjunctiva; *co*, cornea; *cp*, ciliary process; *cs*, conjunctival sac; *ct*, choroid coat; *cv*, central vein and artery of retina; *d*, dura (sheath) of optic nerve; *i*, iris; *on*, optic nerve; *os*, ora serrata (edge of retina); *p*, posterior chamber; *pe*, pigmented epithelium; *r*, retina; *sc*, sclera; *tg*, tarsal gland; *vv*, vorticose vein; *zz*, zonula zinii.

vertebrates the evidence of these organs is simply a foramen in the roof of the skull in ostracoderms, some placoderms, crossopterygians, primitive Amphibia and some of the early orders of reptiles. It usually lies between the parietal bones. The parietal eye of *Sphenodon* is too well covered to be functional, but it contains a retina and lens; sensory nerve fibers go directly down into the stalk from the retinal cells, instead of passing across and between them first as in the paired eyes. In birds these median organs are gone completely, but mammals have a minute ***epiphysis*** or ***pineal organ*** which may prove to be an endocrine gland. (See Figs. 308–314.)

Not related to these, but of entirely different origin, are the ***lateral eyes,*** present and much alike in all vertebrates except a few that dwell in the deep sea, in caverns, or in underground burrows. A brief description of the human eye will provide certain landmarks to use in studying its development and comparative anatomy (Fig. 300). The eye is a nearly spherical ball, movably supported in the ***orbit*** (socket), and connecting

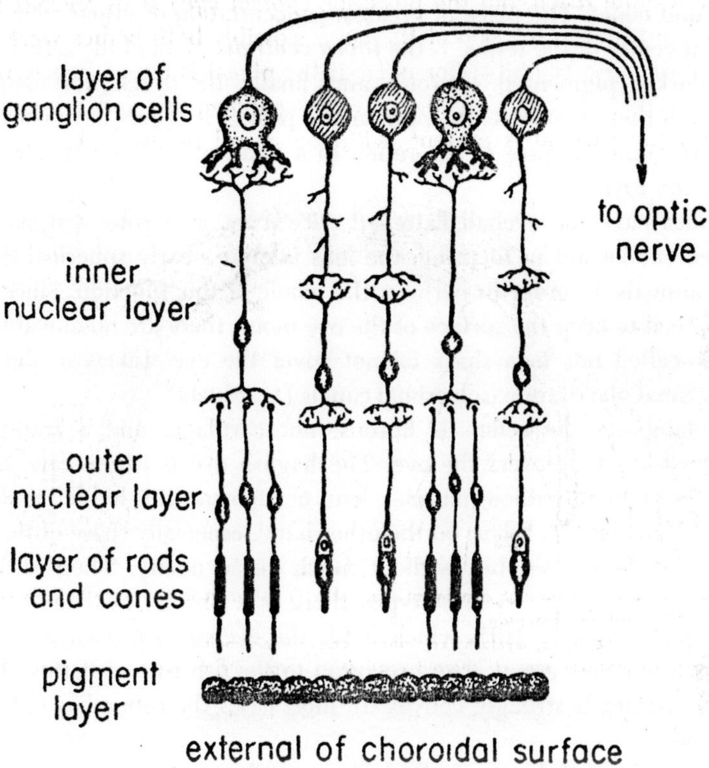

Fig. 301. Diagram of section of retina. (From Best and Taylor, Holt.)

with the diencephalon by the *optic nerve*. A set of six *extraocular muscles* moves the eye (see Chapter VI). Three *tunics* or *coats* form its wall: (1) the outer, which is fibrous and composed of the *sclera* (opaque) and *cornea* (transparent); (2) the middle, which is vascular and made of the *choroid, ciliary body* and *iris,* the last coming partly from the retinal layer; (3) the inner, which is sensory and nervous, the *retina.*

The cornea admits light to the *anterior chamber,* where, by passing through the *pupil* in the center of the iris, it meets the *lens.* Both the cornea and the lens contribute to bending the light rays and hence focusing an image on the retina, but in aquatic animals only the lens can do this, since the cornea has no refractive property when under water. Behind the lens is the *vitreous body,* a clear, slightly viscous fluid transmitting light to the retina. This light has to pass between capillaries on the inner (anterior) surface of the retina, then between nerve fibers and two layers of neurons (Fig. 301) before it reaches the photosensory cells, the

rods and ***cones.*** The place of greatest concentration of visual cells, at the optical center of the retina, is the ***fovea centralis.*** Behind the retina comes the darkly pigmented choroid, and finally the thick fibrous sclera. Through these two coats the optic nerve passes; its fibers have converged over the inner surface of the retina to a point near, but not interfering with, the fovea.

Fishes have the eyeball flattened outwardly, as a rule, and since the cornea cannot aid in focusing, the lens is more nearly spherical than in land animals, in order to perform the whole of this function. Since there is no need to keep the surface of the eye moist, there are no movable lids; the so-called lids in a shark cannot cover the eye. Likewise there are no lacrimal glands, or nasolacrimal canals (tearducts).

In lampreys the sclera is fibrous, not cartilage, and a transparent spectacle of skin covers the eye. The hagfish eye is degenerate, unable to perceive light, without muscles, lens, or pigment; it is hidden under the skin. Elasmobranch fishes, on the other hand, especially those of the deep sea, have large eyes, but without much visual acuity. There is a firm cartilaginous sclera. In most types there is a fibrous cartilage rod, the ***optic pedicel*** (Fig. 210), which holds the eyeball out from the cranial wall within. This group is an exception to the fish rule of flattened eyes, for the cornea is strongly convex. In most cases the retina has only rods, not cones.

Actinopterous fishes typically have a cartilaginous sclera to which are added two or more thin ***sclerotic bones,*** giving additional firmness. This reaches an extreme in some of the big, swift, pelagic fishes, as shown by the bony ring enclosing the eyeball of a tuna or swordfish. On the other hand a few teleosts have neither cartilage nor bone in the sclera. The ***myodome*** is another common feature, a deep funnel-shaped pit in the posterior wall of the orbit, occupied by the elongated rectus muscles. *Amia* and the teleosts have a unique ***choroid gland,*** which is not a gland but a U-shaped mass of capillaries in the back of the eyeball, between the retina and the choroid, supposedly concerned with adjustments of intraocular pressure. Finally, the floor of the eyeball supports a ridge, the ***falciform process,*** which in turn carries the ***retractor lentis*** muscle, used to move the lens inward for accommodation; nothing else like it occurs in other vertebrates.

With Amphibia come movable lids, lacrimal glands, and nasolacrimal canals, but these appear at metamorphosis; the eyes of aquatic stages are essentially fishlike. The modern types have a ***retractor bulbi*** muscle, evidently derived from the external rectus. It serves not only to protect the eye by suddenly withdrawing it, but as an aid in swallowing food, since the eye can press down the roof of the mouth. The eyes of Caecilia are reduced, and a tactile tentacle has partly substituted for them, even using two of the eye muscles for its own movements. Scleral cartilage

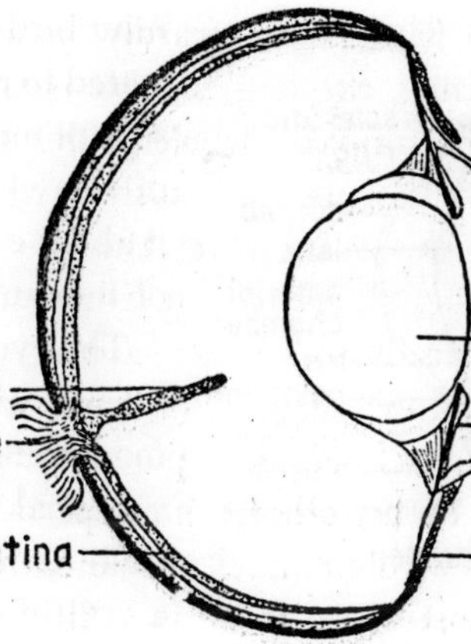

Fig. 302. Lizard eye, showing position of conus, diagrammatic. (Reproduced by permission from *Comparative Anatomy* by L. A. Adams & S. Eddy, published by John Wiley & Sons, Inc., 1949.)

is often absent in frogs; the lens is very large. Accommodation in most land vertebrates is done by marginal pressure on the lens, either to bulge or flatten. In mammals the contraction of the ciliary muscle releases the tension in the marginal fibers which tend to hold the lens a bit flattened, and its own elasticity causes it to become rounder.

A new structure appearing in reptiles, especially lizards, is the ***conus papillaris*** (Fig. 302), a process resting on the head of the optic nerve and pointing forward in the vitreous chamber. It serves as a source of nutriment for the retina, since the latter lacks capillaries. The snakes, derived perhaps from a burrowing type of monitorlike lizard, show in their eyes a remarkable set of reconstructed features, evidently following a period in which the eyes had partly degenerated. Instead of lids, a spectacle of skin covers the surface; the conus has apparently been rebuilt from mesoderm instead of ectoderm, and the lacrimal gland, of course, has disappeared.

The bird's eye is based on that of reptiles (Fig. 303). The conus has become a complex pleated or folded ***pecten***, serving the same function and occupying the same place. A ***nictitating membrane*** supplements the action of the lids, being far more rapid in blinking. The eyes are relatively very large, nearly meeting each other midway in the head, and the eyeball usually has a ring of scleral ossicles besides the cartilage cup. The only bird with a (partly) degenerate eye is the New Zealand kiwi, *Apteryx*. The extraordinary abundance of cones in the retina is responsible for the acute vision of most birds. The cones are estimated to reach one million per square millimeter in the fovea of a hawk, *Buteo*, which gives it about eight times the power of human vision.

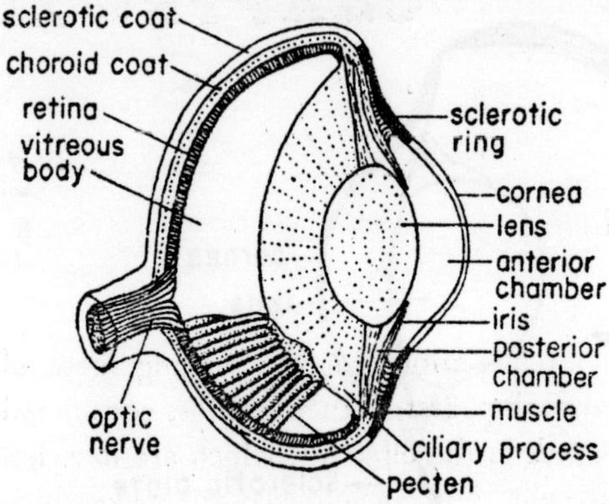

Fig. 303. Eye of chicken, showing pecten. (Redrawn from Adams and Eddy.)

The eye of mammals has no scleral cartilage except in the monotremes, ordinarily is spherical, possesses a retractor bulbi in most types, and has a distinctive pulleylike arrangement for the superior oblique, by which this muscle passes through a ligamentous loop in the anterior wall of the orbit before it goes to the eye. The conus of reptiles persists in a few marsupials, but otherwise is missing. All three major divisions of living mammals show evidence of present or past nocturnal adaptation in the fact that cones are either absent (in most primitive types) or redeveloped in several modern lines independently. In this case they are probably modified rods.

The only mammals in which eyes are vestigial are the burrowing insectivores (moles) and a few Old World rodents of similar habit, such as Spalacidae and Bathyergidae, but the nocturnal types generally have small eyes (bats, shrews, etc.). It is worth noting that the eye, like the inner ear and the brain, does not increase its bulk in proportion to the size of the head and body in very large animals. The human eye is ***relatively*** smaller than that of a mouse, and in a rhinoceros, elephant, or whale it is still smaller.

CHAPTER XI

Nervous System

The nervous system is an incredibly complex arrangement of communicating cells, having relationship, first, to the receptors considered in the previous chapter, and second, to the effectors, which are muscles, glands, and contractile pigment cells. For convenience we make three divisions: the ***central nervous system*** (brain and spinal cord), ***peripheral nervous system*** (cranial and spinal nerves), and ***autonomic system*** (sympathetic and parasympathetic nerves). The central nervous system develops as a tube of ectoderm in the embryo, folded in from the neural plate as we described in Chapter II. The other two divisions grow out more or less directly from this tube, their cells migrating to the final locations. We shall not attempt to consider the many fiber tracts and centers in the cord and brain, for these involve histology and physiological experiment rather than comparative anatomy.

In the ***neuromuscular cells*** of the lower invertebrates, such as *Hydra,* the functions of receptor, conductor, and effector are often combined, and these cells are not localized but generally distributed. Later specialization produces receptors, conductors (neurons) and effectors (Fig. 304). Animals which are sessile (attached to a surface) tend to distribute each of these types radially, and the nervous system becomes a ring with branches. Animals which travel habitually in one direction, and so have a head, tend to group their receptors at the anterior end, and as a result the bulk of the nervous system is also anterior, with nerves going back to the effectors. If the body shows metameric segmentation (annelids and arthropods), there is serial repetition of parts, including the receptors and effectors, and consequently a repetition of both sensory and motor nerves.

But such animals also have a head end (cephalization), especially the advanced arthropods; therefore the series of nerves and nerve centers is not uniform but is strongly developed in the head and diminishes from there back. The body acts increasingly as a unit and less as a chain of organs.

In the chordates the nervous system has a segmental relationship to the muscular system by way of paired nerves, but at the same time it shows pronounced cephalization. The brain of the most primitive known vertebrates is not large, but it is unmistakably a brain and not a series of fused nerve centers. The nervous system of vertebrates does not appear to be derived from segmental nerve chains of invertebrates, and it differs also in being dorsal instead of ventral.

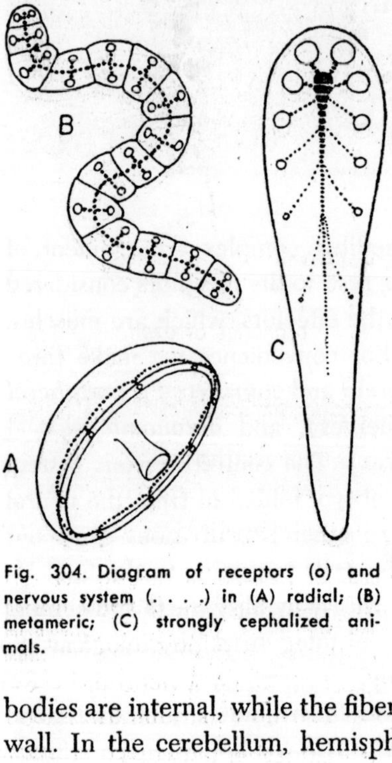

Fig. 304. Diagram of receptors (o) and nervous system (. . . .) in (A) radial; (B) metameric; (C) strongly cephalized animals.

The walls of the embryonic neural tube grow until they are much thicker than the cavity of the tube, and the cells transform into (1) ***neuroglia*** or supporting cells, and (2) ***neurons***, many of which extend their fibers outward in the peripheral nerves. Other fibers grow lengthwise along the tube itself. The cell bodies and the fibers growing from them take up a specific arrangement. In the spinal cord cell bodies are internal, while the fibers lie in tracts along the outer part of the wall. In the cerebellum, hemispheres, and optic lobes of the brain the reverse is true. Since most nerve fibers are enclosed in a white myelin sheath, while the cell bodies are not, it follows that the white matter of the cord and brain is composed of fibers, while the gray matter is the region of cell bodies.

A ***neuron*** (Fig. 305) is peculiar in having its nucleus and most of its cytoplasm confined to a small cell body, out of which run, typically, two or more branching fibers. In one direction goes a ***dendrite*** or a cluster of them; their function is to bring an impulse *to* the cell body. The impulse is carried on, *away* from the cell body, by the other type of fiber, the ***axon.*** Sometimes, as in motor neurons, the axon is much longer than the dendrite (Fig. 306), but the opposite may be true, as with most sensory

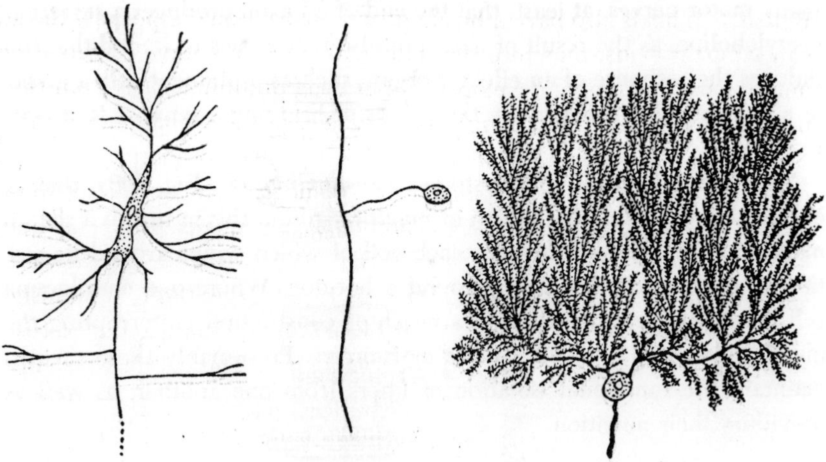

Fig. 305. Examples of neurons.

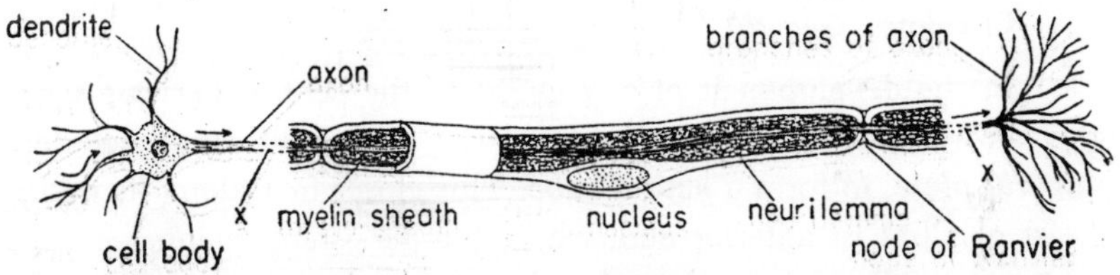

Fig. 306. Neuron with medullated fiber. Direction of impulse indicated by (→); gaps where most of length of fiber is omitted indicated by (X).

neurons. Thus no fundamental difference exists between the two, except in the direction in which they carry an impulse. It follows, too, that normally an impulse can go in only one direction along a neuron, because only the dendrite receives it.

We must be careful to keep in mind a distinction between a ***stimulus,*** which is received by a sense organ or nerve ending, an ***impulse,*** which is a wave of negative charges spreading along a fiber, and a ***sensation,*** which is something perceived, something experienced. The latter is a subject for psychology and has no place, as yet, in functional anatomy.

Since both axons and dendrites may branch at their tips and communicate with the fibers of other neurons, the nature of this connection, the ***synapse,*** is important. The fibers contain cytoplasm, covered by the plasma membrane of the cell, and this covering is complete, even at the end. Therefore a synapse is like the meeting of any two cells, the contact or proximity of two plasma membranes. It has been shown, in the case of many motor nerves, at least, that the end of an axon produces a secretion, acetylcholine, as the result of each impulse that comes to it, and that this induces the response of an effector organ, such as a pigment cell, a gland, or a muscle. That this may be the means of bridging a synapse is at least likely.

Most nerve fibers in gnathostomes are ***myelinated*** (Fig. 306), that is, enclosed in a white fatty sheath of ***myelin.*** Around this in turn is a sheath made of cells, the ***neurilemma,*** each cell of which wraps around its section of fiber much like a roll around a hot-dog. Where one neurilemma cell meets the next one there is a notch or constriction (interrupting the myelin layer also), called a ***node of Ranvier.*** Presumably these sheaths maintain the functional isolation of fibers from one another, as well as providing their nutrition.

CENTRAL NERVOUS SYSTEM

Chordates, General

In the development of any chordate the central nervous system begins, as we have said, by the folding in of an area of dorsal ectoderm, the ***neural plate,*** to form a ***neural tube.*** This closes first along its middle, and last of all at its anterior end, where the persistent opening is called the ***neuropore.*** Posteriorly the tube closes over, but often remains connected temporarily with the archenteron; this connection is the ***neurenteric canal.***

Among protochordates, Amphioxus and the tunicates follow this scheme, and in both of them the neuropore remains open for a while in larval life. When it closes, the cells adjacent to it form, in Amphioxus, the ***"olfactory pit,"*** and in tunicates the ***"neural gland."*** The neural tube of Amphioxus becomes nearly as long as the body, supplying dorsal and ventral segmental nerves to the muscles, but fails to develop a brain. On the other hand, in tunicates the spinal cord is present to supply the body muscles only during the brief free-swimming stage of the larva, and then degenerates when the tail and its muscles disappear, leaving only a ***"dorsal ganglion"*** in the position of a brain.

The wormlike hemichordates have a short neural tube only in the collar. From it extend peripheral nerves, and, unlike any other chordates, a nerve ring around the mouth, with a median ventral nerve going back beneath. The small, sessile Pterobranchia, however, do not develop far enough to have a neural tube, but the neural plate remains external. This is probably not a primitive condition, but the result of an extreme lag in development.

Vertebrates

In all vertebrates the brain forms an enlarged bulb of the central nervous system in the head, while the spinal cord emerges posteriorly and continues down the trunk. Reflex arcs concerned with body and limb muscles are centered, of course, in the cord, but longitudinal fiber tracts run to the brain, where action is coördinated and directed. Besides this, the brain contains the centers of the innumerable sensory and motor functions of the head.

Both brain and cord are wrapped in membranes of connective tissue, the ***meninges,*** lying between the brain and skull and between the cord and the walls of the neural canal in which it lies. Fishes have a single meninx, but amphibians, reptiles, and birds split this into two, an outer ***dura mater*** and inner ***pia mater***, the latter in contact with the brain and cord. A third meninx, the ***arachnoid,*** appears in mammals over the surface of the pia mater. Cerebrospinal fluid occupies the space between the meninges, as well as the cavity of the cord and the ventricles of the brain.

The brain, during its development, grows unevenly, forming at first three divisions: ***prosencephalon, mesencephalon, rhombencephalon.*** (Since the prosencephalon becomes distinct a little before the mes- and rhombencephalon, the latter two are sometimes lumped as ***deuterencephalon.***) These first parts are only temporarily recognizable, for the prosencephalon (forebrain) makes two, the ***telencephalon*** and ***diencephalon***, while the rhombencephalon (hind brain) similarly makes the ***metencephalon*** and ***myelencephalon.*** Thus, with the mesencephalon (midbrain) remaining undivided, there are five major parts. (See Fig. 82.)

The cavity within also expands unevenly (Fig. 307), so that the telencephalon comes to harbor the first two ***ventricles,*** the diencephalon the third ventricle, and the fourth ventricle lies in both met- and myelencephalon, becoming restricted to the latter in animals above fishes; all these connect with one another and with the lumen of the cord. Now the ventricles are spaces inside the brain, but at two places, the roof of the diencephalon and roof of the myelencephalon, they are covered by nothing but a thin loose membrane. The membrane in each case is folded into the ventricle and carries with it a network of small blood vessels, a ***choroid plexus.*** Thus the ***anterior*** choroid plexus hangs into the third ventricle (partly into the first and second), and the ***posterior*** into the fourth.

Connecting right and left walls of the brain, especially in its floor, run crosswise tracts of fibers, the *commissures*, coördinating motor and sensory functions of the two sides. Three parts of the brain roof are especially well developed: the *cerebellum* (roof of the metencephalon, for motor coördination), the *optic lobes* (roof of the mesencephalon, for visual cen-

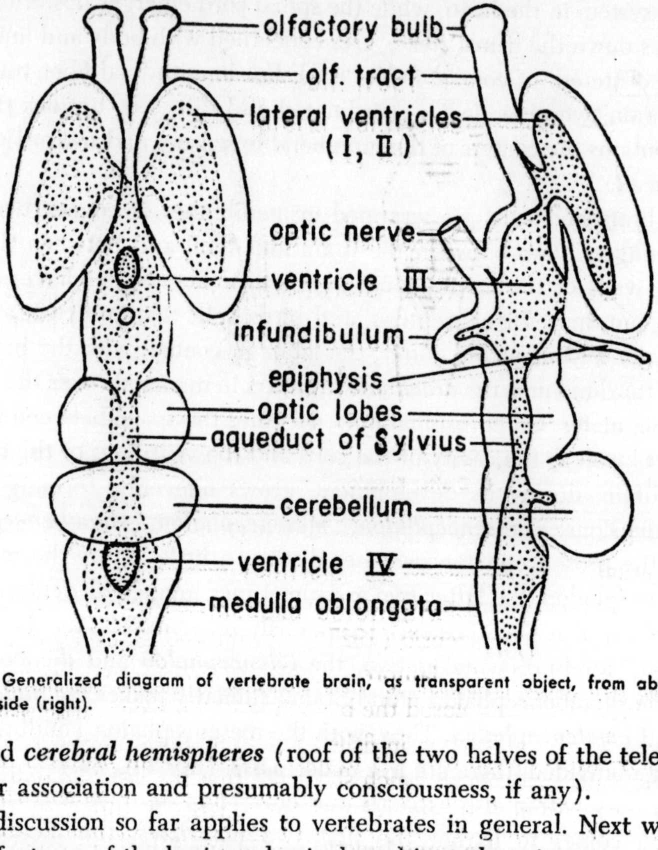

Fig. 307. Generalized diagram of vertebrate brain, as transparent object, from above (left) and from side (right).

ter), and *cerebral hemispheres* (roof of the two halves of the telencephalon, for association and presumably consciousness, if any).

The discussion so far applies to vertebrates in general. Next we shall look at features of the brain and spinal cord in each major group.

BRAIN AND SPINAL CORD

Cyclostomes

The brain in cyclostomes (Fig. 308, A, B) is narrow, elongate, emphasizing the olfactory lobes (ventral part of the telencephalon), optic lobes (dorsal part of the midbrain), and center for equilibrium (metencephalon). The *pallium* or roof of the telencephalon is thin and does not form cerebral hemispheres in any proper sense of the word. Underneath the diencephalon is a ventral pouch, the ***infundibulum,*** which forms the posterior lobe of the ***pituitary gland*** in all vertebrates. But the anterior lobe, the ***hypophysis,*** is at least partly represented in cyclostomes by an open pouch running down under the front of the brain from the nasal sac to the infundibulum; this is not, of course, a part of the nervous system.

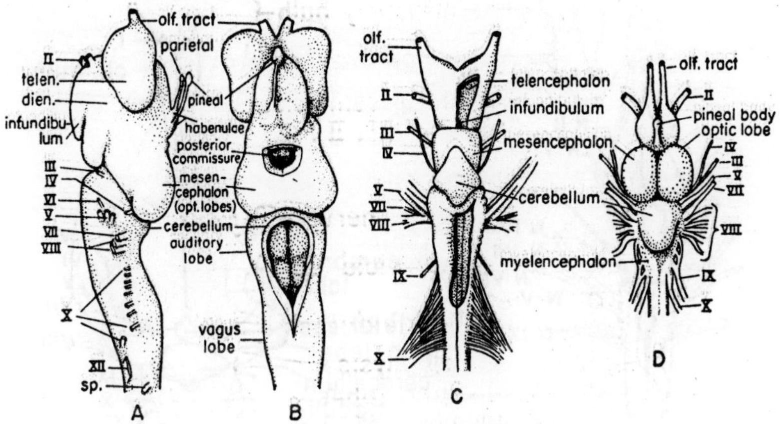

Fig. 308. Brains of fishes. A, lamprey, *Petromyzon,* from side; B, same, dorsal; C, shark, *Heptanchus;* D, pike, *Esox lucius.* (A, B, C after Kingsley; D after Goodrich.)

The cerebellum is very small, as the brain takes no great part in motor coördination.

The brain and cranial nerves of the ancient ostracoderm, *Cephalaspis,* have been restored by Stensiö (1927), who showed that they were much like those of the modern lamprey except for one or two features. The skull, being massive, enclosed the brain and all ten pairs of cranial nerves which are associated with it in fishes. But in modern lampreys and hagfishes the skull has been so much shortened that the last two pairs (glossopharyngeal, IX, and vagus, X) are left outside it, although they are present and perform the same functions as in any fish. Thus some authors refer to cyclostomes as having eight pairs of cranial nerves instead of ten.

The spinal cord in cyclostomes is a flattened, thick-walled tube in which the gray matter occupies the middle part, showing in cross section as an indistinct horizontal band. There are no longitudinal grooves (***fissures***) such as appear in most vertebrates.

Gnathostome Fishes

With fishes above cyclostomes the pallium remains little developed, but there is generally an enlargement of the optic lobes, since the eyes are important (Fig. 308, C, D). With strong development of the lateral line system the lateral lobes of the *medulla* (myelencephalon) may become much swollen. The cerebellum is moderately large in the most active fishes, especially teleosts. The pituitary gland is fundamentally as in higher vertebrates, for the hypophysial pouch closes early in embryonic

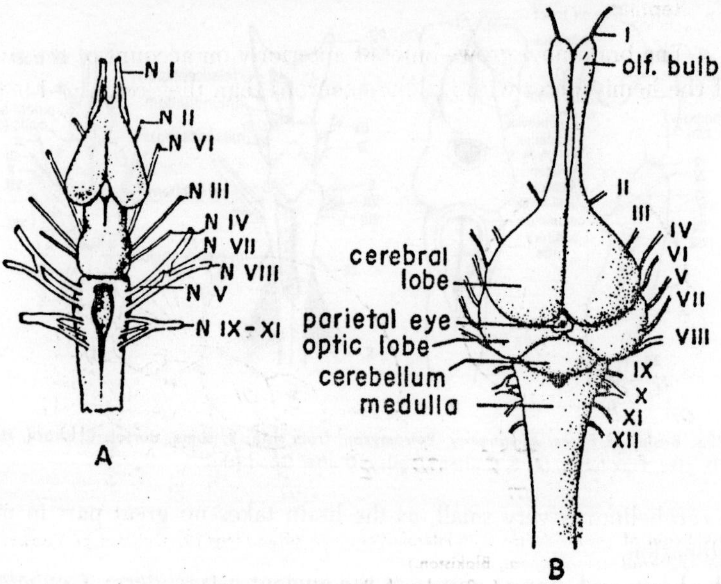

Fig. 309. Brain of (A) a primitive frog, *Pipa pipa*; (B) a lizard, *Sceloporus*. (A from *Biology of the Amphibia* by Noble; B reproduced by permission from *Comparative Anatomy* by L. A. Adams & S. Eddy, published by John Wiley & Sons, Inc., 1949.)

life and becomes the small, round anterior lobe. The spinal cord usually, but not always, has dorsal and ventral fissures.

Amphibia

The most notable change from the fish brain is the thickening of the pallium to make the beginning of cerebral hemispheres. It becomes occupied by neurons and their synapsing fibers, but still lacks the cortical layer of associative fibers. The olfactory and optic lobes are still the most conspicuous parts of the brain, however. The cerebellum is relatively small (Fig. 309A).

We may note here the question of the posterior limits of the brain, as marked by the number of cranial nerves included in the skull. With *modern* Amphibia the number is ten, as in fishes. But they have arrived at this number by reduction of the posterior part of the skull, for in ancestral Amphibia, the labyrinthodonts, 12 pairs were included, just as we find in reptiles, birds, and mammals. Consequently the presence of ten pairs in frogs and salamanders does not indicate relationship to fishes, but a shortening of the skull.

Reptiles

The brain now grows bulbous anteriorly on account of the swelling of the hemispheres with far more neurons than they contained before

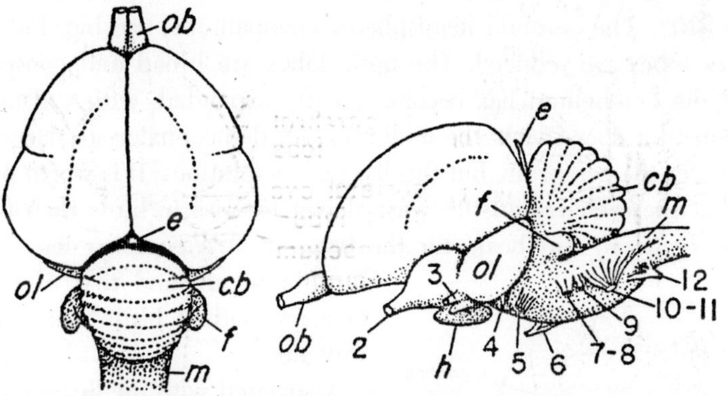

Fig. 310. Brain of goose, dorsal and lateral. (Redrawn after Kingsley, *Outlines of Comparative Anatomy of Vertebrates*, 3rd ed., Blakiston.)
cb, cerebellum; e, epiphysis; f, flocculus; h, hypophysis; m, medulla; ob, olfactory bulb; ol, olfactory lobe; 1–12, cranial nerves.

(Fig. 309B). The cerebral hemispheres are thus well established, although their surfaces remain smooth and they do not conceal the olfactory lobes in front. In some reptiles we find the beginning of a new associative area in the hemispheres, the *neopallium.* The optic lobes are big, for most reptiles have good eyes. The cerebellum remains modest, although larger than in Amphibia. While the reptilian brain in recent types is quite uniform, yet it is known (Edinger, 1949) that "The pterosaurs developed birdlike brains, and mammal-like reptiles had long cerebra with large olfactory bulbs." The reptiles have twelve pairs of cranial nerves, a heritage, as we have said, from the labyrinthodont Amphibia.

It would hardly be correct to regard the modern reptilian brain as ancestral to that of either birds or mammals, as contemporary groups cannot be precisely ancestors or descendants of one another. Rather, all three of these are built upon a previous type met in the earliest reptiles. Therefore the superficial resemblances we may find between bird and mammal brains are quite independent, and go back into the remote past.

There are both dorsal and ventral fissures in the spinal cord of Amphibia and reptiles, but not as deep as in birds and mammals. The gray matter forms an H or X in cross section.

Birds

A bird has big eyes, complex behavior which includes a certain amount of learning, and remarkable motor coördination. But the sense of smell is unimportant. All these facts are reflected in the form of its brain (Fig. 310). The cerebral hemispheres (neopallium) are big, but the olfactory lobes are reduced. The optic lobes are broad and conspicuous, while the cerebellum has become greatly expanded, with a convoluted surface. In a dorsal view the midbrain and diencephalon are largely hidden under the cerebrum, but this has no convolutions. It is worth pointing out that these characteristics were slower to arise in birds than those of wings, feathers, and beak, for the brain of *Archeopteryx* was scarcely birdlike at all, and that of the Cretaceous toothed birds was only part way along.

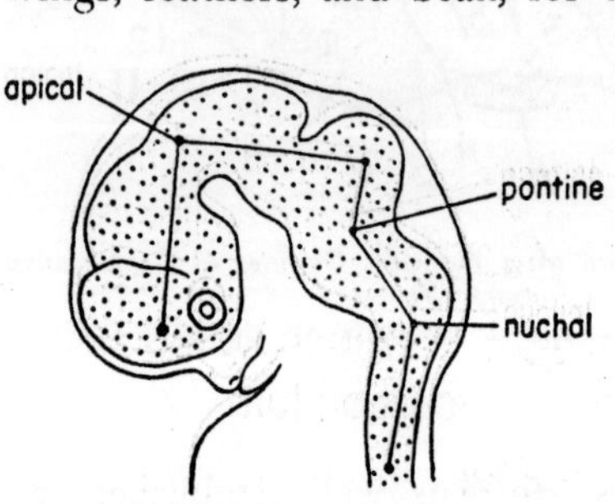

Fig. 311. Flexures of human brain at seven weeks.

Associated with the flexible jointing of the head on the neck, the medulla shows a bend, present also in reptiles and mammals, the ***nuchal flexure.*** In embryonic life, in these three groups, the brain grows relatively fast at the beginning, and to help it fit the cramped space available for it, two other flexures also appear, the ***pontine*** (an upward bend of the metencephalon against the medulla), and the ***apical*** (a downward bend of the anterior parts of the brain, beginning with the midbrain). These flexures become less marked later on. (See Fig. 311.)

Mammals

Although monotremes have a simple, reptilian brain (Fig. 312), the marsupials and especially the placentals show great progress. A characteristic mammalian feature is the ***pons,*** a belt of transverse fibers across the ventral side of the metencephalon. In progressive modern orders the cerebral hemispheres become the largest parts of the brain (Figs. 313C,

314); although they were not so in the early members of these orders. For instance the Eocene horse, *Eohippus,* had a brain much like that of an opossum (Fig. 313 A, B; Edinger, *loc. cit.*). The hemispheres become convoluted in modern ungulates, proboscideans, carnivores, and primates, thus increasing the *cortex* enormously, allowing room for vast numbers of

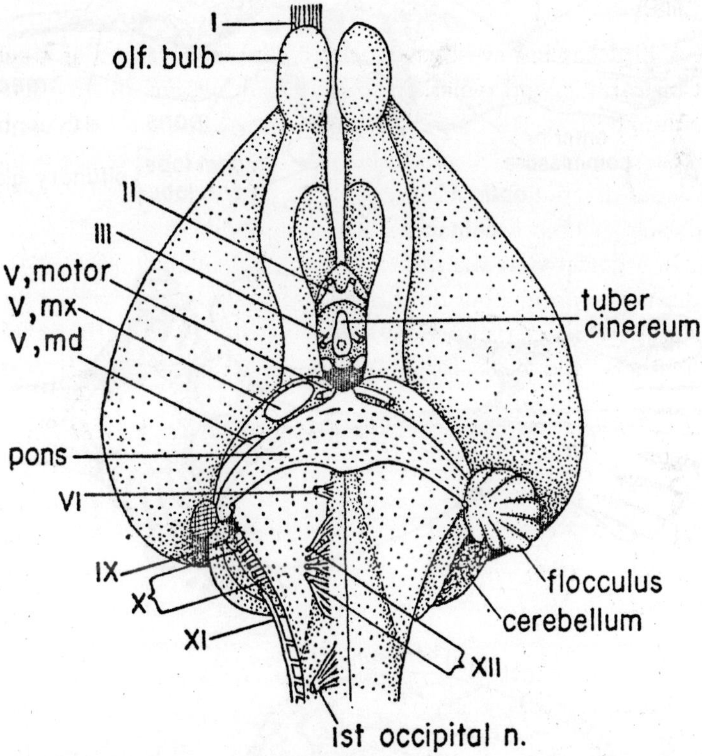

Fig. 312. Brain of monotreme, *Ornithorhynchus,* ventral view. (After Parker and Haswell.)

neurons and connections. The folds (ridges) are *gyri,* and the grooves or creases between folds are *sulci.* Crossing between the hemispheres is a thick, flat commissure, the *corpus callosum.* The optic lobes, in all other classes numbering two, the *corpora bigemina,* now become four (*corpora quadrigemina*). The cerebellum is convoluted, as in birds, but broader, and increased by two lateral lobes; its sagittal section shows a treelike branching of the fibrous interior, the *arbor vitae.*

The spinal cord fails to reach the end of its bony canal in mammals. In fact the spinal nerves emerge from the cord farther anteriorly than the level at which they leave the vertebral column. This is especially evident

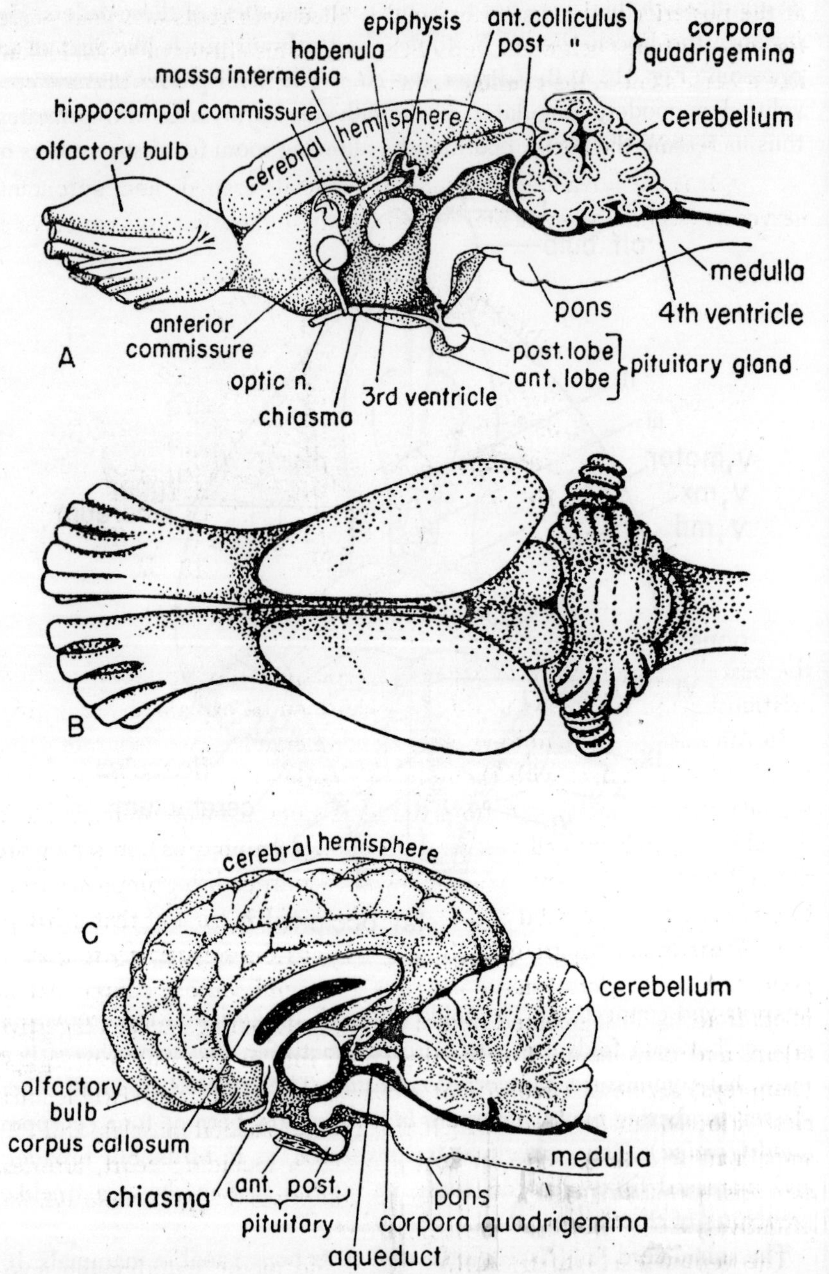

Fig. 313. Brain of opossum. A, sagittal view; B, same, dorsal view. C, brain of horse, sagittal view. (Redrawn from Edinger.)

at the posterior end, where the canal is filled with roots of spinal nerves for some distance below the end of the cord. This cluster of roots, looking like a horse's tail, is the ***cauda equina.***

SPINAL NERVES

It is an advantage to treat the spinal, cranial, and autonomic nerves as essentially segmental nerves. Of these the spinal nerves serve as

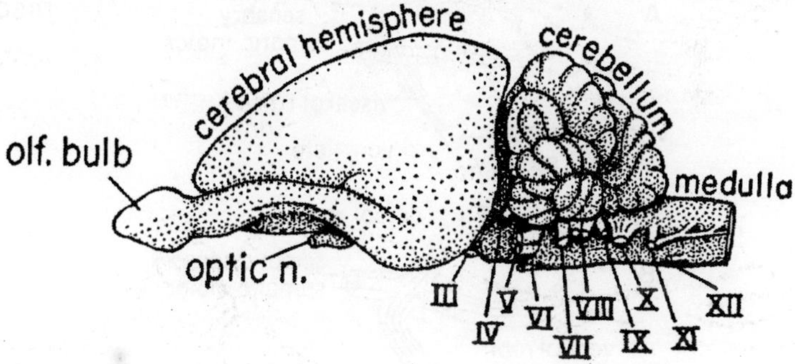

Fig. 314. Brain of hare, *Lepus.* (From Parker and Haswell.)

the best illustration, for they provide a type, probably primitive, and the relationships of the others to this type can then be explained (Fig. 315).

In Amphioxus and lampreys a segmental nerve has two separate parts, which do not connect with each other. The dorsal one is really an intersegmental nerve, going out from the cord along a myoseptum. It has a dorsal root ganglion in all vertebrates (but not Amphioxus), in which are the cell bodies of sensory neurons whose dendrites bring impulses from the skin and the segmental organs (muscles, etc.) in front of that myoseptum. These fibers are grouped, functionally, as the ***somatic sensory component.*** Also the dorsal nerve contains a ***visceral sensory component*** of fibers from the unsegmented part of the body adjoining, such as digestive organs and peritoneum. Finally, the dorsal root in primitive vertebrates (lampreys) carries a ***visceral motor component*** to (1) the visceral muscles (if located in the head) which are striated, skeletal muscles, and (2) smooth muscles of the viscera and elsewhere, including heart, arteries, skin, iris, etc. This latter group of nerves comprises the ***autonomic system,*** primitively connected with the ***dorsal*** root.

The ventral nerve is strictly segmental and motor, entering the myomere in front of the corresponding dorsal nerve. It is said to be a ***somatic motor*** nerve, and its cell bodies are in the gray matter of the cord.

Now in hagfishes and all vertebrates above cyclostomes the dorsal and ventral nerves join to form a common trunk with a dorsal root, bearing the ganglion, and a ventral root, without a ganglion. This does not happen to the segmental nerves in the head. When, as in the spinal nerves, a trunk is formed with two roots, the visceral motor component (auto-

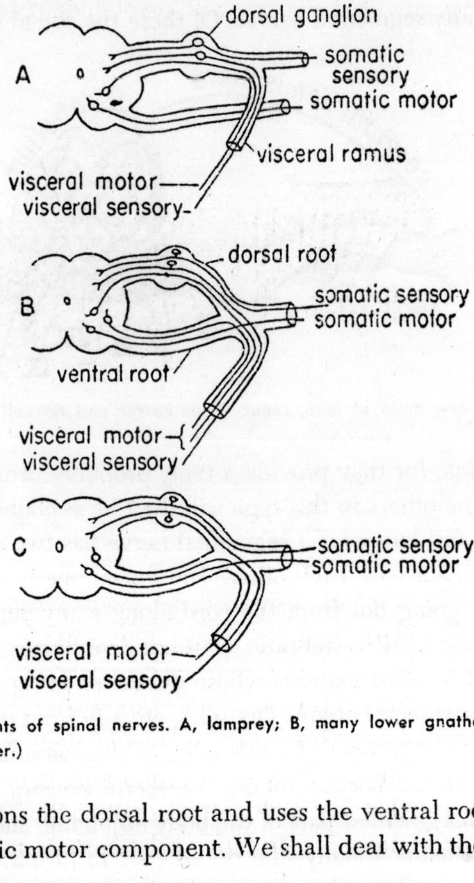

Fig. 315. Components of spinal nerves. A, lamprey; B, many lower gnathostomes; C, mammals. (Adapted from Romer.)

nomic) abandons the dorsal root and uses the ventral root instead, along with the somatic motor component. We shall deal with the autonomic system farther on.

The number of pairs of spinal nerves usually agrees, or nearly so, with the number of vertebrae in the column, even though some of these may be fused. The nerves are virtually identical down the series in fishes, except that those in the region of the gill basket have to bend back and down to reach the pectoral fin, and their lower ends then stretch forward again to supply the hypobranchial muscles. Since the fin muscles in fishes are direct outgrowths of the myotomes, each slip of a fin muscle receives a branch from its appropriate spinal nerve. In the fins of sharks, and even more conspicuously in the pectoral fins of rays (Fig. 316), the nerves gather together into two or three large trunks as they pass through the foramina in the girdle, and then fan out again into the same number of distal divisions as there were nerves involved.

With land animals the same principle holds, but the limb muscles no longer can be associated individually with myotomes. A few spinal nerves

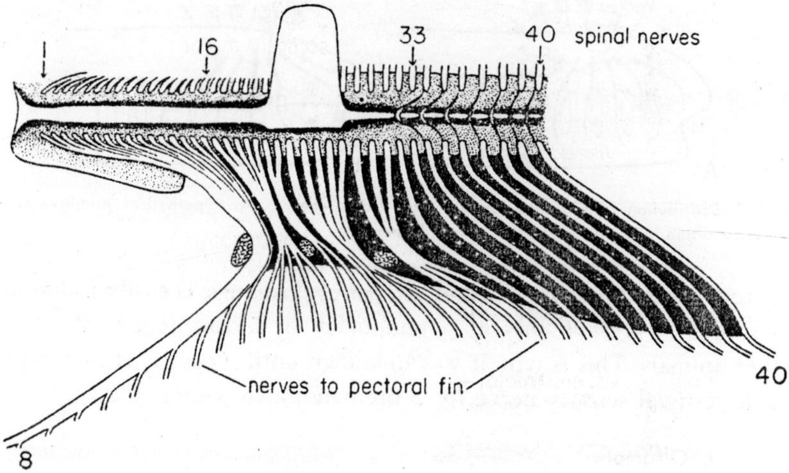

Fig. 316. Spinal nerves entering pectoral fin in ray, *Raja*, left side (anterior end is to left).

combine to form the ***brachial plexus*** (or the ***sacral plexus***) and then enter the fore (or the hind) limb as a large nerve. When the limb is extended laterad with the thumb or big toe forward, the innervation of muscles and skin in it follows this general pattern: the more anterior nerves of the plexus serve the parts nearest the body and along the anterior margin of the limb, while the posterior nerves run to the more distal and posterior parts. The fact that any such pattern exists indicates that the limb architecture keeps some of its primitive relationships to the body segments.

CRANIAL NERVES

There is a great deal of difference, not due entirely to function or position, between cranial nerves and spinal nerves. It is probable that some, but not all, cranial nerves represent what were once separate dorsal (Fig. 317) or ventral nerves belonging to particular head segments. One pair of cranial nerves, the optic, are actually lobes of the brain, hence not comparable with the others. Between one class of vertebrates and another there are few major differences in cranial nerves, so we shall not treat the classes separately (but see Fig. 318).

O. Terminal Nerve

The nervus terminalis is not one of the familiar twelve, but is a pair of small sensory nerves lying along the inner or anterior side of the

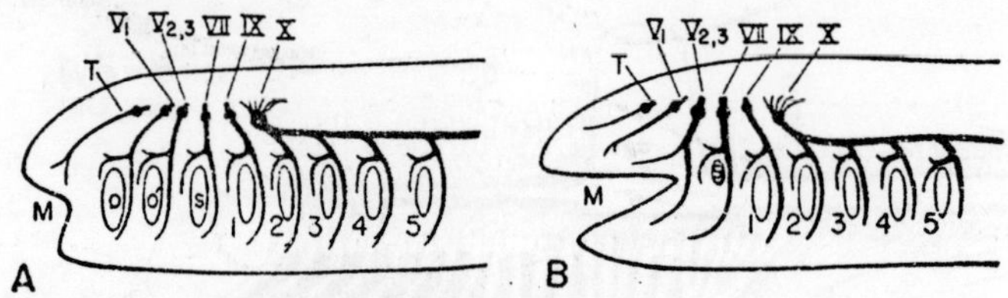

Fig. 317. Distribution of branchial (dorsal root) cranial nerves. A, hypothetical primitive condition; B, condition in jaw-bearing fishes. (From Romer.)

olfactory lobes. It occurs in fishes, except cyclostomes, is easily found in a shark, but is present only as a temporary or inconspicuous thread in higher animals. This is why it was unknown until recently. It may represent a vestigial sensory nerve for a premandibular segment.

I. Olfactory

From the anterior end of each lobe of the telencephalon goes a white *olfactory tract*, ending forward in the *olfactory bulb.* This much, apparently a big nerve, is actually a lobe of the brain, stretching forward almost to the olfactory organ. The true *olfactory nerve* is simply an aggregation of short sensory fibers reaching from the olfactory epithelium back to the bulb.

Ascending the scale of vertebrate life, we find the olfactory tracts prominent in fishes, amphibians, and most reptiles, because the olfactory sense is important in the life of these animals; they are conspicuous because not overhung by the cerebral hemispheres. With birds the sense of smell is much less important, while among mammals it varies from a remarkable development in many carnivores to almost complete atrophy in Cetacea.

II. Optic

The optic nerves grow out from the walls of the embryonic diencephalon, first as hollow vesicles, then becoming narrow stalks; the nerve fibers grow back from the retina to the brain, as described in the previous

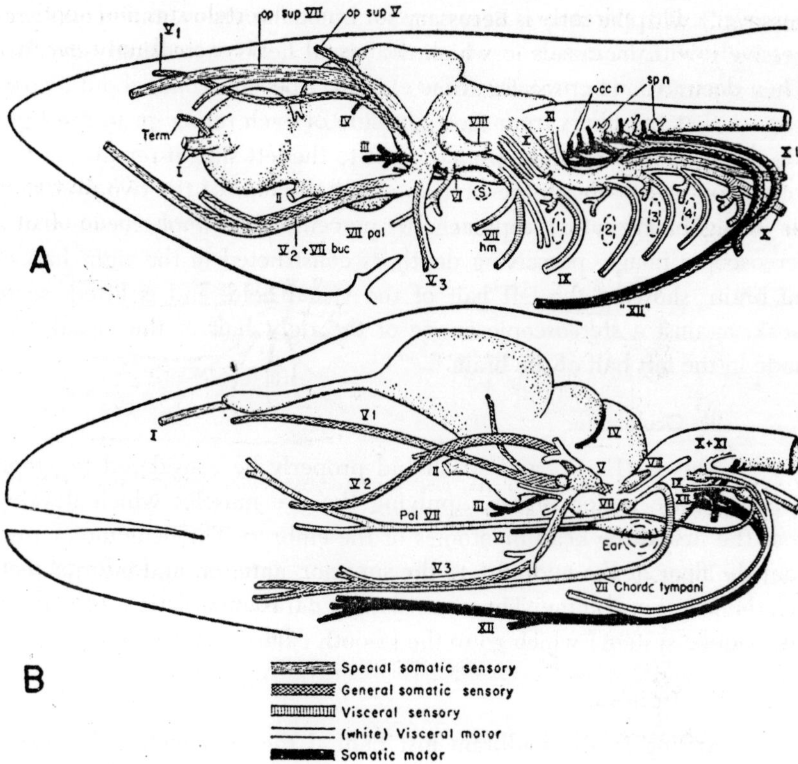

Fig. 318. Distribution and components of cranial nerves. A, a shark, *Squalus;* B, a lizard, *Anolis.* (From Romer.)

chapter. The presence of optic nerves and degree of development depends on the functional importance of the eyes in particular animals, and this we have dealt with. A notable feature is the *optic chiasma,* meeting place of the two optic nerves, seen as a white X on the ventral side of the diencephalon just in front of the pituitary body.

In the chiasma of lower vertebrates, to and including birds, the fibers from each eye cross to the opposite side of the brain before following the internal tracts to the optic lobes or tectum of the mesencephalon. But in mammals the visual centers are in the cerebral hemispheres; perhaps this is related to the psychological complexities of stereoscopic vision, which call for exceedingly intricate associative pathways among the neurons. In mammals, too, some of the fibers from each retina separate from the rest at the chiasma and go to the same side of the brain as that of the eye from which they came. This is called *incomplete decussation* (incomplete crossover). It apparently is necessary for binocular vision, and occurs progressively with mammals in which the visual fields increasingly overlap. When decussation (crossover at the chiasma) becomes only about 50 percent, so that the fibers from the right side of each retina go to the right side of the brain, and those from the left to the left side, it remains at this proportion, even though the overlap of visual fields of the two eyes may (as in higher Primates) approach 100 percent. This simply means that a stereoscopic image, perceiving depth, is constructed in the right half of the brain, showing the left half of the visual field, and is fitted, so to speak, against a stereoscopic image of the right half of the visual field made in the left half of the brain.

III. Oculomotor

Nerves III, IV, and VI should properly be considered together, since they are motor nerves supplying the eye muscles which develop from the first three head myotomes of the embryo. The oculomotor runs from the floor of the midbrain to the superior, anterior, and inferior recti and the inferior oblique. There are with it parasympathetic motor fibers (autonomic system) which go to the smooth ciliary and iris muscles.

IV. Trochlear

Arising in the midbrain just behind the oculomotor, this nerve supplies only the superior oblique muscle, a product of the second myotome.

V. Trigeminal

Just as nerves III, IV, and VI have a close relationship to each other, so have V, VII, IX, and X. It is sometimes difficult to separate V and VII at the base in dissection, and all four are mixed nerves with quite comparable distribution of their branches.

The trigeminal in fishes has four main branches, one of which, the *profundus* (deep ophthalmic, or ramus ophthalmicus profundus), arises a little separately from the rest and is primitively a distinct nerve. On the reasonable assumption that the eye-muscle nerves represent ventral, motor roots of segmental nerves, which also had dorsal roots, it is thought that the profundus is the dorsal (sensory) nerve of the first segment, hence that it belongs with the oculomotor. It arises from the side of the metencephalon, and has its own ganglion, the ***profundus ganglion,*** just outside the brain. It goes to the skin of the snout. Among higher vertebrates the profundus ganglion unites with the Gasserian ganglion of the trigeminal proper, and the profundus nerve combines with the superficial ophthalmic; this is why it was not originally numbered as a separate member of the series (Fig. 319).

The remaining branches of the trigeminal arise from the *Gasserian* (trigeminal, or semilunar) *ganglion,* and probably represent a dorsal root of the second segment, belonging with the trochlear. At the same time

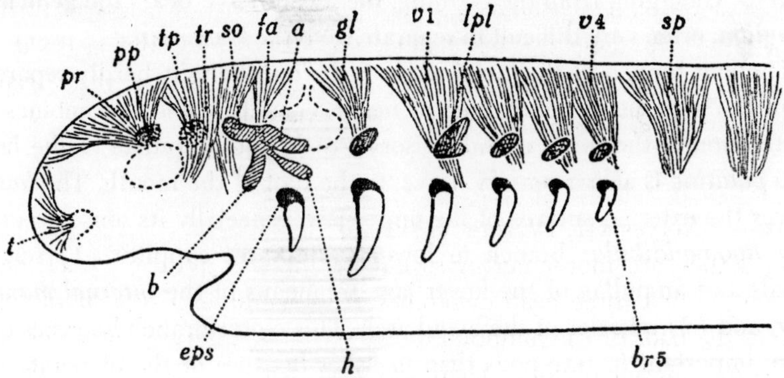

Fig. 319. Diagram of origins of segmental ganglia in gnathostome. (From Goodrich, *Studies on the Structure and Development of the Vertebrates,* Macmillan.)

a, auditory branch; *b,* buccal branch; *br5,* fifth branchial slit; *eps,* epibranchial placode; *fa,* facial; *gl,* glossopharyngeal; *h,* hyoid branch; *lpl,* lateral line placode of vagus; *pp,* profundus placode; *pr,* profundus; *so,* superior ophthalmic.

they are to be regarded as the nerves of the upper and lower jaws, that is, the first visceral arch. The *superficial ophthalmic* (ramus ophthalmicus superficialis) runs through the orbit above the eye, supplying the skin of the upper part of the head. The *maxillary,* also sensory, goes to the upper jaw and its teeth. The *mandibular* nerve supplies (1) the lower jaw and its teeth, as a sensory nerve, and (2) the jaw muscles, as a visceral motor nerve.

How can a dorsal root nerve carry a motor component? We have to remember that primitively the dorsal roots carried the visceral motor fibers for all segments, and that the jaw muscles are visceral, arising from the hypomere, not from myotomes. Thus, as we shall see, cranial nerves V, VII, IX, and X all have such visceral motor divisions, because they represent dorsal segmental roots.

VI. Abducens

This is the motor nerve of the third myotome, innervating the external rectus muscle and, in those animals that have one, the retractor bulbi. The name "abducens" comes from the function of the external rectus in man, to abduct or turn outward the eye.

VII. Facial

As the trigeminal is associated with the first visceral arch, the facial is related to the second, the hyoid arch. It is also thought to be the dorsal root for the third segment, and therefore belongs with the abducens. Emerging from the medulla, the facial nerve bears the ***geniculate ganglion,*** often very difficult to separate from the Gasserian.

The first branch of the facial nerve is the ***ophthalmic,*** hardly separable from the superficial ophthalmic of nerve V, with which it combines. Its destination is the more dorsal sensory canals and ampullae of the head. The ***palatine*** is also a sensory nerve, to the roof of the mouth. The ***buccal*** serves the external surface of the upper jaw, especially its sensory canals. The ***hyomandibular*** branch in lower vertebrates supplies (1) sensory canals and ampullae of the lower jaw, by means of the ***internal mandibular,*** and (2) muscles of the hyoid arch. This motor branch becomes even more important in tetrapods than in fishes because of the elaborate specialization of muscles formerly connected with the hyoid arch. Thus in mammals the facial muscles, the platysma, part of the digastric, and others are served by the facial nerve. A branch running through the middle ear, the ***chorda tympani,*** carries sensory fibers from the taste buds and motor (parasympathetic) fibers to the salivary glands. This chorda tympani, found in reptiles, birds, and mammals, is derived from the internal mandibular branch in fishes.

VIII. Auditory

This nerve is sensory only, going to the end-organs of the inner ear. It has three branches to the ampullae of the semicircular canals, and a larger division in higher vertebrates to the organ of Corti in the cochlea. Thus it is a nerve of both equilibrium and hearing. We cannot say definitely that it is a part of any primitive segmental nerve. But it is a member of a group of lateral-line nerves, which may originally have been independent of the others, for we have seen already how the inner ear develops as a part of the sensory canal system.

IX. Glossopharyngeal

This nerve is associated with the third visceral arch (first branchial), the one behind the hyoid. It has a sensory palatine branch serving the same area as the palatine branch of the facial, and usually combining with it. The motor part supplies the visceral muscles of the third arch. Its ganglion is the *petrosal,* and it is thought to represent a dorsal root for the fourth head segment, for which no myotome and no ventral motor root develop in vertebrates.

X. Vagus (Pneumogastric)

This is the most important and most problematical of all the cranial nerves. In fishes it appears to be a combination of several visceral-arch nerves (all those behind the third arch), supplying to each of its arches a visceral motor component, for the muscles, and two sensory branches. But then it continues into the trunk, as a general visceral nerve with parasympathetic fibers to the digestive organs, lungs, and heart. It has also, in fishes and aquatic Amphibia, a lateralis branch serving the entire lateral line except for the anterior parts already supplied by nerves VII and IX. At its origin there are several roots and a ***vagus ganglion.*** Embryonically it seems that the various roots have their own ganglia, but the majority of these disappear when the developing nerve comes to depend on the more anterior ganglion.

Probably the vagus represents a dorsal root of the fifth head segment, for which no myotome or ventral root is present, plus the visceral branches from the dorsal roots of about four more segments, whose ventral roots are represented by nerve XII. It has, so to speak, stolen these visceral branches by linking them with its own ganglion, thus taking responsibility for all the visceral arches behind the third. It is generally held that two myotomes are missing between the three which make the eye muscles and the first ones to make segmental muscles behind the head. The dorsal nerves related to these missing segments would then be, as we have said, IX and the first root of X.

In tetrapods, of course, the lateralis division disappears and the visceral arch branches are much reduced, so that the vagus of higher animals is simply a nerve to the viscera, but to a physiologist it is the nerve of greatest interest in the whole body.

XI. Spinal Accessory

The eleventh cranial nerve is a visceral motor nerve, hence dorsal. It has a center in the medulla, and supplies muscles developed from the most posterior branchial arches. This means that in fishes it is really included in the vagus, but in tetrapods it innervates the trapezius and sternomastoid (or their equivalents), for these muscles are derived from the branchial group. In tetrapods, except modern Amphibia, it emerges separately from a foramen in the skull. In mammals parasympathetic fibers from it still accompany the vagus to the viscera.

XII. Hypoglossal

Again, this nerve is "cranial" only in tetrapods, except modern Amphibia. It corresponds to the motor divisions of at least three spinal nerves and has its center in the medulla. It runs to hypobranchial muscles in the fishes and to corresponding muscles of the tongue in mammals.

To summarize the relationships of cranial nerves:

1. Nerve I is a group of neuroepithelial cells connected to a lobe of the brain, and II is itself a lobe of the brain.
2. Nerves III, IV, and VI may be considered ventral, motor divisions of primitive segmental nerves going to the first three myotomes. This does not imply that they were ever connected with the dorsal divisions.
3. The profundus branch of nerve V may be the dorsal, sensory nerve of segment 1, while nerve O may be similarly related to a still more anterior segment, which no longer appears.
4. Nerves V and VII are probably the dorsal, sensory and visceral motor nerves of segments 2 and 3, respectively.
5. For segments 4 and 5 there are no ventral, motor nerves, since no myotomes develop.
6. For segment 4 the dorsal nerve is IX.
7. Nerve X is the dorsal nerve for segment 5 plus at least 4 more segments. It is primarily visceral.
8. The visceral motor nerve for the most posterior branchial muscles, or their derivatives, is XI, although it is not separate in lower vertebrates.
9. Nerve XII is a combination of motor nerves for at least three segments.
10. Nerve VIII and the sensory-canal branches of VII, IX, and X may perhaps not be related to primitive segmental nerves but belong to a group originally distinct, the lateralis nerves; this question is unsettled.

AUTONOMIC SYSTEM

We generally call "autonomic" those ***visceral motor nerves*** which go to smooth and cardiac muscles, to glands, contractile pigment cells, etc., but not to the visceral arch muscles of the head. As we saw, in lampreys the autonomic nerves connect with dorsal divisions of segmental nerves in both head and trunk, but in gnathostomes this connection remains only in the case of cranial nerves VII, IX, X, and XI. Otherwise the visceral motor fibers leave the spinal cord by way of the ventral root. (The oculomotor nerve, incidentally, comes in the same category, as a ventral motor nerve.)

Why this happened is not understood, but presumably the lamprey condition is primitive, and certainly it gives us an anatomical basis for arranging these nerves which is quite different from the functional basis commonly used. To escape confusion let us finish with the anatomy and then come to the function.

In a typical spinal nerve with dorsal and ventral root, then, visceral motor fibers run out the ventral root and then turn off ventrad to a segmental ***sympathetic ganglion*** lying against the dorsal wall of the body cavity. These fibers are ***preganglionic*** and white, since they are myelinated. So they connect the ganglion to the spinal nerve by a ***white ramus.*** In the sympathetic ganglion some of them end, while others go on through, to reach an unpaired ***collateral ganglion*** among the viscera, or still farther in the case of parasympathetic fibers, to the viscera directly. But in *all* these cases there is a synapse with a second neuron, called a ***postganglionic neuron***, whose fiber is not myelinated, hence is gray. Thus autonomic impulses pass through two linked neurons, preganglionic and postganglionic, before ending in a muscle or gland. Postganglionic fibers beginning in the paired sympathetic ganglia nearest the spinal cord may go in three directions: (1) up to the trunk of the adjacent spinal nerve, by a ***gray ramus,*** and out with the peripheral branches to the skin, or in lower vertebrates out to the skin separately; (2) down to the viscera, passing through collateral ganglia without stopping, or sometimes by-passing collateral ganglia; (3) through longitudinal trunks connecting successive sympathetic ganglia, but in this case they go on out to their destination anyway. (We should add here that visceral sensory fibers may use the same nerves, returning to cell bodies in the dorsal root ganglia, but the autonomic system is usually thought of as simply an outflow or motor system, in which these sensory fibers are only incidental, not differing from any other sensory fibers.)

Thus in the thoracolumbar region the autonomic system shows as a double chain of ganglia linked by two parallel trunks, and from each of the paired ganglia two short rami, gray and white, go to the adjacent spinal nerve.

The more important unpaired, collateral ganglia in the trunk (Fig.

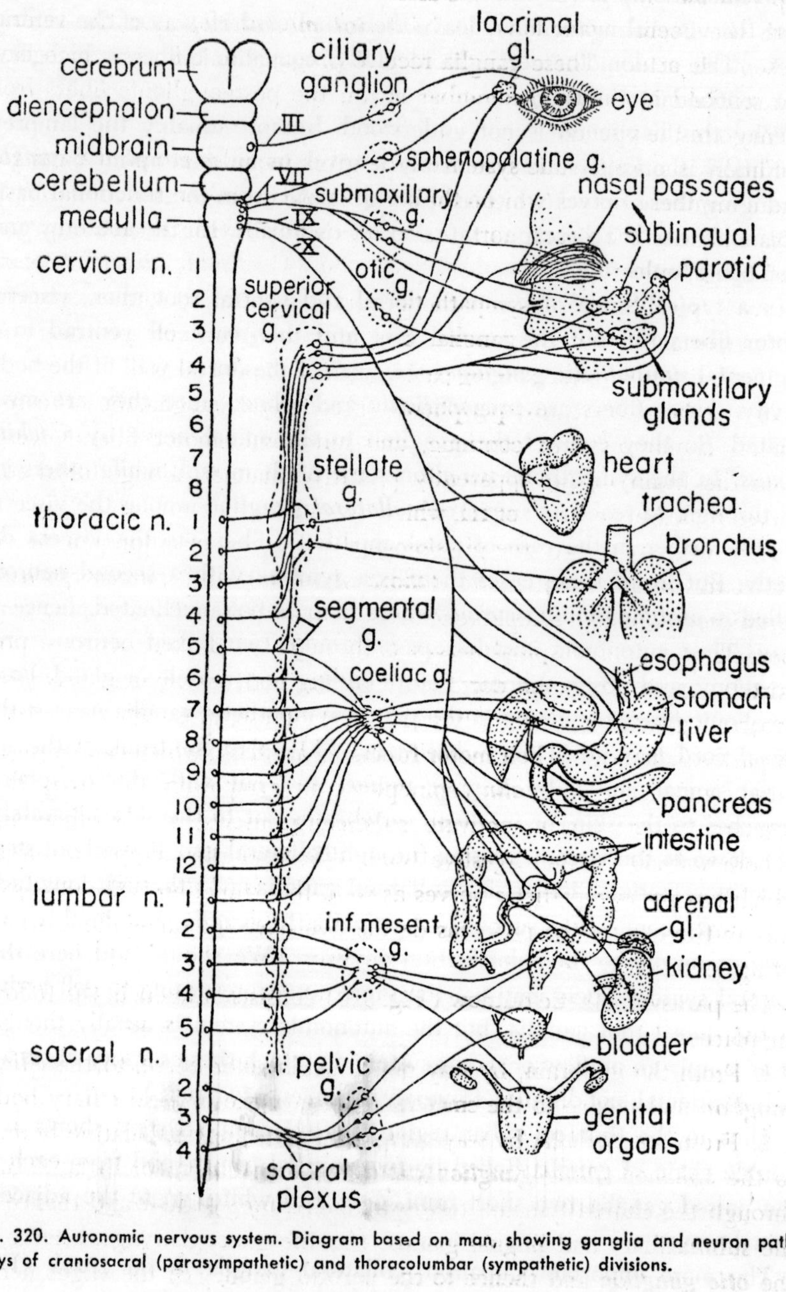

Fig. 320. Autonomic nervous system. Diagram based on man, showing ganglia and neuron pathways of craniosacral (parasympathetic) and thoracolumbar (sympathetic) divisions.

320), are the ***coeliac*** (near where the coeliac artery leaves the aorta), ***superior mesenteric*** (next to the artery of the same name, but often fused with the coeliac ganglion), and ***inferior mesenteric***, near the inferior mesenteric artery. These ganglia receive preganglionic fibers which have not stopped in the thoracolumbar chain; the postganglionic fibers from them go to the viscera.

Finally, the autonomic system in the trunk includes ***peripheral ganglia*** and a number of ***plexi*** in or on the surfaces of visceral organs and arteries. This includes the heart, aorta, and arterial walls, the digestive organs and lungs, spleen, gonads and kidneys, etc. The "***solar plexus***" (coeliac plexus) is a cluster of sympathetic nerves associated with the coeliac and superior mesenteric ganglia. The plexi of ***Auerbach*** and ***Meissner*** lie in the intestinal walls, taking part in control of peristalsis.

In the head there are peripheral or collateral ganglia, but the segmental chain does not continue, and autonomic motor fibers emerge from the brain in the apparently primitive manner, along dorsal root nerves with the exception of III, which is ventral.

Here we may turn to the physiological separation of autonomic nerves into two types: ***sympathetic*** (***thoracolumbar***) and ***parasympathetic*** (***craniosacral***). The sympathetic division includes the fibers coming from thoracic and lumbar spinal nerves (Fig. 320), as just described, to the paired ganglia along the dorsal wall of the body cavity and thence to unpaired ganglia among the viscera. From these, fibers go to virtually all parts of the body. Their motor functions are those which can be duplicated by the drug ***adrenalin*** (epinephrin, or "sympathin"). That is, they accelerate heart action, activate sphincter muscles, and inhibit most smooth muscles of the digestive tract. Wherever they supply an organ which has parasympathetic nerves as well, their function is the opposite; only in the case of the pancreas do the two have the same effect, stimulating secretion.

The parasympathetic outflow (Fig. 320) consists of fibers in the following places:

1. From the midbrain, by way of the oculomotor nerve, to the ***ciliary ganglion*** and thence to the smooth muscles of the iris and ciliary body.
2. From the medulla, by way of: (a) the facial nerve, palatine branch, to the ***sphenopalatine ganglion*** and thence to the lacrimal gland; also through the chorda tympani to the ***submaxillary ganglion*** and thence to the submaxillary and lingual glands; (b) the glossopharyngeal nerve to the ***otic ganglion*** and thence to the parotid gland; (c) the vagus nerve

to the heart and arteries, digestive organs (except colon) and lungs.

3. From the first three sacral nerves to the colon, rectum, bladder, and external genital organs.

This parasympathetic division has, in general, functions which can be duplicated by use of the drug acetylcholine, namely to inhibit the action of the heart and sphincters, and to stimulate smooth muscle action and glandular secretion in most of the other parts named.

CHAPTER XII

Panorama

Today the distinctions between one biological "subject" and another are breaking down rapidly, and a student needs to see why this is so. The study of ***form*** (morphology), of ***function*** (physiology), of ***development*** (embryology), of ***classification*** (taxonomy, or systematics) and of the ***fossil record*** (paleontology) all contributes to the "subject" of comparative anatomy. No field of biology is richer in ideas or carries wider significance.

THE RISE OF COMPARATIVE ANATOMY

First, none of these studies sprang up over night. The curiosity of countless men during more than two thousand years has brought into being, a little at a time, the orderly, reasonable system of modern science. Not wishing merely to catalog names or dates, we ought to meet a few of these people and find out what they thought, and what difference their thinking made in the world. We need not consider their contributions to other branches of biology, but only those that have a bearing on comparative anatomy as it is now. It is misleading to say as little as we must here, because many of these men were far bigger in accomplishment and wider in understanding than we give them credit for being. Our view is broader, perhaps, than theirs simply because we stand on their shoulders.

One need not go back to Aristotle (384–322 B.C.) to learn about the developing parts of an embryo chick, although he describes them, nor do we need today his accounts of some 540 kinds of animals known to him (Fig. 321). But it is a measure of his stature that more than eighteen hundred years after his death he was still the leading authority on natural science, as well as on much else in human thought. The idea at the center of morphology today, that form develops to its highest perfection out of matter which is formless, was his; he related it both to the development of an individual (***ontogeny***) and to that of the world of living things (***phylogeny***). That something without form can of itself accomplish this involves another of his ideas, ***potentiality.***

By gathering all contemporary knowledge of animal life and arranging animals in a system according to their ways of living, actions, habits, and bodily parts, and finally by considering that this system showed the rise of life toward its ultimate perfection in man, Aristotle can be said to have produced a workable theory of evolution, which is thus far from being an upstart notion of our day.

Although anatomical dissection was done in Italian universities from the thirteenth century on, it was not with the idea of learning anything new. Medical professors pointed out the organs as surgeons toiled with the dissection; they used the writings of Galen (131–?210 A.D.), who had dissected animals but not human bodies, for authority, while the ever-present philosophy professors quoted Aristotle. It is true that even the use of ancient authorities represented an advance in Europe, where virtually no activity of this sort had taken place before.

Like a fresh breeze through this murky medieval atmosphere came the work of two great men, Leonardo da Vinci (1452–1519) and Andreas Vesalius (1514– ?). Totally unlike and unknown to each other, both held the novel belief that experience, not classical authority, is the source of knowledge. Leonardo began his study of anatomy because of its relationship to painting, but he made numerous comparisons between human anatomy and that of animals. He realized that fossils were the remains of animals once living, and that marine shells found embedded in the rock of mountains proved the former submergence of land beneath the sea. (This was also held by Xenophanes, about 500 B.C., but has been considered heresy by almost all thinkers up to the nineteenth century.) Because most of Leonardo's studies remained in private notes and drawings, unpublished almost to the present time, he had relatively little effect on the scientific practices of his day.

Vesalius, on the other hand, was a medical man, remarkable for his skill in dissecting and teaching (Fig. 322). While still in his twenties he saw that he must get away from Galen's authority and describe the human body as he found it, with correct illustrations instead of fanciful

Fig. 321. Aristotle.

Fig. 322. Vesalius.

Fig. 323. Harvey.

Fig. 324. Lamarck.

diagrams. With the strength of his convictions he published the first thorough-going anatomy in the modern spirit Eventually this work won its place against tremendous opposition from the old-timers.

Still, this was only a beginning. William Harvey (1578–1657), by years of the most careful work, proved that blood circulates through the body in arteries and veins, that the heart is muscular and pumps actively, not merely filling passively, and that the valves control the direction of flow (Fig. 323). He did not see the capillary connections because his work was done just a little before microscopes became available. In his studies of embryology he arrived at the view that the source of every living creature is an egg, yet this was really a confirmation of Aristotle's theory that life perfects its form, so far as it can, out of a formless beginning.

With the advent of the Dutch school of lens makers in the seventeenth century a large number of workers began to explore the fine details of animals, plants, and the human body. Antony van Leeuwenhoek (1632–1723) made out the capillaries and saw blood cells pouring through them. He studied the spermatozoa of several kinds of animals. Marcello Malpighi (1628–1694) likewise applied the lens to everything he could find. He, too, saw capillaries, and he distinguished the kidney tubules. Many other men like these were making discoveries to which their names are still given. Probably more fundamentally new details were seen in the seventeenth century than at any earlier or later time. But with this came a good deal of fantastic speculation and dogmatism, which only more recently has disappeared. The work of Emmanuel Swedenborg (1688–1772) on the brain shows both the progress and the fallacies of the age in ample measure.

In the eighteenth century men worked, still in a pioneering way and still on somewhat unfamiliar ground, but with an attitude more like that of the present. Georges L. L. de Buffon (1707–1788) had probably the greatest influence on later biology. He succeeded in uniting biological knowledge in a comprehensive "natural history," which treated man also as a natural object, and showed that animals and plants are subject to physical laws, to be learned by experiment and observation. Realizing what fossils are, and seeing in part the meaning of geological strata, Buffon rejected the biblical story of creation and used instead a series of epochs to cover the past history of the earth. His most direct relationship to comparative anatomy comes in the work of his collaborator, Louis Daubenton (1716–1800), who carried out the anatomical part of the natural history under Buffon's plan. Anatomies of many different animals were compared for the first time.

The studies of a few other eighteenth-century men will illustrate certain advances in thought which took place then. Charles Bonnet (1720–1793) originated a theory of preformation based in part on his work on aphids; he discovered parthenogenesis (reproduction without males) in these insects, and inferred that the females carry within themselves, minute but already formed, the young of succeeding generations. This belief, even though extended to other animals, was not so unreasonable as it seems now to us, and it served to stimulate much other work. Bonnet carried further Buffon's idea of prehistoric epochs, which he thought had ended with violent catastrophes such as world-wide floods, new life developing each time out of eggs or seeds that somehow escaped destruction.

Caspar Friedrich Wolff (1733–1794) by his embryological work put an end to preformation theories, demonstrating (as Aristotle and Harvey had believed) that development makes new structures where none had been before. Petrus Camper (1722–1789) made special anatomical studies of some of the larger animals—rhinoceros, elephant, reindeer, orangutan; he discovered air sacs and their connection with the hollow limb bones in birds. Felix Vicq D'Azyr (1748–1794), studying mammals more extensively, pointed out the functional correlation between parts of the body as used in a particular way of life; for instance sharp claws, cutting teeth and a simple kind of stomach go together in meat eaters.

The famous German poet, Johann Wolfgang von Goethe (1749–1832) has a place in eighteenth-century anatomy because of his studies on the skeleton of mammals. He supported (but did not originate) the theory that the skull resulted from a fusion of separate vertebrae. He also saw in a human skull the unfused premaxillary bones, at a time when their absence was thought to be an absolute distinction between man and animals.

Every biology student knows the name of Lamarck (Jean B. P. A. de M. de Lamarck, 1744–1829), and the relationship between his theory of the inheritance of acquired characters and the natural selection theory of Darwin and Wallace (Fig. 324). But here we should notice his arrangement of the animal kingdom in a linear series according to the degree of specialization, that is, the presence or absence of organs, and their complexity, etc. He attributed these characters to the habits of animals and the conditions in which they and their ancestors lived. Although this arrangement implied evolutionary descent, we may note that Lamarck was the last to use a single line to express the relationships of all animals.

To Marie François Xavier Bichat (1771–1802) we owe the very important recognition of tissue types, and the categories of organs and systems. Considering that he worked in human anatomy without the help of a microscope, and that he died before he was 31, his contributions were among the foremost of his time.

Georges L. C. F. D. Cuvier (1769–1832) was a man of extraordinary ability, to whom we owe the creation of both comparative anatomy and paleontology, for no one before him had tied together anatomical studies in so comprehensive a manner (Fig. 325). We have seen, up to this time, either isolated, disconnected work or a submergence of anatomy in efforts which were primarily something else. Cuvier began in a new way, dissecting marine animals, which had scarcely been investigated before. He went on to some special work on large mammals, and was the first to compare in detail the bones of fossil vertebrates with those of their modern relatives. He developed the correlation principle beyond the place where Vicq D'Azyr left it, to the point where it was thought possible to reconstruct the whole of an unknown animal from the structure of one part. Thus he could "restore" an extinct animal from an incomplete skeleton.

Cuvier denied, however, that any species of animal evolved from, or into, different ones. He objected to the linear-series scheme of Lamarck and others, and used instead a system of four primary types of animals (vertebrates, molluscs, articulates, and radiates), four major themes upon which all animal structure could be based. To account for the obvious differences between extinct and living species he used the catastrophe theory of Bonnet, supposing that after each destructive event a new population arose from some remote part of the world that had been spared.

Thus the nineteenth century began with controversy over the principle of evolution, in which the ablest man, who had himself contributed more evidence than any other, stood on the opposing side. Perhaps the resulting delay of general acceptance was not actually time lost, because Cuvier set an example of critical thoroughness and accuracy which was badly needed.

Richard Owen (1804–1892) was the best-known English anatomist of the nineteenth century. His comparative studies of teeth, of the anatomy

Fig. 325. Cuvier.

Fig. 326. Huxley.

of lungfishes, of *Archeopteryx*, of the extinct moas in New Zealand, and the apes, have permanent value. The concepts of homology and analogy, as usually explained, are due to him. He meant by ***homology*** a relationship between corresponding organs of different animals, whether superficially alike or not, the relationship being due to derivation from a single "archetype;" for example, a bird's wing, a whale's flipper and a horse's foreleg are homologous. ***Analogy***, of course, refers to a similarity of form and function between parts which are not derived from one archetype, as the analogy between a bird's wing and an insect's wing. Naturally these ideas have undergone a great deal of discussion and some change, especially within the last twenty years.

Fig. 327. Cope.

Mentioning two or three more eminent European figures will bring us close to our own period. T. H. Huxley (1825–1895), known for his mastery of written and spoken English and his powerful support of Darwin, was also important in comparative anatomy and paleontology (Fig. 326). Among other things he was able to show that the skull does not develop by a fusion of vertebrae, and thus to lay a sound basis for the embryological studies made shortly after. Carl Gegenbaur (1826–1903) was typical of the German morphological school, thorough and single minded. Taking part in the general enthusiasm over Darwinism that followed publication of *The Origin of Species* (1859), Gegenbaur made phylogenetic study his sole aim, especially basing it on the skeleton. Thus he compared limb bones of vertebrates until he found what he thought was the primitive type, the "archipterygial" fin of lungfishes. He believed that gill arches are comparable with ribs. His work, although not free from errors of interpretation, was essential at the time.

Two of the notable generalizations made by Ernst Haeckel (1834–1919) were his principle of recapitulation, or "biogenetic law," and his theory of the germ layers. Both of these have needed qualification in recent years, but the original idea of recapitulation, that embryos repeat in abbreviated manner the ancestral history of the race, is sometimes very nearly true. The most striking examples may not be found among vertebrates, and we can see that usually immature stages possess their own adaptations without reference to whether these occurred in the adults of remote ancestors, but for many anatomical details the concept has value. In these days of "debunking" earlier work it is a common fault to throw overboard *in toto* a great generalization when we find that newer evidence calls for partial revision, and then to dismiss all thinking on the subject as "mere speculation."

The germ-layer theory, that primitive animals possessed ectoderm and endoderm, presently enclosing mesoderm between, and that organs in all higher animals develop only from the germ layers with which they were originally associated, has borne abundant fruit in the field of comparative embryology, even if recent students are finding it impossible to keep the original definitions intact.

In the United States the most productive work of the nineteenth century in our field was probably that with a slant toward paleontology, because here, more than in the Old World, a vast store of undiscovered fossil vertebrates awaited the explorer. First to exploit these riches were Joseph Leidy, Edward D. Cope (Fig. 327), Othniel C. Marsh, Henry Fairfield Osborn, and their colleagues. All were able scholars (most opinion puts Cope foremost), and all described quantities of hitherto unknown forms, primarily among Mesozoic reptiles and early Cenozoic mammals, as well as encouraging numerous other students. Theories of the evolution of mammalian teeth, the specialization of limbs and toes in ungulates, and the detailed family trees of modern orders and families come largely from them.

It would serve no useful purpose to try to evaluate the labors of men still living, for any student may consult their books and papers. But it is worth emphasizing that the ***method of evolution,*** as well as its course, is under intense investigation by men who, trained in systematics, genetics, anatomy, or paleontology, see over the boundaries that formerly separated these fields. They are trying, with notable success, to reach agreement on principles, for the documentary groundwork is now pretty well laid.

ORIGIN OF THE VERTEBRATES

The question of the origin of vertebrates is also a question of the relationship between Chordata and the invertebrate phyla. We shall present here certain evidence which must be considered, and some of the difficulties in the way, adding that the matter is still obscure.

Now quite generally zoologists lean toward the echinoderms as the nearest relatives of Chordata. Resemblances which have been shown between Chordata and Echinodermata, as probably indicating relationship, include these:

1. The microscopic, ciliated tornaria larva of the acorn worms (Hemichordata) very strikingly resembles the corresponding stage of many echinoderms (Fig. 328).

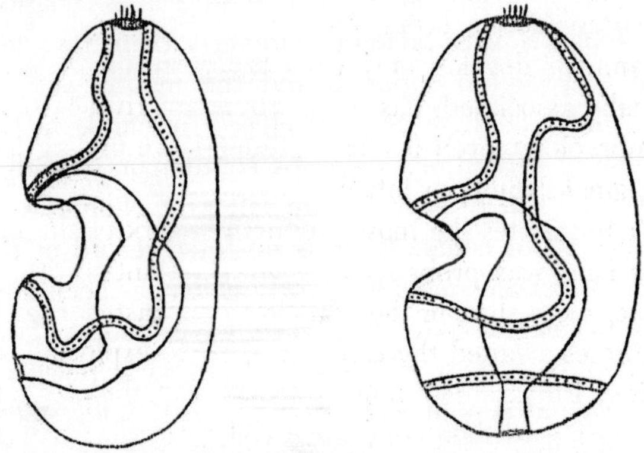

Fig. 328. Resemblance of the tornaria larva of a hemichordate (right) to an echinoid larva (left).

2. The embryos of hemichordates and of Amphioxus produce mesoderm in the form of paired coelomic pouches from the archenteron, just as in echinoderms.

3. The early history of cleavage, gastrulation, and the formation of the anus in the position of the blastopore is essentially the same in the two phyla.

4. Between the most primitive echinoderms (the early Paleozoic cystoids) and the oldest known vertebrates (the ostracoderms, also early Paleozoic) there are surprising similarities (Fig. 329), which of course cannot be found between *modern* echinoderms and *modern* vertebrates.

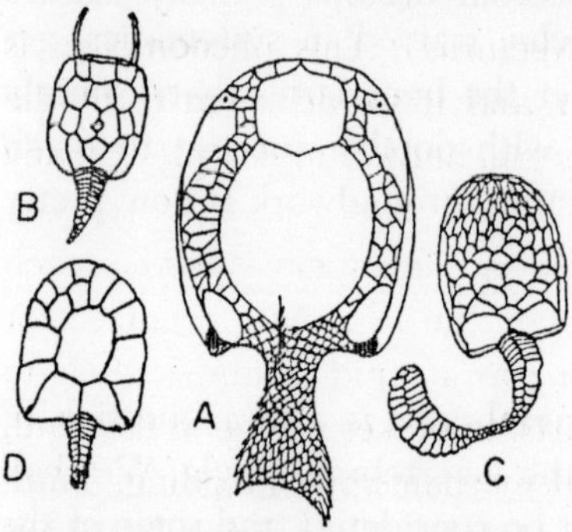

Fig. 329. Resemblance of an early ostracoderm, *Drepanaspis* (A), to several early cystoid echinoderms (B, C, D). (Redrawn from Gregory, *Quarterly Review of Biology*, 1946, vol. 21 (4), Williams & Wilkins.)

5. Serological comparisons show a quantitative likeness between the two phyla; such measurements, of course, simply indicate a degree of chemical similarity.

Now merely to enumerate resemblances and differences will get us nowhere unless we see whether they lead us to a possible picture of the origin of chordates. We know from the fossils that most cystoid echinoderms were sessile, attached to a surface by a stalk, that their feeding mechanism was a series of food grooves on the surface of the body which brought microscopic food to the mouth, that a few types became detached and were able to move themselves about by a tail-like stalk, and that in correlation with this freedom they took on more bilateral symmetry than the attached kinds, having dorsal and ventral surfaces, and the mouth was ventral. The dermal armor plates of these cystoids closely resemble in form and arrangement those of certain ostracoderms contemporary with them.

It has been our habit to consider the soft protochordates as somehow primitive types, and of course there is no fossil record of their history. But with all the immense amount of work which has been done on them, it is impossible to imagine how any of them gave rise to an armored primitive fish. The tendency now is to see the process in reverse; the acornworms, tunicates, and Amphioxus are more easily understood as having diverged from the line ancestral to vertebrates, and survived by becoming specialized in their own peculiar ways. And only by supposing this could a connection between cystoids and ostracoderms be admitted.

De Beer (1940) has described many instances of a process which he was among the first to recognize, ***pedogenesis,*** by which a general retarding of the rate of bodily development in the course of evolution results in adults of later types showing features of the larvae or young of their ancestors. Thus among Amphibia several species of salamanders fail to complete the usual metamorphosis, but become sexually mature while keeping a larval structure (e.g., *Necturus*). This phenomenon is very widespread in the animal kingdom, and it would seem to fit the problem of the protochordates, for they begin, but fail to complete, many of the structures of vertebrates.

The hemichordates, starting out like larval echinoderms, never reach the point of having a stalk or external armor, but become specialized for burrowing instead. Amphioxus, developing as a rapid swimmer, fails to put on any armor or produce any hard parts, but takes refuge in the sand, and depends on an elaborate pharyngeal mechanism for feeding. Tunicates, even in their most generalized form, make only a brief, hasty approach to the form of Amphioxus, and then, as it were, change their minds and become sessile, completely making over their bodies to fit the new situation; and this obviously has no connection with the sessile habit of echinoderms.

While we cannot deal here with the matter in enough detail, it will be clear that the approach involves several different methods of study. At every step one must be prepared to meet exacting criticism; for instance, those who study modern fishes do not, in general, agree that armored ostracoderms really were primitive, but believe there must have been soft-bodied, "normal" types from which the armored ones were derived. This, if true, would push the connections of echinoderms and chordates back beyond reach of any direct evidence, which might, after all, be correct.

Two other theories are that the ancestors of chordates were ***annelid worms*** (Dohrn, 1875; Minot, 1897), and that they were ***arthropods*** (Gaskell, 1896; Patten, 1912). A remarkable accumulation of evidence was presented with these views, especially the second one, and on reading Patten's work, at least, a student feels attracted and almost persuaded. Patten's connecting links were, in arthropods, the arachnids (such as *Limulus*, the horseshoe crab), and, in chordates, the early placoderm fishes, such as *Bothriolepis;* he did not make the recent distinction between placoderms and ostracoderms. Both the annelid and the arthropod theories require a broad assumption which cannot be supported by evidence, namely the inversion of the animal as a whole (Fig. 330). For in those two phyla the central nervous system is ventral (except the brain), and blood is pumped forward on the dorsal side. So if a vertebrate is descended from an arthropod or annelid, it must have turned upside down. Moreover it must have abandoned its old mouth (as well as eyes) and developed a new one, because otherwise the brain would be below the mouth. These changes were granted by the authors of the theories, but few others are willing to do so, and it is doubtful, with the present

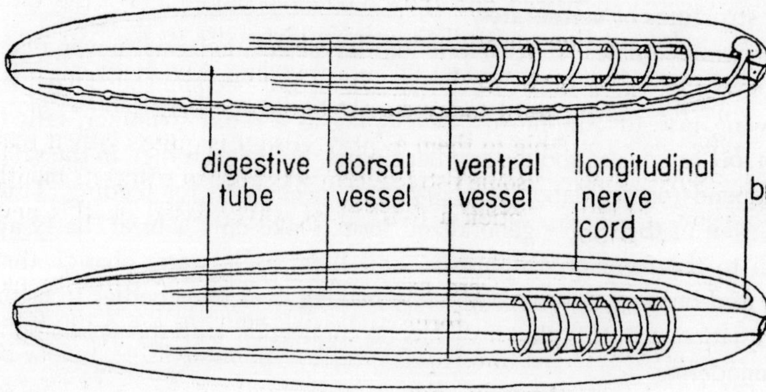

Fig. 330. Comparison of the annelid-arthropod body plan (above) with the vertebrate plan (below).

knowledge of early fishes, that even they would still hold such a view.

Other groups which perhaps should be mentioned as possible ancestors are ***Sagitta*** (the arrow worm) and ***Bryozoa.*** E. B. Wilson once commented that every invertebrate phylum except the molluscs has been regarded as ancestral to the chordates—"and I think I could make out a case for the molluscs!" The student should not take this to mean that the problem is insoluble, even at the present time, or that it does not need to be solved.

PANORAMA OF THE VERTEBRATES

On the grounds of ignorance, we shall say no more about the first wave of chordate evolution, which established the characters of the phylum, but go on to what looks like a second wave. In the remote ages when no fish swam in the sea, and the dry land was virtually barren, and no insect had ever yet buzzed or crawled, there lived in fresh water early types of jawless armored fishes, the ostracoderms. Their unknown larval stages must have been like those of modern lampreys, feeding on microscopic food by a mechanism of ciliated bands in the mouth and pharynx. To help counteract the tendency of their body fluids to absorb water from outside, these fishes developed a pronephric kidney, which maintained a fairly rapid outflow of water from the coelom.

Evidently, for their time, they had produced a successful "basic patent," for they persisted in four or five orders through the Silurian and Devonian periods, spread into the sea, and made, apparently, more than one experiment with paired fins and speedier locomotion. By the time their wave subsided they must already have given rise to the two parasitic naked orders, Petromyzontia (lampreys) and Myxinoidea (hagfishes), and these lived until today simply because other fishes of more modern type were available to them as prey. Also it is quite evident that during the Silurian period some ostracoderm was able to equip its mouth with jaws and so set in motion a third great wave, based on this new patent.

Again, it is as if nature were experimenting with the earliest gnathostomes, for curious armored forms appeared. Some of the arthrodires were of enormous size and built quite unlike any other fishes before or since. In one of the Antiarchi, *Bothriolepis,* there is an indication of paired lungs and a spiral valve type of intestine. It seems as if the first step in producing jaws is shown by the little fresh-water acanthodians, in which the conversion of a gill arch into an upper and lower jaw had not yet gone far enough to eliminate the gill which it still carried.

By the middle Devonian period, fishes representing the modern classes had appeared, and seemingly their success spelled the doom of the two major groups that preceded them. Sharks, having made their way down from the fresh water, flourished during the next few million years in the sea, producing many new families and orders. While they came from bony ancestors, their skeletons were considerably reduced and all trace of bone was lost. Their ability to meet the new physiological environment of sea water was at least partly due to retention of urea in the blood, to the point where it brought an osmotic balance between the body and the water outside. The Carboniferous seems to have been the golden age of elasmobranchs, for since then they have remained a minority among fishes.

The really progressive vertebrates of the Devonian period, however, were the modern bony fishes, Osteichthyes; one class of these, the Actinopteri, took as its fundamental innovation the freely movable fin with jointed fin rays and a thin membrane between. Starting with a scaly armor much like that of acanthodians, they thinned it gradually, made innumerable changes in bodily proportions, and eventually occupied almost every conceivable situation in both fresh and salt water. The whole story of the rise of teleosts and their relatives would take another book to tell, but certainly we should mention the conversion of air-breathing lungs into a hydrostatic swim bladder, the reduction of kidney glomeruli in marine types to conserve water in the body, and the diversity of feeding and reproductive habits, as among the essentials of that history.

Turning to the other class, Choanichthyes, we can justly regard it as the forerunner of a fourth great wave of evolution, the terrestrial vertebrates. This was a transformation more important to us than any previous one, for it brought to the animal a new set of characters fitting it to live out of water, breathing air, supporting the weight of the body by its own structure for the first time, and withstanding physical conditions far more rigorous than an aquatic animal is likely to meet. Therefore we need to picture this transformation as a whole, which in previous chapters we mentioned only in relation to particular systems.

Fossil evidence shows that a fresh-water lobefin (crossopterygian) of the late Devonian was in all probability the type from which amphibians arose. Although this lobefin was a fish, with all the equipment it needed for remaining a fish, it already possessed certain features that could be turned to advantage in venturing out of water. One, of course, was a pair of lungs, the common heritage of lobefins, lungfishes, and the primitive ray-fins as well. But this lobefin also had a pair of nostrils, barely raised above the rim of the mouth, so that evidently it was a part-time air breather as an adult, although its gills still bore the brunt of respiration. For leisurely swimming it used the curious appendages that give the name "lobefin" to the group; the paired fins were like paddles, having a narrow, bony, muscular base, and broadening out distally to a flat undulating lobe.

Whatever crisis provoked the original attempt to leave the water, we may be sure that only the adults made it, and that the immature stages for a very long time remained aquatic. A life history then was obliged to repeat, each time, the whole sequence of changes that had been accomplished, and to a large degree the modern amphibians still do so. These changes were: loss of gills, reduction of gill-bearing arches and their accompanying muscles and arteries, closing the external gill openings with loss of the operculum which covered them, closure of the spiracle (cleft between jaw and hyoid arch), with the very remarkable development of a middle ear in the same location, using the hyomandibular bone now as the stapes for sound transmission, loss of fin membranes and of all the median fins because they were useless out of water, conversion of the basal parts of the paired fins into fore and hind legs with appropriate digits and joint mechanisms, development of dorsal processes of the pelvis to connect with one or two pairs of sacral ribs in order that the belly should be lifted from the ground during locomotion, reduction of the dermal shoulder girdle so that the head became freely movable, and finally the gradual replacement of scales by a glandular, naked skin.

In rivers and lakes of the late Paleozoic era, and around their swampy shores, there lived a diversity of amphibians, some of them apparently reluctant to leave the water at all, others venturing farther away, to places from which they might not return even for breeding. These labyrinthodonts were ancestors of reptiles; sometime in the Pennsylvanian period this break was made. Again a new class, emancipated from certain shortcomings of its forefathers, outweighed the latter in competition for a place in the world, and the last labyrinthodonts died in the Triassic. Only the little, specialized frogs, salamanders, and caecilians remained to hold the Amphibian niche.

Producing a reptile (to judge by those still living) required certain fundamental changes, but not as many as took place in the origin of Amphibia. First was the problem of reproduction without going back to water, solved by (1) internal fertilization, using copulatory organs, (2) depositing eggs in the ground, each enclosed in a tough skin which would not easily let water evaporate, (3) provision of a large yolk in the egg, allowing the embryo to develop far beyond the stage corresponding to the amphibian larva, so that when hatched the young reptile was ready for terrestrial life, (4) formation by the embryo of a sac, the amnion, to hold a watery reservoir around its body, and thus free it, during its delicate early stages, from pressures or contacts hampering growth in any direction. From this primitive requirement of a fluid medium no vertebrate has ever wholly escaped.

A scaly epidermis then made it possible for the animal to come and go with little regard for relative humidity. A new type of kidney was introduced, the metanephros, which carried on filtration of fluid from the plasma and its reabsorption into the blood more rapidly than in aquatic animals, leaving a smaller proportion to be lost, and the formation of a urinary bladder (also present in adult terrestrial Amphibia) permitted control of the release of urine. The effect of temperature, however, could not be directly overcome; the rate of nervous and muscular activity remained high in a warm climate or during warm seasons, but cold brought torpor.

Climate seems not to have stood in the way of the rise of reptile life during the mild Mesozoic ages. Little ones and giants, herbivores and carnivores, soft delicate bodies and armored fortresses, runners, jumpers, swimmers and fliers, all found opportunity during more than a hundred million years of reptile supremacy.

But, unknown to any of them, two new classes meanwhile came into being, both provided with an answer to the temperature problem, and both with improved brains, enabling them to profit a little by experience. Birds had, in addition to this, a spectacular new locomotor mechanism, the feathered wing. They must have been remarkably agile even in the beginning, back in the Jurassic jungles, for they were undoubtedly arboreal reptiles before they were birds. The circulatory system made progress, for the right aortic arch came to be used alone, and the left disappeared, as the heart became four-chambered. In females only the left ovary and oviduct remained. All bones became thin and light, the sacrum stiffened by a fusion of many vertebrae, and the hollow limb bones in most cases received extensions from the air sacs branching out of the lungs. The eyes grew relatively large, and more acute than in any other vertebrates.

With wings, then, and an increasingly effective motor coördination, birds found much of the world free. We know little of the Mesozoic kinds, but enough to show that they did not wait for extinction of the dinosaurs before undertaking their adaptive radiation. Finally they dispensed with teeth, and during the Cenozoic era helped themselves to every environmental situation in the world except under ground and in deep water. Thus today they outnumber all classes of vertebrates except the bony fishes.

We may now review, in this setting of the whole vertebrate story, the arrival of mammals, and last consider man's place among them. We find the first mammals already appearing at the end of the Triassic. They remained of minor importance during a full two-thirds of their entire history. The transformation of a line of therapsid reptiles into authentic mammals involved a series of details which we may summarize here:

Already present was the heterodont dentition, the single temporal fossa, the beginning of a hard palate, division of the occipital condyle into two, and elevation of the body off the ground by vertical legs. The lower jaw bones became reduced to a dentary alone, as a part of the process of remodeling the middle ear, for when the reptilian *quadrate* bone (upper jaw) withdrew from its share in jaw articulation to become the ***incus,*** the ***articular*** (lower jaw) came along with it as the ***malleus,*** and finally the ***angular,*** just below, got involved in the auditory canal as the ***tympanic.*** This left the dentary alone, hinging against the squamosal. (The latter, in man and many other mammals, fuses with the bones of the ear capsule to constitute the ***temporal*** bone.) The middle ear now had three sound-transmitting bones, the stapes, incus, and malleus.

Soft parts, whose evolution cannot be traced directly in the fossils, also underwent certain changes. The heart, as in birds, completed its division into four chambers, but pumped systemic blood into the left aorta, abandoning the right. The transverse septum took on a layer of muscle to make the diaphragm, as an aid in respiration. The nasal passages developed folds supported on the turbinal bones, increasing the area of olfactory epithelium. The skin became hairy instead of scaly, and sweat glands began to help in the control of temperature. Mammary glands appeared, probably before the early mammals gave up egg laying.

Remains of Mesozoic mammals are rare and fragmentary. But by the close of the Cretaceous, when dinosaurs disappeared, there were several orders of mammals, including two still living, the marsupials and insectivores. Therefore the reproductive habits of modern mammals must have been initiated by then. Evidently egg-laying monotremes lived somewhere in the world, since they are still with us, but so far none of their fossil progenitors have been found. Now, just above the last Cretaceous rocks, we come to those of the Paleocene, and at once we step into the age of mammals. From then on, order after order appears, and they deploy into all available habitats. We select from them, as we could from any large group of animals, illustrations of two or three principles not specifically mentioned before.

1. Comparing the structure of a full series of horses, from the little Eocene "dawn horse," *Hyracotherium* (*Eohippus*), to the modern *Equus* (horses, donkeys, zebras), one is impressed with the ***regularity*** of their change of proportions (Fig. 331). As if plotted on a straight-line graph, three principal modifications keep step with each other: the increase of total size, the gradual reduction in number of toes (*Hyracotherium,* four in front and three behind; *Equus,* one), and the development of com-

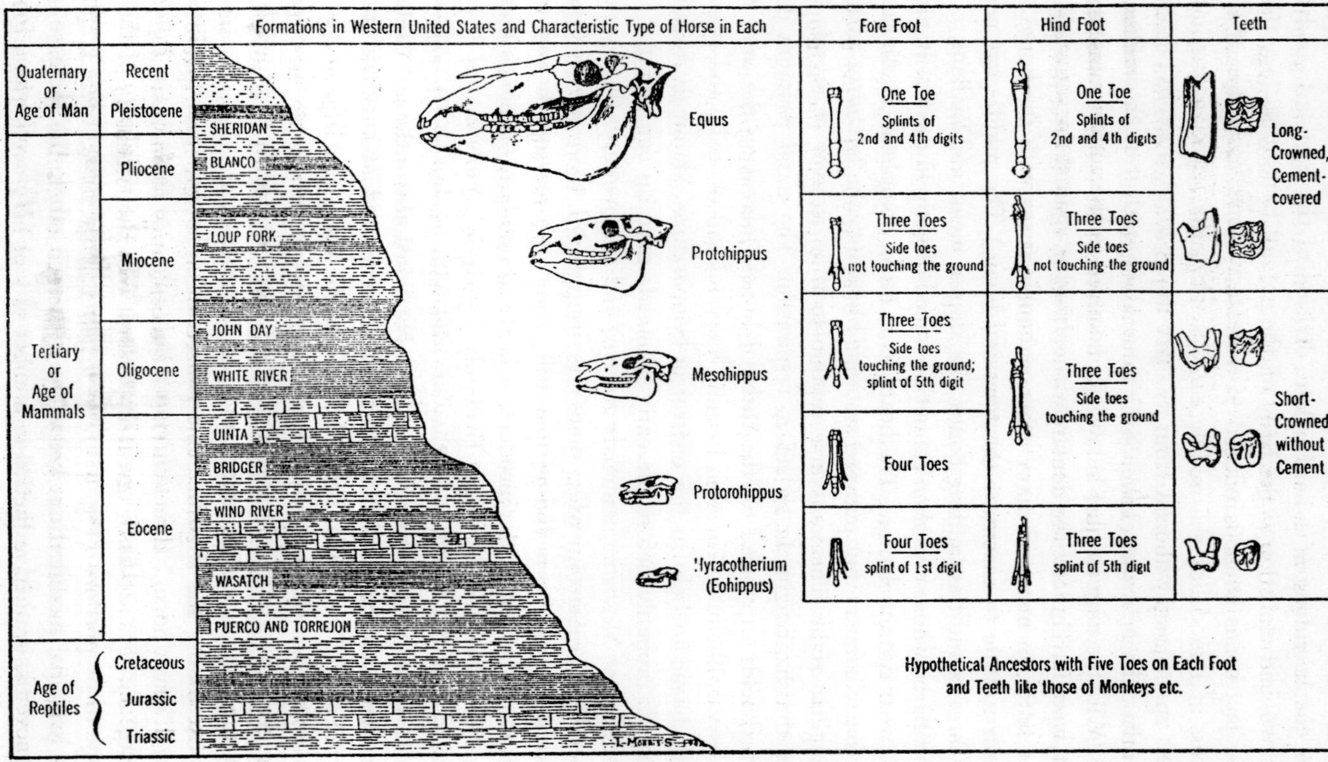

Fig. 331. Progressive changes in skulls, limbs, and teeth of horses, often cited as an example of orthogenesis. (From Parker and Haswell, after W. D. Matthew.)

plex enamel ridges on the molar teeth. This apparent straight-line evolution has been called ***orthogenesis;*** Osborn and others have supposed that it indicates a continuous uncontrollable change in the genetic make-up of the animals, leading in some cases (as titanotheres) to a maladjustment that brought about extinction. But the prevalent opinion now is that the changes can be explained by natural selection.

2. With a great number of different families and orders going their own ways, fitting into new situations from time to time, it often happened that similar adaptations arose independently. If two or more different lines go through a corresponding series of changes, arriving at quite similar results, this is called ***parallelism;*** for instance, the history of camels has been much like that of horses, and even more so has the history of the litopterns, an extinct South American group which became one-toed. If from very different ancestors nearly identical forms evolve, we call it ***convergence;*** e.g., the true moles (Insectivora), the marsupial "mole," and certain molelike rodents.

In conclusion we shall try to picture the origin of man, for one of the prime values of comparative anatomy is to give a modern perspective of man as a part of nature. His order, the Primates, arose from insectivores in the Paleocene, some sixty million years ago. Indeed, one of the foremost students of mammals (Simpson, 1945) considers the "tree-shrews," generally called insectivores, to be Primates. Within the order the basic habit is tree-climbing. The grasping structure of hands and feet, the flat nails instead of claws, the binocular vision, and enlargement of the brain, all probably came as contributions to the perfection of arboreal life.

Apes (family Pongidae) and man (Hominidae) come next to the Old World monkeys (Cercopithecidae) in the zoological scheme. Modern apes stand nearer to man in their structure than do any other living animals, but they cannot represent exactly the form which gave rise to man some two or three million years ago; the human family has been divergent at least that long. Now among the few known fossil apes of that age are some whose tooth and skull characters are distinctly more manlike than those of gorilla, chimpanzee, orang, or gibbon. Apparently man diverged in one direction and the recent apes in another. But the divergence of man was the greater; in fact it was phenomenal.

In all known types of fossil and modern men the body is erect and the legs long, for sustained upright walking instead of climbing. The feet, whenever known, have the modern characters of heel, arch, and a great toe parallel to the others, for these are part of the walking mechanism.

But the arrangement of muscles in the foot points to an arboreal ancestry, when the great toe, as in apes, stood out like a thumb and was used for grasping.

In an ancient fossil man of Java, *Pithecanthropus erectus,* probably dating from the early Pleistocene, the capacity of the braincase was well ahead of that in the largest modern ape (gorilla, about 600 cc.), varying from 775 to 900 cc. The skull in this Java ape-man must have hung a bit forward on the neck; eyebrow ridges were large and the forehead low; the jaws jutted forward but the chin receded. Another species, *P. robustus* (Fig. 332B), has also been found in Java.

The very similar "Peking Man," described as *Sinanthropus,* is properly referred to *Pithecanthropus,* so that its name now stands as *P. pekingensis.* The cranial capacity of this species, found in caves near Peiping, China, averaged about 1050 cc. Lately two kinds of giants, one in Java and one in southeastern China, have become known from jaw fragments and molar teeth a great deal larger than in any modern man or ape; they are *Meganthropus* and *Gigantopithecus.* The "Piltdown Man" found in England in 1912, *Eoanthropus dawsoni,* must be added to the list, but with a question mark, for there is disagreement about whether its skull is authentic or, perhaps, is a human cranium associated with an ape's jaw; the two were not in just the same spot.

Very recently several remarkable types of fossil man-apes have been found in South African caves. They are *Australopithecus africanus* and *A. prometheus* (Fig. 332A), with subhuman brains but in many respects manlike, *Plesianthropus transvaalensis, Paranthropus robustus* and *P. crassidens,* the latter now one of the best known of all prehuman remains. The jaws were immense, adult skulls had a sagittal crest, yet the cranial capacity approached or exceeded 1000 cc.

While all living men belong to one species, *Homo sapiens* (Fig. 332D), the Old World has given up fossil remains of other species of *Homo,* one from Java (*H. soloensis*), one from South Africa (*H. rhodesiensis*), and two from Europe (*H. heidelbergensis* and *H. neanderthalensis,* Fig. 332C). The student must be prepared for continuing changes in the classification of early men, as our knowledge of them is growing more rapidly now than at any time before.

The history of fossil men cannot yet be expressed in correct sequence, but it has involved growth of the forebrain, producing a higher forehead, reduction (finally) of eyebrow ridges, shortening of jaws so that the teeth recede and leave the chin more prominent, and a balancing of the

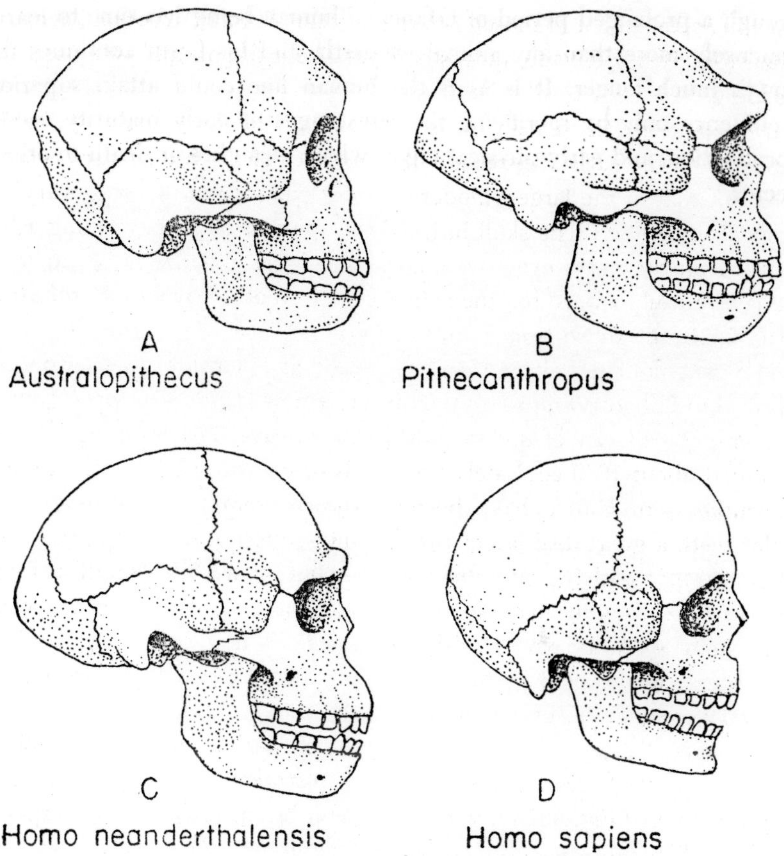

Fig. 332. Skulls of (A, B, C) early, and (D) modern man.

skull more nearly upon its center of gravity. The modern average brain capacity, some 1500 cc., was probably reached about 100,000 years ago by Neanderthal man, and was exceeded by the Cro-Magnon race of *Homo sapiens* living in caves of France and Spain during the last glacial age, twenty to thirty thousand years ago. It is not likely that we shall ever know when the body lost its hairy covering, or when speech became possible, for bones do not tell such things.

It seems probable that the round cranium, vertical face, reduced jaws, and naked body are associated with the equally remarkable delay in human development. It takes a man far longer to reach maturity than any other animal, even whales and elephants. The physical features just mentioned can be found in a fetal ape. By drawing them out, as it were, through a prolonged period of infancy, a human being has time to learn immensely more than any animal on earth, and his brain continues its growth much longer. It is as if the human line could attain superior intelligence only by sacrificing the advantages of early maturity, early reproduction, and early physical vigor, which mean life or death to other species.

Bibliography

Adams, L. A., and Eddy, Samuel, *Comparative Anatomy. An Introduction to the Vertebrates,* John Wiley and Sons, 2nd ed., 1949.

Arey, L. B., *Developmental Anatomy,* P. Blakiston's Son & Co., 5th ed., 1946.

Barrington, E. J. W., *Observations on feeding and digestion of Glossobalanus minutus,* Quarterly Journal of Microscopical Science, Vol. 82 (2), 1940, pp. 227 ff.

Bolk, L., et al., *Handbuch der vergleichende Anatomie der Wirbeltiere,* Berlin, Vienna, 1931–1938, 6 vols.

Cott, Hugh B., *Adaptive Coloration in Animals,* Macmillan, 1940.

Daniel, J. Frank, *The Elasmobranch Fishes,* University of California Press, 3rd ed., 1934.

Davison, A., and Stromsten, F. A., *Mammalian Anatomy, with Special Reference to the Cat,* P. Blakiston's Son & Co., 7th ed., 1947.

De Beer, G. R., and Moy-Thomas, J. A., *On the skull of Holocephali,* Philosophical Transactions Royal Society, B, Vol. 224, 1935, pp. 287 ff.

De Beer, G. R., *The Development of the Vertebrate Skull,* Oxford University Press, 1937.

De Beer, G. R., *Embryos and Ancestors,* Oxford, Clarendon Press, 1940.

Denison, R. H., *The soft anatomy of Bothriolepis,* Journal of Paleontology, Vol. 15 (5), 1941, pp. 553 ff.

Dohrn, Anton, *Der Ursprung der Wirbelthiere,* 1875.

Eaton, Theodore H., Jr., *Skeletal supports of the median fins of fishes,* Journal of Morphology, Vol. 76 (3), 1945, pp. 193 ff.

Edinger, Tilly, *Evolution of the Horse Brain,* Geological Society of America Memoir 25, 1948.

Evans, F. Gaynor, *The morphology and functional evolution of the atlas-axis complex from fish to mammals,* Annals New York Academy of Science, Vol. 39 (2), 1939, pp. 29 ff.

Evans, Harold M., *Brain and Body of Fish; a Study of Brain Pattern in Relation to Hunting and Feeding in Fish,* P. Blakiston's Son & Co., 1940.

Francis, E. T. B., *The Anatomy of the Salamander,* Oxford University Press, 1934.

Gaskell, W. H., *The Origin of Vertebrates,* Proceedings of the Cambridge *Philosophical Society,* Vol. 9, 1896, pp. 19 ff.

Goodrich, E. S., *Vertebrata Craniata. First Fascicle. Cyclostomes and Fishes.* Part IX of A Treatise on Zoology, ed. by E. R. Lankester. A. and C. Black, 1909.

Goodrich, E. S., *Studies on the structure and development of Vertebrates,* Macmillan, 1930.

Gregory, W. K., *Fish Skulls: a Study of the Evolution of Natural Mechanisms,* Transactions American Philosophical Society, Vol. 23, 1933, pp. 75 ff.

Gregory, W. K. and Raven, H. C., *Studies on the origin and early evolution of paired fins and limbs,* Annals New York Academy of Science, Vol. 42 (3), 1941, pp. 273 ff.

Gregory, W. K., *Evolution Emerging,* Macmillan, 1951.

Hartman, C. G., and Straus, W. L., Jr., *The Anatomy of the Rhesus Monkey,* Williams & Wilkins Co., 1933.

Howell, A. B., *Anatomy of the Wood Rat,* Williams & Wilkins Co., 1926.

Huxley, T. H., *Classification of Birds,* Proceedings of the Zoological Society of London, 1867.

Hyman, L. H., *Comparative Vertebrate Anatomy,* University of Chicago Press, 1942.

Jordan, D. S., *Fishes,* D. Appleton Co., 1925.

Kerr, J. G., *Text-book of Embryology, Vol. II, Vertebrata with the Exception of Mammals,* Macmillan, 1919.

Kingsley, J. S., *Outlines of Comparative Anatomy of Vertebrates,* P. Blakiston's Son & Co., 3rd ed., 1926.

Leach, W. J., *The archetypal position of Amphioxus and Ammocoetes and the roles of the endocrines in evolution,* American Naturalist, Vol. 78, 1944.

Lull, R. S. *Organic Evolution,* Macmillan, 2nd ed., 1940.

Maximow, A., and Bloom, W., *A Textbook of Histology,* W. B. Saunders, 5th ed., 1948.

Minot, C. S., *Cephalic Homologies: a Contribution to the Determination of the Ancestry of Vertebrates,* American Naturalist, Vol. 31, 1897, pp. 927 ff.

Moy-Thomas, J. A., *Palaeozoic Fishes,* Chemical Publishing Co., 1939.

Neal, H. V., and Rand, H. W., *Comparative Anatomy,* P. Blakiston's Son & Co., 1936.

Newman, H. H., *The Phylum Chordata,* Macmillan, 1939.

Noble, G. K., *Biology of the Amphibia,* McGraw-Hill, 1931.

Nordenskiöld, Eric, *The History of Biology,* Tudor Publishing Co., 1935.

Norman, J. R., *A History of Fishes,* Ernest Benn, Ltd., 2nd ed., 1936.

Parker, T. J., and Haswell, W. A., *A Textbook of Zoology,* Vol. II, Macmillan, 6th ed., 1940.

Patten, William, *The Evolution of the Vertebrates and their Kin,* P. Blakiston's Son & Co., 1912.

Quiring, Daniel P., *Functional Anatomy of the Vertebrates,* McGraw-Hill, 1950.

Reighard, J. E., and Jennings, H. S., *Anatomy of the Cat,* Henry Holt & Co., 3rd ed., 1935.

Reynolds, T. E., *Hydrostatics of the Suctorial Mouth of the Lamprey,* University of California Publications in Zoology, Vol. 37 (2), 1931, pp. 15 ff.

Romer, Alfred S., *The braincase of the Carboniferous Crossopterygian Megalichthys nitidus,* Bulletin of Museum of Comparative Zoology, Harvard, Vol. 82 (1), 1937, pp. 1–73.

Romer, Alfred S., *Cartilage an Embryonic Adaptation,* American Naturalist, Vol. 76, 1942.

Romer, Alfred S., and Witter, Robert V., *Edops, a primitive Rhachitomous Amphibian from the Texas Red Beds,* Journal of Geology, Vol. 50 (8), 1942, pp. 925 ff.

Romer, Alfred S., *Notes on Certain American Paleozoic Fishes,* American Journal of Science, Vol. 240, 1942, pp. 216 ff.

Romer, Alfred S., *The development of tetrapod limb musculature—the thigh of Lacerta,* Journal of Morphology, Vol. 71 (2), 1942, pp. 251 ff.

Romer, Alfred S., *The development of tetrapod limb musculature. The shoulder region of Lacerta,* Journal of Morphology, Vol. 74 (1), 1944, pp. 1–41.

Romer, Alfred S., *Vertebrate Paleontology,* University of Chicago Press, 2nd ed., 1945.

Romer, Alfred S., *The Vertebrate Body,* W. B. Saunders Co., 1949.

Sawin, H. J., *The Cranial Anatomy of Eryops megacephalus,* Bulletin of the Museum of Comparative Zoology, Vol. 88, 1941, pp. 407 ff.

Schaeffer, Bobb, *Evolution of the Tarsus in Amphibians and Reptiles,* Bulletin American Museum of Natural History, Vol. 78, 1941, pp. 395 ff.

Scott, W. B., *A History of Land Mammals in the Western Hemisphere,* Macmillan, 2nd ed., 1937.

Simpson, G. G., *The principles of classification and a classification of the mammals,* Bulletin American Museum of Natural History, Vol. 85, 1945.

Sisson, S., and Grossman, J. D., *The Anatomy of the Domestic Animals,* W. B. Saunders Co., 3rd ed., 1938.

Smith, Hobart M., *Handbook of Lizards,* Comstock, 1946.

Smith, Homer W., *Physiology of the Kidney,* Oxford, 1937.

Stensiö, E. A., *The Downtonian and Devonian Vertebrates of Spitzbergen. I. Family Cephalaspidae,* Skrifter om Svalbard og Nordishavet, No. 12, 1927.

Walls, Gordon L., *The Vertebrate Eye and its Adaptive Radiation,* Cranbrook Institute of Science, Bulletin 19, 1942.

Watson, D. M. S., *The Evolution and Origin of the Amphibia,* Philosophical Transactions Royal Society, B, Vol. 214, 1926.

Watson, D. M. S., *The Acanthodian Fishes,* Philosophical Transactions Royal Society, B, Vol. 228, 1937.

Weber, M., and others, *Die Säugetiere,* Jena, 2 vols., 2nd ed., 1927–1928.

Westoll, T. S., *The origin of the tetrapods,* Biological Reviews, Vol. 18, 1943, pp. 78 ff.

Westoll, T. S., *The origin of the primitive tetrapod limb,* Proceedings of the Royal Society of London, B, Vol. 131, 1943, pp. 373 ff.

Zangerl, Rainer, *Contributions to the Osteology of the Skull of the Amphisbaenidae,* American Midland Naturalist, Vol. 31 (2), 1944, pp. 417 ff.

Zittel, Karl A. von, *Text-book of Palaeontology,* Vol. II, 2nd English ed., Macmillan, 1932.

INDEX

Italicized page numbers mean that the item is illustrated on the page, although not necessarily labeled in every case.